Manual of Chronic Total Occlusion Interventions

Manual of Chronic Total Occlusion Interventions

A Step-by-Step Approach

Second Edition

Emmanouil Brilakis

ACADEMIC PRESS

An imprint of Elsevier

Academic Press is an imprint of Elsevier
125 London Wall, London EC2Y 5AS, United Kingdom
525 B Street, Suite 1800, San Diego, CA 92101-4495, United States
50 Hampshire Street, 5th Floor, Cambridge, MA 02139, United States
The Boulevard, Langford Lane, Kidlington, Oxford OX5 1GB, United Kingdom

Notices
Knowledge and best practice in this field are constantly changing. As new research and experience
broaden our understanding, changes in research methods, professional practices, or medical
treatment may become necessary.

Practitioners and researchers must always rely on their own experience and knowledge in
evaluating and using any information, methods, compounds, or experiments described herein. In
using such information or methods they should be mindful of their own safety and the safety of
others, including parties for whom they have a professional responsibility.

To the fullest extent of the law, neither the Publisher nor the authors, contributors, or editors,
assume any liability for any injury and/or damage to persons or property as a matter of products
liability, negligence or otherwise, or from any use or operation of any methods, products,
instructions, or ideas contained in the material herein.

Library of Congress Cataloging-in-Publication Data
A catalog record for this book is available from the Library of Congress

British Library Cataloguing-in-Publication Data
A catalogue record for this book is available from the British Library

ISBN: 978-0-12-809929-2

For information on all Academic Press publications visit our website at
https://www.elsevier.com/books-and-journals

  Working together
to grow libraries in
developing countries

www.elsevier.com • www.bookaid.org

Publisher: Mica Haley
Acquisition Editor: Stacy Masucci
Editorial Project Manager: Sam Young
Production Project Manager: Mohanambal Natarajan
Designer: Miles Hitchen

Typeset by TNQ Books and Journals

To Nicole, Stelios, and Thomas

Contents

For any questions, comments, or suggestions regarding this manual please contact Emmanouil Brilakis at esbrilakis@gmail.com

Contributors

Khaldoon Alaswad, Wayne State University, Detroit, MI, United States

Ehrin J. Armstrong, University of Colorado School of Medicine, Denver, CO, United States

Alexandre Avran, Arnault Tzanck Institut St. Laurent du Var Nice Saint Laurent du Var, France

Lorenzo Azzalini, San Rafaelle Scientific Institute, Milan, Italy

Subhash Banerjee, University of Texas Southwestern Medical Center, Dallas, TX, United States

Nicolas Boudou, Rangueil University Hospital, Toulouse, France

Marouane Boukhris, University of Tunis El Manar, Tunis, Tunisia; Abderrahmen Mami Hospital, Ariana, Tunisia

Leszek Bryniarski, Jagiellonian University Medical College, Krakow, Poland

M.N. Burke, Minneapolis Heart Institute, Minneapolis, MN, United States

Mauro Carlino, San Raffaele Scientific Institute, Milan, Italy

Charles E. Chambers, Penn State University School of Medicine, Hershey, PA, United States

Konstantinos Charitakis, McGovern Medical School at Texas Medical Center, Houston, TX, United States

James W. Choi, Baylor Heart and Vascular Hospital at Baylor University Medical Center, Dallas, TX, United States

Antonio Colombo, San Raffaele Hospital and Columbus Hospital, Milan, Italy

Stephen L. Cook, Oregon Heart & Vascular Institute, Springfield, OR, United States

Kevin J. Croce, Harvard Medical School, Boston, MA, United States

Tony J. DeMartini, Interventional Cardiologist Advocate Heart Institute, Downers Grove, IL, United States

Ali E. Denktas, UT McGovern Medical School, Houston, TX, United States; Michael E. DeBakey VA Medical Center, Houston, TX, United States

Joseph Dens, Ziekenhuis Oost Limburg, Genk, Belgium

Anthony H. Doing, UC Health Cardiology Medical Center of the Rockies, Loveland, CO, United States

Parag Doshi, Chicago Cardiology Institute, Schaumburg, IL, United States

Mohaned Egred, Freeman Hospital & Newcastle University, Newcastle upon Tyne, United Kingdom

Stephen Ellis, Cleveland Clinic, Cleveland, OH, United States

Javier Escaned, Hospital Clínico San Carlos and Universidad Complutense de Madrid, Madrid, Spain

Alfredo R. Galassi, University of Catania, Catania, Italy

Santiago Garcia, University of Minnesota, Minneapolis, MN, United States

Gabriele L. Gasparini, Humanitas Research Hospital, Milan, Italy

Omer Goktekin, Bezmialem Vakif University, Istanbul, Turkey

J. Aaron Grantham, University of Missouri Kansas City, Kansas City, MO, United States

Luis A. Guzman, Virginia Commonwealth University Pauley Heart Center, Richmond, VA, United States

Sean Halligan, Avera Heart Hospital, Sioux Falls, SD, United States

Scott Harding, Wellington Hospital, Wellington, New Zealand

Tarek Helmy, St. Louis University School of Medicine, Saint Louis, MO, United States

Jose P.S. Henriques, Academic Medical Centre of the University of Amsterdam, Amsterdam, The Netherlands

Elizabeth M. Holper, The Heart Hospital Baylor Plano, Plano, TX, United States

Wissam Jaber, Emory University School of Medicine, Atlanta, Georgia

Farouc Jaffer, Harvard Medical School, Boston, MA, United States; Massachusetts General Hospital, Boston, MA, United States

Yangsoo Jang, YUHS, Seoul, Republic of Korea

Risto Jussila, Helsinki Heart Hospital, Helsinki, Finland

Sanjog Kalra, Albert Einstein Medical Center, Philadelphia, PA, United States

Arun Kalyanasundaram, Seattle Heart and Vascular Institute, Seattle, WA, United States

David E. Kandzari, Piedmont Heart Institute, Atlanta, GA, United States

Judit Karacsonyi, University of Texas Southwestern Medical Center and VA North Texas Healthcare System, Dallas, TX, United States; University of Szeged, Szeged, Hungary

Dimitri Karmpaliotis, Columbia University Medical Center, New York, NY, United States

Jaikirshan Khatri, Cleveland Clinic, Cleveland, OH, United States

Ajay J. Kirtane, Columbia University Medical Center, New York, NY, United States

Michalis J. Koutouzis, Hellenic Red Cross Hospital, Athens, Greece

Thierry Lefèvre, Hopital Privé Jacques Cartier, Massy, France

Nicholas J. Lembo, Columbia University Medical Center, New York, NY, United States

Martin B. Leon, Columbia University Medical Center, New York, NY, United States

John R. Lesser, Minneapolis Heart Institute, Minneapolis, MN, United States

William Lombardi, University of Washington, Seattle, WA, United States

Michael Luna, UT Southwestern Medical Center/Parkland Memorial Hospital, Dallas, TX, United States

Ehtisham Mahmud, University of California, San Diego, CA, United States

Kambis Mashayekhi, University Heartcenter Freiburg, Bad Krozingen, Germany

Lampros K. Michalis, University of Ioannina, Ioannina, Greece

Jeffrey Moses, Columbia University Medical Center, Roslyn, NY, United States; Columbia University Medical Center, New York, NY, United States

Bilal Murad, United Heart and Vascular Clinic, St. Paul, MN, United States

William J. Nicholson, Wellspan Health System, York, PA, United States

Göran Olivecrona, Skåne University Hospital-Lund/ University of Lund, Lund, Sweden

Mitul P. Patel, UC San Diego Sulpizio Cardiovascular Center, La Jolla, CA, United States

Stylianos A. Pyxaras, Coburg-Clinic, Coburg, Germany

Mark J. Ricciardi, Northwestern University Feinberg School of Medicine, Chicago, IL, United States

Stéphane Rinfret, McGill University, Montreal, QC, Canada

Habib Samady, Emory University School of Medicine, Atlanta, GA, United States

James Sapontis, Monash University, Melbourne, VIC, Australia

Kendrick Shunk, University of California, San Francisco, CA, United States

George Sianos, AHEPA University Hospital, Thessaloniki, Greece

Anthony J. Spaedy, Missouri Heart Center, Columbia, MO, United States

Bradley H. Strauss, Sunnybrook Health Sciences Center, Toronto, ON, Canada

Peter Tajti, University of Szeged, Szeged, Hungary; Minneapolis Heart Institute, Minneapolis, MN, United States

Craig Thompson, Boston Scientific, Natick, MA, United States

Catalin Toma, University of Pittsburgh Medical Center, Pittsburgh, PA, United States

Thomas T. Tsai, University of Colorado, Denver, CO, United States

Etsuo Tsuchikane, Toyohashi Heart Center, Toyohashi, Japan

Imre Ungi, University of Szeged, Szeged, Hungary

Barry F. Uretsky, University of Arkansas for Medical Sciences and Central Arkansas Veterans Health System, Little Rock, AR, United States

Minh N. Vo, Mazankowksi Alberta Heart Institute, Edmonton, AB, Canada

Gerald S. Werner, Medizinische Klinik I, Klinikum Darmstadt GmbH, Darmstadt, Germany

William Wilson, Royal Melbourne Hospital, Parkville, VIC, Australia

R.M. Wyman, Torrance Memorial Medical Center, Torrance, CA, United States

Masahisa Yamane, Saitama St Luke's International Hospital, Tokyo, Japan

Robert W. Yeh, Harvard Medical School, Boston, MA, United States

Foreword

Chronic total occlusion (CTO) percutaneous coronary intervention (PCI) was once viewed as the last barrier to full percutaneous revascularization of patients. During my 12 years of involvement, I have often been asked how I got here and what advice I have for someone learning. First I accept that I am not as good as I need to be at my profession. Second, I have never accepted right or wrong—I learn from everyone, focusing on efficiency, success, and teaching. I am always evolving and trying to get better. I have watched many of those I taught become leaders in this arena. I have also seen many of those early adapters failing to follow simple rules. Treat patients not angiograms, teach others to be better than you. You should always be self-critical to improve your techniques and teaching. Do the right things for patients. There has been too much focus on success and showing off. That era has now ended. With PROGRESS, UK Hybrid, Euro CTO, Japanese Expert Registry, and OPEN CTO all showing success rates between 88% and 92%, it is clear that if you put in the time and effort you can be highly successful at CTO PCI.

I offer you a different thought as you read through the following pages. Think less about success and more about how to improve. A successful procedure is different than a successful therapy. When you have a technical success but it took more than 2 h, reflect on why. When you watch live case demonstrations, follow the example of operators with clear plans, algorithms, and simplified approaches to solve problems efficiently and safely. Understand that although there may be more than 30 different CTO guidewires and 10 different OTW microcatheters, many of them are just "me-too" devices. Ask those teaching whether the guidewire or device is unique or whether it mirrors one that already exists. Understand the nuances of your equipment so that you maximize the strength of a product and minimize its weakness.

As you perform procedures, accept failure of a technique or device and know the next step to avoid getting stuck or being slow or inefficient. Arteries do not care how they are opened nor which device or operator performs the procedure. All the myocardium knows is that it is receiving oxygenated blood. While reading the chapters, don't get overwhelmed by the minutia or unclear plans. Look for algorithms and simplified problem solving, and focus on tried and true techniques to learn fundamental skills before you start to experiment.

The main premise of hybrid PCI was to use a thoughtful algorithm for preprocedure planning and to maintain intraprocedural efficiency. Those who follow this premise also use a minimum number of devices, have intraprocedural solutions for problems such as the impenetrable cap by device, wire across collateral microcatheter that won't follow, how to deal with ambiguity antegrade and retrograde, and so on. While reading, make notes and simplify your approach; minimize choices and embrace evolution and change both within the procedure and as you grow your skill set. Remember in the end there are only two techniques (wires and dissection reentry) from two directions (antegrade and retrograde).

As I send you on your path to improvement I will leave you with one last thought. Please remember that there is no impossible case. I have heard many experts opine that x, y, or z can't be done. There is no impossible case, just those unwilling to learn or accept failure. Help our specialty grow and help it become one where we find solutions and treat patients. Those who do will find a way. Those who don't will find excuses. Let's stop making excuses and find a way to achieve complete percutaneous revascularization for all who need it.

William Lombardi, MD

Chapter 1

When to Perform Chronic Total Occlusion Interventions

1.1 CHRONIC TOTAL OCCLUSION DEFINITION

A coronary chronic total occlusion (CTO) is defined as 100% occlusion in a coronary artery with noncollateral thrombolysis in myocardial infarction (TIMI) 0 flow of at least 3 month duration.[1] The duration of occlusion may be difficult to determine if there has been no prior angiogram demonstrating presence of the CTO. In such cases estimation of the occlusion duration is based upon first onset of symptoms and/or prior history of myocardial infarction in the target vessel territory.

Occluded arteries discovered within 30 days of a myocardial infarction, such as those included in the Open Artery Trial,[2] are not considered to be CTOs. Hence, the lack of benefit observed with percutaneous coronary intervention (PCI) in these subacute lesions should not be extrapolated to CTO PCI.

1.2 PREVALENCE OF CHRONIC TOTAL OCCLUSIONS

Coronary CTOs are common, found in approximately one in three patients undergoing diagnostic coronary angiography (Table 1.1).[3–10]

Among 14,439 patients undergoing coronary angiography at three Canadian centers, at least one CTO was present in 18.4% of patients with coronary artery disease (CAD).[6] The CTO prevalence was higher (54%) among patients with prior coronary artery bypass graft surgery (CABG) and lower among patients undergoing primary PCI for acute ST-segment elevation myocardial infarction (10%) (Fig. 1.1). Left ventricular function was normal in >50% of patients with CTO and half of the CTOs were located in the right coronary artery. In a Swedish nationwide study[10] and an Italian multicenter registry[9] the prevalence of CTOs among patients with coronary artery disease was 16% and 13%, respectively.

1.3 SHOULD CHRONIC TOTAL OCCLUSION PERCUTANEOUS CORONARY INTERVENTION BE PERFORMED IN THIS PATIENT?

CTO PCI is a tool in the armamentarium for the treatment of CAD. As with every patient with CAD, treatment of patients with coronary CTOs should

Manual of Chronic Total Occlusion Interventions. http://dx.doi.org/10.1016/B978-0-12-809929-2.00001-6

TABLE 1.1 Prevalence of Coronary Chronic Total Occlusions

First Author	Country	Year	Number of Sites	n	CTO Prevalence (%)	CTO Prevalence Among Prior CABG Patients (%)
Kahn[3]	United States	1993	1	287	35	–
Christofferson[4]	United States	2005	1	8,004	52	–
Werner[5]	Germany	2009	64	2,002	35	–
Fefer[6]	Canada	2012	3	14,439	18	54
Jeroudi[7]	United States	2013	1	1,669	31	89
Azzalini[8]	Canada	2015	1	2,514	20	87
Tomasello[9]	Italy	2015	12	13,423	13	–
Ramunddal[10]	Sweden	2016	30	89,872	16	–

CABG, coronary artery bypass graft surgery; CTO, chronic total occlusion.

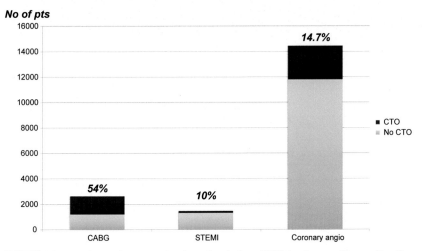

No of pts

FIGURE 1.1   Prevalence of coronary chronic total occlusions (CTO) in a large multicenter Canadian registry. *Reproduced from Fefer P, Knudtson ML, Cheema AN, et al. Current perspectives on coronary chronic total occlusions: the Canadian multicenter chronic total occlusions registry. J Am Coll Cardiol 2012;59:991–7, Elsevier.*

include optimal medical therapy (OMT) (every patient should receive aspirin and a statin unless they have a contraindication) and possibly coronary revascularization, with either PCI or CABG.

Revascularization is indicated in patients with angina or other symptoms due to ischemia, such as dyspnea, and possibly patients with ischemia on noninvasive testing, or left ventricular dysfunction (Fig. 1.2).[8,9,11] Percutaneous revascularization is preferred in case of single-vessel disease and in post-CABG patients (especially those with patent left internal mammary artery grafts to the left anterior descending artery) due to the high risk and technical challenges associated with redo CABG. In case of multivessel disease, CABG is generally preferred in patients with complex disease (especially if they are diabetic) whereas PCI is preferred in patients with simpler disease.[11–13]

The decision on whether to perform CTO PCI depends on (1) anticipated benefit and (2) estimated risk.

1.3.1 Estimating Benefit

Potential benefits of CTO PCI depend on (1) clinical presentation (i.e., symptoms, extent of ischemia, myocardial function) and (2) the likelihood of success.[14] Risks include both acute (procedural)[15] and long-term adverse events (Fig. 1.3).

The main demonstrated benefit of CTO PCI is symptomatic improvement. In patients who are truly asymptomatic, CTO PCI is generally not indicated, except possibly in patients with a large area of ischemia (>10%).

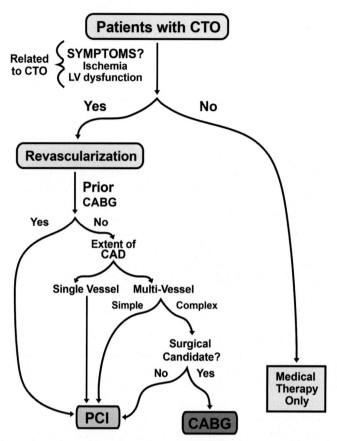

FIGURE 1.2 **Revascularization options for patients with coronary chronic total occlusions (CTOs).** Algorithm for determining the need for coronary revascularization in patients with coronary CTOs. Revascularization is indicated in patients with symptoms, and possibly those with significant ischemia or left ventricular dysfunction attributable to the CTO(s). Patients with prior coronary bypass graft surgery (CABG) are almost always treated with percutaneous coronary intervention (PCI) given the increased risk of redo CABG. In patients without prior CABG, CTO PCI and CABG are both treatment options, with CABG preferred for patients with multivessel complex disease, and PCI (including CTO PCI) preferred for patients with simpler multivessel or single-vessel disease or patients who are poor candidates for CABG. *Modified with permission from Azzalini L, Torregrossa G, Puskas JD, et al. Percutaneous revascularization of chronic total occlusions: rationale, indications, techniques, and the cardiac surgeon's point of view.* Int J Cardiol *2017, Elsevier.*

CTO patients may have to progressively limit physical activity to prevent symptoms, and often present with atypical symptoms such as dyspnea rather than angina.[16] Therefore, objective evaluation (e.g., with a treadmill stress test, 6-min walking test, standardized quality of life questionnaires, etc.) may be particularly useful in patients with limited or no symptoms.

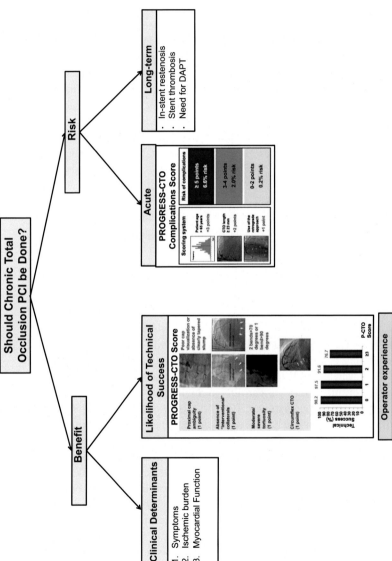

FIGURE 1.3 **Deciding whether chronic total occlusion percutaneous coronary intervention (CTO PCI) should be performed.** Deciding on whether CTO PCI should be performed depends on the anticipated risk/benefit ratio. The anticipated benefit depends on the patient's baseline clinical characteristics as well as the likelihood of technical success. Potential risks include periprocedural and long-term risks. Assessment of the likelihood for CTO PCI success and the risk for periprocedural complications can be performed using various scores, such as the Prospective Global Registry for the Study of Chronic Total Occlusion Intervention (PROGRESS-CTO)[14] and the PROGRESS-CTO Complications[15] score.

1.3.1.1 Clinical Determinants of Chronic Total Occlusion Percutaneous Coronary Intervention Benefit

CTO PCI can provide the following benefits:

1. **Improve quality of life**

    For patients with medically refractory angina caused by a CTO, successful CTO recanalization can reduce or eliminate the angina and the need for antianginal medications and improve exercise capacity.[17–20]

    Many patients with coronary CTOs may present with dyspnea and/or fatigue instead of classic angina.[16] Patients with such types of symptoms are frequently miscategorized as asymptomatic, as they can get accustomed to these symptoms and may not report them, or may minimize their severity. Many patients may also substantially curtail their physical activities and misattribute these adverse lifestyle changes to normal aging or other factors. Bruckel et al. also demonstrated that several patients with CTOs also suffer from undiagnosed major depression, and such patients derived the most benefit from successful CTO PCI.[21]

    Patients who undergo successful CTO PCI usually require fewer or no antianginal medications, obviating the medication-related cost and side effects. Eliminating nitrate intake can also allow patients to take phosphodiesterase inhibitors (e.g., sildenafil, vardenafil, tadalafil) for erectile dysfunction, which is common in CAD patients. The Drug-Eluting Stent Implantation Versus Optimal Medical Treatment in Patients With Chronic Total Occlusion (DECISION-CTO) trial (presented at the 2017 American College of Cardiology meeting) randomized 834 patients with coronary CTOs to OMT alone versus OMT + CTO-PCI. Patients in the OMT + CTO PCI group had similar clinical outcomes and quality of life during a median follow-up of 3.1 years. However, the study has several limitations, including early termination before achievement of target enrollment, high cross-over rates (18% in the OMT alone group underwent CTO PCI), revascularization of non-CTO lesions in most patients in both groups, and mild baseline symptoms in both study groups. The EuroCTO (A Randomized Multicentre Trial to Evaluate the Utilization of Revascularization or Optimal Medical Therapy for the Treatment of Chronic Total Coronary Occlusions; NCT01760083) trial (presented at the 2017 EuroPCR meeting) was also stopped early because of slow enrollment after randomizing 407 patients to OMT alone versus OMT + CTO-PCI. Compared with patients randomized to OMT only, patients randomized to CTO PCI had more improvement in angina frequency at 12 months, as assessed by the Seattle Angina Questionnaire.

2. **Reduce the need for CABG (and offer revascularization options to patients who are poor candidates for CABG)**

    In patients with complex stable coronary disease, CABG is the preferred revascularization modality, as it can reduce mortality and the risk of myocardial infarction, whereas outcomes are similar with PCI and CABG in patients with less complex disease (such as those with Syntax score

≤22)[22] (Fig. 1.3).[13] However, many patients decline CABG for nonmedical reasons or because of concerns regarding complications and recovery. Other patients are poor candidates for CABG due to high surgical risk (for example patients with multiple comorbidities or patients who require redo CABG). In such cases CTO PCI provides an attractive alternative treatment option. Examples of patients in whom CTO PCI is preferable over CABG include those with single vessel right or circumflex coronary artery CTO and intractable, medically-refractory angina and those with prior CABG, especially if they have a patent left internal mammary artery graft to the left anterior descending artery (Fig. 1.3).[13]

3. **Reduce ischemia**

   Studies using fractional flow reserve measurement after CTO crossing but before stent implantation showed that myocardial territories supplied by a CTO are ischemic, even when extensive collateral circulation is present. When fractional flow reserve (FFR) was performed in 92 patients immediately after CTO crossing with a microcatheter but before balloon angioplasty and stenting, resting ischemia was observed in 78% of patients and with hyperemia FFR was <0.80 in all patients.[23] Similar findings were observed in a study of 50 CTO patients, all of whom were ischemic regardless of the presence and extent of collateral circulation.[24]

   The hemodynamic significance of a lesion in the donor vessel that collateralizes the CTO vessel may change after successful CTO recanalization. In a study by Sachdeva et al., six of nine donor vessels that had baseline ischemia, as assessed by FFR measurement (FFR ≤ 0.80) reverted to nonischemic FFR after successful CTO recanalization.[25] The mean increase in donor vessel FFR after CTO recanalization was 0.098 ± 0.04. In another study the mean increase in donor vessel FFR after CTO recanalization was 0.03.[26] Therefore, evaluation of possible ischemia in a donor vessel supplying collaterals to a CTO territory by FFR should be viewed with caution due to risk of false positive results.

   In a study of 301 patients who underwent myocardial perfusion imaging before and after CTO PCI, a baseline ischemic burden of >12.5% was optimal in identifying patients most likely to have a significant decrease in ischemic burden post-CTO PCI. CTO PCI is, therefore, more likely to benefit patients with significant baseline myocardial ischemia.[27]

   The presence of a CTO was the strongest independent predictor of incomplete revascularization in the PCI arm of patients treated for multivessel CAD in the SYNTAX trial. Irrespective of surgical or percutaneous revascularization strategy, incomplete revascularization and consequent ischemic burden was associated with significantly higher 4-year clinical event rates including mortality.[28] Successful CTO PCI was relatively low in this cohort compared with current standards.

4. **Improve myocardial function**

   Successful CTO revascularization can improve left ventricular systolic function,[29–37] provided that the CTO-supplied myocardium is viable[34,35] and the vessel remains patent during follow-up.[32,33] In patients with systolic heart failure, CTO revascularization was associated with improvement in left ventricular ejection fraction and improvement in New York Heart Association functional class, angina, and brain natriuretic peptide levels.[38] Three-year follow-up after successful CTO PCI suggested a beneficial effect on left ventricular remodeling, as well as tendency toward improvement in left ventricular ejection fraction.[35]

Note

Viability can be assessed using several techniques. However, if the affected myocardial segment is hypokinetic but not akinetic and if there are no Q-waves in the corresponding region of the electrocardiogram,[39] then viability is highly likely.

In the Evaluating Xience and Left Ventricular Function in Percutaneous Coronary Intervention on Occlusions After ST-Elevation Myocardial Infarction (EXPLORE) trial, patients who underwent primary PCI for ST-segment elevation acute myocardial infarction and were found to have a concomitant CTO in a non–infarct-related artery were randomized to CTO PCI or medical therapy alone within 7 days.[40] Core laboratory adjudicated procedural success was 73%. At 4 months left ventricular ejection fraction and left ventricular end-diastolic volume were similar in the two study groups (Fig. 1.4).

5. **Improve long-term survival**
   Several observational studies[41–43] and metaanalyses[18,44,45] have reported better long-term survival after successful versus failed CTO PCI, even among patients with well-developed collateral circulation.[46] Also in a large registry, patients with a CTO had higher mortality than patients without a CTO.[10] A potential beneficial effect of CTO recanalization on long-term survival may be due to protection from future coronary events in vessels supplying collateral perfusion to the ischemic CTO territory, improved myocardial contractility, and reduction in the risk for ischemic arrhythmias. All studies performed to date are unfortunately limited by their retrospective, observational designs and were not compared with a control group that was only receiving medical therapy. In addition, patients who have unsuccessful CTO PCI are more likely to have complications at the time of their procedure, potentially biasing the results in favor of successful CTO PCI. As described above, the DECISION CTO and EuroCTO trial did not show differences in the incidence of adverse cardiac events during follow-up, but were underpowered and had multiple other limitations.

6. **Improve tolerance of future coronary events**
   Patients with CTO who develop an acute coronary syndrome (ACS) have significantly worse acute and long-term outcomes as compared with those without a CTO, including patients with multivessel CAD[47–50] (Fig. 1.5) ("double jeopardy").

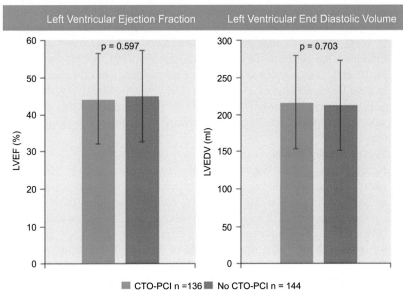

FIGURE 1.4   Left ventricular function and size at 4-month follow-up in ST-segment elevation acute myocardial infarction patients undergoing chronic total occlusion percutaneous coronary intervention (CTO PCI) versus no CTO PCI. There was no difference in left ventricular ejection fraction or left ventricular end-diastolic volume between the two groups, although a subanalysis suggested improvement in left ventricular function after CTO PCI of the left anterior descending coronary artery. *Reproduced with permission from Henriques JP, Hoebers LP, Ramunddal T, et al. Percutaneous intervention for concurrent chronic total occlusions in patients with STEMI: the EXPLORE trial.* J Am Coll Cardiol *2016;68:1622–32, Elsevier.*

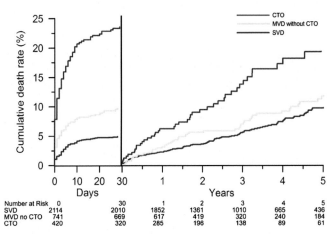

FIGURE 1.5   Impact of the presence of chronic total occlusion (CTO) on outcomes of patients presenting with ST-segment elevation acute myocardial infarction. *Reproduced with permission from Claessen BE, van der Schaaf RJ, Verouden NJ, et al. Evaluation of the effect of a concurrent chronic total occlusion on long-term mortality and left ventricular function in patients after primary percutaneous coronary intervention.* JACC Cardiovasc Interv *2009;2:1128–34, Elsevier.*

There are no prospective studies showing that prophylactic CTO PCI can improve the outcomes of future ACSs.[51] However, in a metaanalysis of seven studies of patients with ST-segment elevation acute myocardial infarction, the presence of a non–infarct-related artery CTO (present in 11.7% of patients) was associated with increased incidence of all-cause mortality at a median follow-up of 25.2 months (OR 2.90, $P<.0001$).[52] Although this does not prove that prophylactic CTO PCI improves outcomes in patients who subsequently develop acute coronary syndromes, it is suggestive of benefit.

7. **Prevent arrhythmias**
Ischemia may predispose to ventricular arrhythmias. Among 162 patients with ischemic cardiomyopathy who received an implantable cardioverter defibrillator in the VACTO study, 44% had at least one CTO.[53] During a median follow-up of 26 months, the presence of CTO was associated with higher rates of ventricular arrhythmia and death ($P<.01$),[53] although a subsequent study failed to confirm these findings.[54] Infarct-related artery CTO was independently associated with ventricular tachycardia recurrence after successful catheter ablation.[55] Patients with ischemia-induced arrhythmias could benefit from CTO recanalization.[56]

1.3.1.2 Estimating Likelihood of Success

Reported success rates of CTO PCI in the past were approximately 70%–80%.[57] With development of novel equipment, techniques, and treatment strategies CTO PCI success rates have significantly improved, with experienced centers around the world consistently achieving success rates of 85%–90%.[58–62] Less experienced centers tend to have less favorable results: in an analysis from National Cardiovascular Data Registry (NCDR) procedural success of CTO PCI between 2009 and 2013 was 59%.[63]

CTO PCI success depends both on operator experience and lesion complexity. Many angiographic and clinical parameters have been included in various scoring systems (that were developed in diverse CTO PCI cohorts) and can be used to estimate the likelihood of successful CTO PCI (Table 1.2).[14,60,64,65]

The first such score was the J-CTO score (Multicenter Chronic Total Occlusion [CTO] Registry in Japan) that uses five variables (occlusion length ≥ 20 mm, blunt stump, CTO calcification, CTO tortuosity, and prior failed attempt) to create a 5-point score that predicts successful guidewire crossing within the first 30 min (Fig. 1.6).[64]

Another commonly used scoring system is the Prospective Global Registry for the Study of Chronic Total Occlusion Intervention (PROGRESS-CTO) score, which uses four variables (proximal cap ambiguity, moderate/severe tortuosity, circumflex artery CTO, and absence of interventional collaterals) to create a 4-point score that predicts technical success (Fig. 1.7).[14]

An online calculator for several CTO PCI scores is available at http://www.progresscto.org/cto-scores.

TABLE 1.2 Comparison of Various Scores for Chronic Total Occlusion Percutaneous Coronary Intervention

	J-CTO[64]	CL[65]	PROGRESS-CTO[14]	ORA[60]
Number of variables	5	6	4	3
Number of cases	494	1657	781	1073
Setup	12 Japanese centers	2 French centers	7 US Centers	Single expert operator
Dates	2006–2007	2004–2013	2012–2015	2005–2014
Overall success	88.6% (guidewire crossing)	72.5% (procedural success)	92.9% (technical success)	91.9% (technical success)
Clinical				
Age ≥75 years				+
Prior CABG		+		
Prior MI		+		
Prior CTO PCI failure	+			
Angiographic				
Blunt stump	+	+	+[a]	
Ostial location				+
Severe calcification	+	+		
Severe tortuosity	+		+	
CTO length >20 mm	+	+		
CTO target vessel		+ (non-LAD)	+ (circumflex)	
Collaterals			+ (interventional)	+ (Rentrop <2)

CABG, coronary artery bypass graft surgery; CL, Clinical and Lesion-related score; CTO, chronic total occlusion; J-CTO, Multicenter CTO Registry in Japan score; LAD, left anterior descending artery; MI, myocardial infarction; ORA score, Ostial location, Rentrop <2, Age ≥75 years score; PCI, percutaneous coronary intervention; PROGRESS-CTO, Prospective Global Registry for the Study of Chronic Total Occlusion Intervention.
[a]Proximal cap ambiguity.

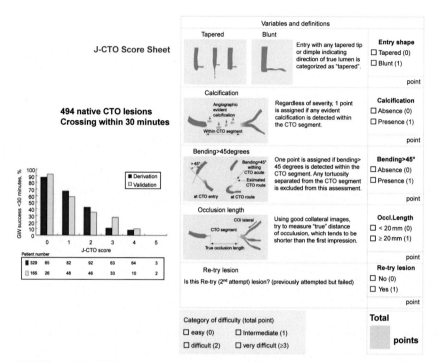

FIGURE 1.6 The J-CTO score. Description of the components of the Multicenter Chronic Total Occlusion Registry in Japan (J-CTO) score that was developed to predict the likelihood of successful guidewire crossing of the occlusion within 30 min. *Reproduced with permission from Morino Y, Abe M, Morimoto T, et al. Predicting successful guidewire crossing through chronic total occlusion of native coronary lesions within 30 minutes: the J-CTO (Multicenter CTO Registry in Japan) score as a difficulty grading and time assessment tool.* JACC Cardiovasc Interv *2011;4:213–21, Elsevier.*

1.3.2 Estimating Risk

1.3.2.1 Complications Score

CTO PCI carries increased risk for complications as compared with standard PCI. In NCDR the risk for major adverse cardiac events (MACE) was 1.6% for CTO PCI versus 0.8% for non-CTO PCI ($P<0.001$).[63] In the PROGRESS-CTO registry, the risk for MACE was 2.8%, and was associated with age >65, occlusion length ≥23 mm, and use of the retrograde approach.[15,66] Similar to the scores developed to determine the likelihood of procedural success, a score has been developed to predict the risk for MACE during CTO PCI (PROGRESS-CTO Complications score) (Fig. 1.8).[15]

1.3.2.2 Contraindications to Chronic Total Occlusion Percutaneous Coronary Intervention

Absolute Contraindications

1. Inability to receive a P2Y12 inhibitor (for example due to high risk for bleeding). Patients with contraindications to P2Y12 inhibitors are best treated with CABG surgery.

Relative Contraindications

1. Inability to receive prolonged dual antiplatelet therapy required after drug-eluting stent implantation, due to noncompliance or high bleeding risk. Due to high restenosis rates of bare metal stents in CTO PCI, drug-eluting stents are commonly used and require extended periods of dual antiplatelet therapy, as described in Chapter 11.
2. Prior radiation skin injury to the torso.
3. Chronic kidney disease, as high contrast volume may be needed during the procedure.
4. Prior (especially recent and high) radiation exposure, or multiple and prolonged prior cardiac procedures requiring fluoroscopy, given the increased risk for radiation skin injury with repeat X-ray exposure.
5. Heparin-induced thrombocytopenia (although bivalirudin could potentially be used in such cases).
6. Anaphylactic reaction to contrast media administration.

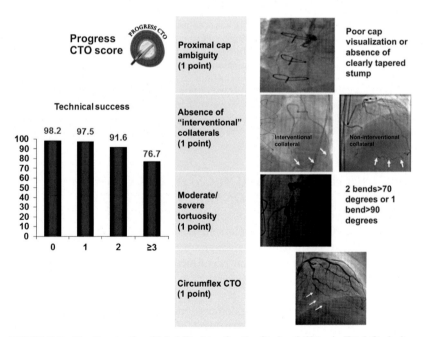

FIGURE 1.7   **The Prospective Global Registry for the Study of Chronic Total Occlusion Intervention (PROGRESS-CTO) score.** Description of the components of the PROGRESS-CTO score that was developed to predict technical success of CTO percutaneous coronary intervention. *Reproduced with permission from Christopoulos G, Kandzari DE, Yeh RW, et al. Development and Validation of a Novel Scoring System for Predicting Technical Success of Chronic Total Occlusion Percutaneous Coronary Interventions: the PROGRESS CTO (Prospective Global Registry for the Study of Chronic Total Occlusion Intervention) Score. JACC Cardiovas Interv 2016;9:1–9, Elsevier.*

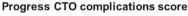

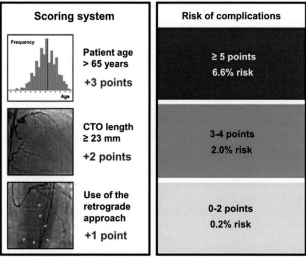

FIGURE 1.8   **The Prospective Global Registry for the Study of Chronic Total Occlusion Intervention (PROGRESS-CTO) Complications score.** Description of the components of the PROGRESS-CTO Complications score that was developed to predict periprocedural complications during CTO percutaneous coronary intervention. Periprocedural complications included any of the following adverse events prior to hospital discharge: death, myocardial infarction, recurrent symptoms requiring urgent repeat target vessel revascularization with PCI or coronary artery bypass graft surgery, tamponade requiring either pericardiocentesis or surgery, and stroke. *Reproduced with permission from Danek BA, Karatasakis A, Karmpaliotis D, et al. Development and Validation of a Scoring System for Predicting Periprocedural Complications during Percutaneous Coronary Interventions of Chronic Total Occlusions: the Prospective Global Registry for the Study of Chronic Total Occlusion Intervention (PROGRESS CTO) Complications Score. J Am Heart Assoc 2016:5 (open access article).*

1.4   GUIDELINES FOR CHRONIC TOTAL OCCLUSION PERCUTANEOUS CORONARY INTERVENTION

In the 2011 American College of Cardiology/American Heart Association PCI guidelines, CTO PCI carries a class IIA/level of evidence B recommendation: "PCI of a CTO in patients with appropriate clinical indications and suitable anatomy is reasonable when performed by operators with appropriate expertise."[67]

In the 2014 European Society of Cardiology/European Association of Cardiothoracic Surgery guidelines on myocardial revascularization CTO PCI also carries a class IIA/level of evidence B recommendation: "Percutaneous recanalization of CTOs should be considered in patients with expected ischemia reduction in a corresponding myocardial territory and/or angina relief."[68]

Despite these guidelines, there is significant variability in the treatment of patients with coronary CTOs between different physicians and institutions.[6]

1.5 SUMMARY AND CONCLUSIONS

In summary, successful CTO PCI can provide patients with significant health benefits when clinically indicated. In symptomatic patients without other coronary artery disease, the myocardium supplied by a CTO is always an ischemic zone, even with well-developed collateral circulation. Ongoing ischemia is associated with worse clinical outcomes and successful CTO PCI is important for relieving ischemia and helping achieve complete coronary revascularization. Deciding when to proceed with CTO PCI depends on the balance of the anticipated risks and benefits of the procedure, should be individualized, and should always be patient-centered. How to successfully and safely perform CTO PCI will be discussed in more detail in the following chapters.

REFERENCES

1. Stone GW, Kandzari DE, Mehran R, et al. Percutaneous recanalization of chronically occluded coronary arteries: a consensus document: part I. *Circulation* 2005;**112**:2364–72.
2. Hochman JS, Lamas GA, Buller CE, et al. Coronary intervention for persistent occlusion after myocardial infarction. *N Engl J Med* 2006;**355**:2395–407.
3. Kahn JK. Angiographic suitability for catheter revascularization of total coronary occlusions in patients from a community hospital setting. *Am Heart J* 1993;**126**:561–4.
4. Christofferson RD, Lehmann KG, Martin GV, Every N, Caldwell JH, Kapadia SR. Effect of chronic total coronary occlusion on treatment strategy. *Am J Cardiol* 2005;**95**:1088–91.
5. Werner GS, Gitt AK, Zeymer U, et al. Chronic total coronary occlusions in patients with stable angina pectoris: impact on therapy and outcome in present day clinical practice. *Clin Res Cardiol* 2009;**98**:435–41.
6. Fefer P, Knudtson ML, Cheema AN, et al. Current perspectives on coronary chronic total occlusions: the Canadian multicenter chronic total occlusions registry. *J Am Coll Cardiol* 2012;**59**:991–7.
7. Jeroudi OM, Alomar ME, Michael TT, et al. Prevalence and management of coronary chronic total occlusions in a tertiary veterans affairs hospital. *Catheter Cardiovasc Interv* 2014;**84**:637–43.
8. Azzalini L, Jolicoeur EM, Pighi M, et al. Epidemiology, management strategies, and outcomes of patients with chronic total coronary occlusion. *Am J Cardiol* 2016;**118**:1128–35.
9. Tomasello SD, Boukhris M, Giubilato S, et al. Management strategies in patients affected by chronic total occlusions: results from the Italian Registry of Chronic Total Occlusions. *Eur Heart J* 2015;**36**:3189–98.
10. Ramunddal T, Hoebers LP, Henriques JP, et al. Prognostic impact of chronic total occlusions: a report from SCAAR (Swedish coronary angiography and angioplasty registry). *JACC Cardiovasc Interv* 2016;**9**:1535–44.
11. Brilakis ES, Abdullah SM, Banerjee S. Who should undergo chronic total occlusion percutaneous coronary Intervention?: the EXPLORation continues. *J Am Coll Cardiol* 2016;**68**:1633–6.
12. Azzalini L. The clinical significance and management implications of chronic total occlusion associated with surgical coronary artery revascularization. *Can J Cardiol* 2016;**32**:1286–9.
13. Azzalini L, Torregrossa G, Puskas JD, et al. Percutaneous revascularization of chronic total occlusions: rationale, indications, techniques, and the cardiac surgeon's point of view. *Int J Cardiol* 2017;**231**:90–6.

14. Christopoulos G, Kandzari DE, Yeh RW, et al. Development and Validation of a Novel Scoring System for Predicting Technical Success of Chronic Total Occlusion Percutaneous Coronary Interventions: the PROGRESS CTO (Prospective Global Registry for the Study of Chronic Total Occlusion Intervention) Score. *JACC Cardiovasc Interv* 2016;**9**:1–9.

15. Danek BA, Karatasakis A, Karmpaliotis D, et al. Development and Validation of a Scoring System for Predicting Periprocedural Complications during Percutaneous Coronary Interventions of Chronic Total Occlusions: the Prospective Global Registry for the Study of Chronic Total Occlusion Intervention (PROGRESS CTO) Complications Score. *J Am Heart Assoc* 2016:5.

16. Safley DM, Grantham J, Jones PG, Spertus J. Heatlh Status benefits of angioplasty for chronic total occlusions – an analysis from the OPS/PRISM studies. *J Am Coll Cardiol* 2012;**59**:E101.

17. Olivari Z, Rubartelli P, Piscione F, et al. Immediate results and one-year clinical outcome after percutaneous coronary interventions in chronic total occlusions: data from a multicenter, prospective, observational study (TOAST-GISE). *J Am Coll Cardiol* 2003;**41**:1672–8.

18. Christakopoulos GE, Christopoulos G, Carlino M, et al. Meta-analysis of clinical outcomes of patients who underwent percutaneous coronary interventions for chronic total occlusions. *Am J Cardiol* 2015;**115**:1367–75.

19. Joyal D, Afilalo J, Rinfret S. Effectiveness of recanalization of chronic total occlusions: a systematic review and meta-analysis. *Am Heart J* 2010;**160**:179–87.

20. Rossello X, Pujadas S, Serra A, et al. Assessment of inducible myocardial ischemia, quality of life, and functional status after successful percutaneous revascularization in patients with chronic total coronary occlusion. *Am J Cardiol* 2016;**117**:720–6.

21. Bruckel JT, Jaffer FA, O'Brien C, Stone L, Pomerantsev E, Yeh RW. Angina severity, depression, and response to percutaneous revascularization in patients with chronic total occlusion of coronary arteries. *J Invasive Cardiol* 2016;**28**:44–51.

22. Mohr FW, Morice MC, Kappetein AP, et al. Coronary artery bypass graft surgery versus percutaneous coronary intervention in patients with three-vessel disease and left main coronary disease: 5-year follow-up of the randomised, clinical SYNTAX trial. *Lancet* 2013;**381**:629–38.

23. Werner GS, Surber R, Ferrari M, Fritzenwanger M, Figulla HR. The functional reserve of collaterals supplying long-term chronic total coronary occlusions in patients without prior myocardial infarction. *Eur Heart J* 2006;**27**:2406–12.

24. Sachdeva R, Agrawal M, Flynn SE, Werner GS, Uretsky BF. The myocardium supplied by a chronic total occlusion is a persistently ischemic zone. *Catheter Cardiovasc Interv* 2014;**83**:9–16.

25. Sachdeva R, Agrawal M, Flynn SE, Werner GS, Uretsky BF. Reversal of ischemia of donor artery myocardium after recanalization of a chronic total occlusion. *Catheter Cardiovasc Interv* 2013;**82**:E453–8.

26. Ladwiniec A, Cunnington MS, Rossington J, et al. Collateral donor artery physiology and the influence of a chronic total occlusion on fractional flow reserve. *Circ Cardiovasc Interv* 2015:8.

27. Safley DM, Koshy S, Grantham JA, et al. Changes in myocardial ischemic burden following percutaneous coronary intervention of chronic total occlusions. *Catheter Cardiovasc Interv* 2011;**78**:337–43.

28. Farooq V, Serruys PW, Garcia-Garcia HM, et al. The negative impact of incomplete angiographic revascularization on clinical outcomes and its association with total occlusions: the SYNTAX (Synergy between Percutaneous Coronary Intervention with Taxus and Cardiac Surgery) trial. *J Am Coll Cardiol* 2013;**61**:282–94.

29. Melchior JP, Doriot PA, Chatelain P, et al. Improvement of left ventricular contraction and relaxation synchronism after recanalization of chronic total coronary occlusion by angioplasty. *J Am Coll Cardiol* 1987;**9**:763–8.

30. Danchin N, Angioi M, Cador R, et al. Effect of late percutaneous angioplastic recanalization of total coronary artery occlusion on left ventricular remodeling, ejection fraction, and regional wall motion. *Am J Cardiol* 1996;**78**:729–35.

31. Van Belle E, Blouard P, McFadden EP, Lablanche JM, Bauters C, Bertrand ME. Effects of stenting of recent or chronic coronary occlusions on late vessel patency and left ventricular function. *Am J Cardiol* 1997;**80**:1150–4.

32. Sirnes PA, Myreng Y, Molstad P, Bonarjee V, Golf S. Improvement in left ventricular ejection fraction and wall motion after successful recanalization of chronic coronary occlusions. *Eur Heart J* 1998;**19**:273–81.

33. Piscione F, Galasso G, De Luca G, et al. Late reopening of an occluded infarct related artery improves left ventricular function and long term clinical outcome. *Heart* 2005;**91**:646–51.

34. Baks T, van Geuns RJ, Duncker DJ, et al. Prediction of left ventricular function after drug-eluting stent implantation for chronic total coronary occlusions. *J Am Coll Cardiol* 2006;**47**:721–5.

35. Kirschbaum SW, Baks T, van den Ent M, et al. Evaluation of left ventricular function three years after percutaneous recanalization of chronic total coronary occlusions. *Am J Cardiol* 2008;**101**:179–85.

36. Cheng AS, Selvanayagam JB, Jerosch-Herold M, et al. Percutaneous treatment of chronic total coronary occlusions improves regional hyperemic myocardial blood flow and contractility: insights from quantitative cardiovascular magnetic resonance imaging. *JACC Cardiovasc Interv* 2008;**1**:44–53.

37. Werner GS, Surber R, Kuethe F, et al. Collaterals and the recovery of left ventricular function after recanalization of a chronic total coronary occlusion. *Am Heart J* 2005;**149**:129–37.

38. Cardona M, Martin V, Prat-Gonzalez S, et al. Benefits of chronic total coronary occlusion percutaneous intervention in patients with heart failure and reduced ejection fraction: insights from a cardiovascular magnetic resonance study. *J Cardiovasc Magn Reson* 2016;**18**:78.

39. Surber R, Schwarz G, Figulla HR, Werner GS. Resting 12-lead electrocardiogram as a reliable predictor of functional recovery after recanalization of chronic total coronary occlusions. *Clin Cardiol* 2005;**28**:293–7.

40. Henriques JP, Hoebers LP, Ramunddal T, et al. Percutaneous intervention for concurrent chronic total occlusions in patients with STEMI: the EXPLORE trial. *J Am Coll Cardiol* 2016;**68**:1622–32.

41. Mehran R, Claessen BE, Godino C, et al. Long-term outcome of percutaneous coronary intervention for chronic total occlusions. *JACC Cardiovasc Interv* 2011;**4**:952–61.

42. Jones DA, Weerackody R, Rathod K, et al. Successful recanalization of chronic total occlusions is associated with improved long-term survival. *JACC Cardiovas Interv* 2012;**5**:380–8.

43. George S, Cockburn J, Clayton TC, et al. Long-term follow-up of elective chronic total coronary occlusion angioplasty: analysis from the U.K. Central Cardiac Audit Database. *J Am Coll Cardiol* 2014;**64**:235–43.

44. Khan MF, Wendel CS, Thai HM, Movahed MR. Effects of percutaneous revascularization of chronic total occlusions on clinical outcomes: a meta-analysis comparing successful versus failed percutaneous intervention for chronic total occlusion. *Catheter Cardiovasc Interv* 2013;**82**:95–107.

45. Hoebers LP, Claessen BE, Elias J, Dangas GD, Mehran R, Henriques JP. Meta-analysis on the impact of percutaneous coronary intervention of chronic total occlusions on left ventricular function and clinical outcome. *Int J Cardiol* 2015;**187**:90–6.

46. Jang WJ, Yang JH, Choi SH, et al. Long-term survival benefit of revascularization compared with medical therapy in patients with coronary chronic total occlusion and well-developed collateral circulation. *JACC Cardiovasc Interv* 2015;**8**:271–9.

47. Claessen BE, Dangas GD, Weisz G, et al. Prognostic impact of a chronic total occlusion in a non-infarct-related artery in patients with ST-segment elevation myocardial infarction: 3-year results from the HORIZONS-AMI trial. *Eur Heart J* 2012;**33**:768–75.

48. Claessen BE, van der Schaaf RJ, Verouden NJ, et al. Evaluation of the effect of a concurrent chronic total occlusion on long-term mortality and left ventricular function in patients after primary percutaneous coronary intervention. *JACC Cardiovasc Interv* 2009;**2**:1128–34.

49. Hoebers LP, Vis MM, Claessen BE, et al. The impact of multivessel disease with and without a co-existing chronic total occlusion on short- and long-term mortality in ST-elevation myocardial infarction patients with and without cardiogenic shock. *Eur J Heart Fail* 2013;**15**:425–32.

50. Lexis CP, van der Horst IC, Rahel BM, et al. Impact of chronic total occlusions on markers of reperfusion, infarct size, and long-term mortality: a substudy from the TAPAS-trial. *Catheter Cardiovasc Interv* 2011;**77**:484–91.

51. Yang ZK, Zhang RY, Hu J, Zhang Q, Ding FH, Shen WF. Impact of successful staged revascularization of a chronic total occlusion in the non-infarct-related artery on long-term outcome in patients with acute ST-segment elevation myocardial infarction. *Int J Cardiol* 2013;**165**:76–9.

52. O'Connor SA, Garot P, Sanguineti F, et al. Meta-analysis of the impact on mortality of non-infarct-related artery coronary chronic total occlusion in patients presenting with st-segment elevation myocardial infarction. *Am J Cardiol* 2015;**116**:8–14.

53. Nombela-Franco L, Mitroi CD, Fernandez-Lozano I, et al. Ventricular arrhythmias among implantable cardioverter-defibrillator recipients for primary prevention: impact of chronic total coronary occlusion (VACTO Primary Study). *Circ Arrhythm Electrophysiol* 2012;**5**:147–54.

54. Raja V, Wiegn P, Obel O, et al. Impact of chronic total occlusions and coronary revascularization on all-cause mortality and the incidence of ventricular arrhythmias in patients with ischemic cardiomyopathy. *Am J Cardiol* 2015;**116**:1358–62.

55. Di Marco A, Paglino G, Oloriz T, et al. Impact of a chronic total occlusion in an infarct-related artery on the long-term outcome of ventricular tachycardia ablation. *J Cardiovasc Electrophysiol* 2015;**26**:532–9.

56. Mixon TA. Ventricular tachycardic storm with a chronic total coronary artery occlusion treated with percutaneous coronary intervention. *Proc (Bayl Univ Med Cent)* 2015;**28**:196–9.

57. Patel VG, Brayton KM, Tamayo A, et al. Angiographic success and procedural complications in patients undergoing percutaneous coronary chronic total occlusion interventions: a weighted meta-analysis of 18,061 patients from 65 studies. *JACC Cardiovasc Interv* 2013;**6**:128–36.

58. Christopoulos G, Karmpaliotis D, Alaswad K, et al. Application and outcomes of a hybrid approach to chronic total occlusion percutaneous coronary intervention in a contemporary multicenter US registry. *Int J Cardiol* 2015;**198**:222–8.

59. Maeremans J, Walsh S, Knaapen P, et al. The hybrid Algorithm for treating chronic total occlusions in Europe: the recharge registry. *J Am Coll Cardiol* 2016;**68**:1958–70.

60. Galassi AR, Boukhris M, Azzarelli S, Castaing M, Marza F, Tomasello SD. Percutaneous coronary revascularization for chronic total occlusions: a novel predictive score of technical failure using advanced technologies. *JACC Cardiovasc Interv* 2016;**9**:911–22.

61. Wilson WM, Walsh SJ, Yan AT, et al. Hybrid approach improves success of chronic total occlusion angioplasty. *Heart* 2016;**102**:1486–93.

62. Habara M, Tsuchikane E, Muramatsu T, et al. Comparison of percutaneous coronary intervention for chronic total occlusion outcome according to operator experience from the Japanese retrograde summit registry. *Catheter Cardiovasc Interv* 2016;**87**:1027–35.

63. Brilakis ES, Banerjee S, Karmpaliotis D, et al. Procedural outcomes of chronic total occlusion percutaneous coronary intervention: a report from the NCDR (national cardiovascular data registry). *JACC Cardiovasc Interv* 2015;**8**:245–53.

64. Morino Y, Abe M, Morimoto T, et al. Predicting successful guidewire crossing through chronic total occlusion of native coronary lesions within 30 minutes: the J-CTO (Multicenter CTO Registry in Japan) score as a difficulty grading and time assessment tool. *JACC Cardiovasc Interv* 2011;**4**:213–21.

65. Alessandrino G, Chevalier B, Lefevre T, et al. A clinical and angiographic scoring system to predict the probability of successful first-attempt percutaneous coronary intervention in patients with total chronic coronary occlusion. *JACC Cardiovasc Interv* 2015;**8**:1540–8.

66. Karatasakis A, Iwnetu R, Danek BA, et al. The impact of age and sex on in-hospital outcomes of chronic total occlusion percutaneous coronary intervention. *J Invasive Cardiol* 2017;**29**:116–22.

67. Levine GN, Bates ER, Blankenship JC, et al. 2011 ACCF/AHA/SCAI guideline for percutaneous coronary intervention. A report of the American College of Cardiology Foundation/American heart association Task Force on practice guidelines and the Society for cardiovascular angiography and interventions. *J Am Coll Cardiol* 2011;**58**:e44–122.

68. Windecker S, Kolh P, Alfonso F, et al. 2014 ESC/EACTS guidelines on myocardial revascularization: the Task Force on myocardial revascularization of the European Society of Cardiology (ESC) and the European association for Cardio-Thoracic surgery (EACTS)Developed with the special contribution of the European association of percutaneous cardiovascular interventions (EAPCI). *Eur Heart J* 2014;**35**:2541–619.

Chapter 2

Equipment

2.1 INTRODUCTION

One of the most frequently asked questions about chronic total occlusion (CTO) percutaneous coronary intervention (PCI), especially from programs early in the learning curve, is, "what equipment do I really need?"[1]

Although many operators would like to have everything available, the reality is that equipment cost and space limitations require prioritization. Here are some criteria to use when deciding the must-haves for CTO PCI:

1. At least one item that fulfills each of the requisite steps in CTO PCI (e.g., septal crossing, wire externalization, snaring, etc.) should be available.
2. The operator should be familiar with the equipment, understand its strengths and limitations, and be willing to actually use it when required (otherwise it will expire on the shelf). In some cases, such as covered stents and coils, equipment expiration is to some extent expected given the low frequency of complications requiring their use (Chapter 12).

Table 2.1 shows a must-have and a good-to-have checklist for CTO PCI, classifying equipment into 12 categories.[1–4]

2.2 SHEATHS

Many high-volume hybrid CTO operators routinely use bilateral femoral 45 cm long sheaths, which provide better guide catheter support and torquability compared with shorter sheaths. Long sheaths straighten iliac vessel tortuosity, facilitating guide catheter and wire manipulation and reducing the risk for guide catheter kinking. The 45 cm length usually allows the tip of the sheath to reach the level of the diaphragm (Fig. 2.1). Although there is increased risk of thrombus formation within longer sheaths, this is rarely an issue, especially for retrograde CTO PCI, given the high ACTs (>350 s) achieved for this procedure.

If radial access is obtained, 6 Fr is the most commonly used sheath, although 7 or 8 Fr can often be used in larger radial arteries. The 7 Fr Slender sheath (Terumo) has 2.79 mm outer diameter, which is not much larger than the outer diameter of a 6 Fr standard sheath (2.62 mm), and allows use of 7 Fr guide catheters. Although radial access and smaller sheath size can reduce the risk

Manual of Chronic Total Occlusion Interventions. http://dx.doi.org/10.1016/B978-0-12-809929-2.00002-8

TABLE 2.1 Checklist of Equipment Needed for Chronic Total Occlusion Interventions

Category No.	Equipment	Must Have	Good to Have
1.	Sheaths	6–8 French standard sheaths	• 45-cm long sheaths • 7 French Slender sheaths for transradial approach
2.	Guides—guide catheter extensions	• XB/EBU 3.0, 3.5, 3.75, 4.0 • AL1, AL0.75 • JR4 • Y-connector with hemostatic valve (such as Co-pilot or Guardian) • Guide catheter extensions (GuideLiner, Trapliner, Guidezilla, Guidion)	• 90-cm long • Side-hole guides, especially AL1 • Sheathless guide catheters
3.	Microcatheters/support catheters	• Corsair, Corsair Pro, and Caravel or Turnpike and Turnpike LP (150 cm for retrograde; 135 cm for antegrade) or NHancer ProX (155 cm for retrograde;135 cm for antegrade) • Finecross (150 cm for retrograde; 135 cm for antegrade) or Micro 14 (155 cm) • Small (1.20, 1.25, or 1.5 mm diameter) 20 mm long over-the-wire balloons of 145 cm or longer total length • TwinPass or other dual lumen microcatheter (NHancer Rx, FineDuo, Crusade)	• Venture • Turnpike Spiral • SuperCross • MultiCross • CenterCross • Prodigy • NovaCross
4.	Guidewires[a]	• Fielder XT, XT-A and XT-R or Fighter • Confianza Pro 12 • Pilot 200 • Gaia 2 and 3 • Sion • Sion black • Suoh 03 • Fielder FC • RG3 or R350 (for externalization)	• Ultimate 3 • Hornet 14 and 10 • Astato 20 • Miracle 6 and 12 • Wiggle • Extra support wires (IronMan, Grand Slam, BHW)

5.	Dissection/reentry equipment	• CrossBoss catheter • Stingray LP balloon and wire	
6.	Snares	Ensnare or Atrieve 18–30mm or 27–45 mm	Amplatz Gooseneck snares
7.	Balloon uncrossable/undilatable lesion equipment	• Small 20 mm long over-the-wire and rapid-exchange balloons • Threader • Tornus or Turnpike Gold • Laser • Atherectomy catheters (rotational, orbital)	Angiosculpt
8.	Intravascular imaging	IVUS (any)	• IVUS (solid state), especially with short-tip • OCT
9.	Complication management	• Covered stents • Coils (ideally compatible with 0.014 inch microcatheter, such as Axium coils); if only 0.018 coils are available larger microcatheters are needed (such as Progreat or Renegade)	• Pericardiocentesis tray • Particles for embolization • Thrombin
10.	Radiation protection		• Radiation scatter shields • X-ray machine with radiation-reduction protocols • Zero Gravity system
11.	Balloons and Stents	• Noncompliant, long balloons • Trapper balloon • Drug-eluting stents	• Ostial Flash • Cutting balloon
12.	Hemodynamic support	• Intraaortic balloon pump • Impella CP	ECMO

aFor radial operators, 300-cm wires may be required because the trapping technique (Section 3.7) cannot be used in a 6 Fr guide catheter for trapping over-the-wire balloons, the CrossBoss catheter, and the Stingray LP balloon. In such cases, using a 6 Fr Trapliner can be useful. Alternatively, guidewire extensions (for the Asahi and Abbott guidewires) are needed. However, trapping can be performed for the Finecross, SuperCross, TwinPass, Turnpike (including Turnpike LP, Spiral, and Gold), Corsair, and the Tornus 2.1 microcatheter through a 6 Fr guide catheter with 0.071 inch internal diameter (such as Medtronic Launcher).
Another solution available to the radialist is the introduction of a 7 Fr guide catheter through a 7 Fr Slender sheath (Terumo). In addition, there are 6.5 Fr and 7.5 Fr sheathless radial guides available (Asahi) which result in a smaller diameter radial arteriotomy than a traditional 6 Fr sheath yet afford the requisite inner diameter for all trapping techniques (Section 2.3).

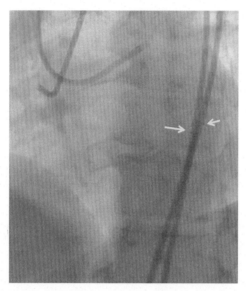

FIGURE 2.1   Location of the distal tip of 45-long femoral sheaths.

for vascular access complications, disadvantages of 6 Fr guides for CTO PCI include:

1. Weaker support as compared with larger guide catheters.
2. Inability to use the trapping technique (Section 3.7) for various key CTO PCI equipment, such as over-the-wire balloons and the CrossBoss catheter.
3. Inability to use simultaneously a balloon and a covered stent (block and deliver technique, Section 12.1.1.2.3) in case of perforation.

A sheathless guide system (Eaucath, Asahi Intecc) allows CTO PCI with 7.5 Fr guides through an arterial puncture equivalent to that created by a 5 Fr sheath. An alternative approach is using regular 8 Fr guides delivered with a sheathless approach or using a short 8 Fr sheath. It offers all the advantages of regular 8 Fr catheters and the safety of the transradial approach. The inner and outer diameters of an 8 Fr guide are similar to a 6 Fr sheath. The technique involves introducing a long 110 cm dilator that comes with a 6 Fr 90 cm long Cook Shuttle sheath into a regular 8 Fr guide. Once the guide is inserted in the radial artery, the 0.035″ guidewire is removed, Rotaglide or a radial "cocktail" (60 mg lidocaine and 5 mg verapamil) is injected through the dilator, the wire is reintroduced, and the guide is advanced to the ascending aorta. Alternatively the balloon-assisted tracking technique can be used for sheathless placement of an 8 Fr guide catheter.[5]

2.3 GUIDE CATHETERS, Y-CONNECTORS, GUIDE CATHETER EXTENSIONS

2.3.1 Diameter, Length, and Shapes

Dual 8 Fr guides are most commonly used for CTO PCI in the United States, whereas 7 Fr guides are most commonly used in Europe. Compared with smaller caliber guide catheters, 8 Fr guides provide enhanced support and improve vessel visualization.

In the donor vessel, 6 Fr guides can usually provide adequate support for delivering retrograde gear and may reduce the risk of donor artery dissection. Therefore, many operators use 6 Fr catheters from either the femoral or radial approach for retrograde access. This combination of a 6 Fr radial guide for the retrograde side and an 8 Fr femoral guide for the antegrade side is becoming increasingly popular. To maintain consistency and minimize the chance of accidental injections made through the wrong guide catheter during a case, many operators employ a standard approach of placing the right coronary artery guide catheter in the right femoral artery and the left coronary artery guide catheter in the left femoral artery.

Using at least one 90 cm long guide is often helpful for the retrograde approach, as it shortens the distance that a retrograde wire needs to cover to be externalized. It can also facilitate intraprocedural strategy change to a retrograde approach and allow the retrograde microcatheter to reach the antegrade guide for subsequent wire externalization, although 100 cm guides can also be used in most cases if the RG3 or R350 guidewires are available. Shorter (80 cm long) guides may not reach the coronary ostia in some patients and are not commonly used. The must-have guides are those with supportive shapes, such as the XB and EBU for the left coronary artery and AL for the right coronary artery.

Occasionally a vessel may not be able to be engaged despite using various guide catheters, requiring primary retrograde crossing (see Online Case 18).

2.3.2 Shortening the Guide Catheter

With availability of long externalization guidewires, such as the RG3 and R350, shortening of the guide catheter is seldom required, but may sometimes be needed for retrograde cases via bypass grafts or apical collaterals.

If manufactured short guide catheters are not available, a 100 cm long guide can be shortened using the following technique (Fig. 2.2)[6] (see online video: "How to shorten a coronary guide catheter").

How to Shorten a Coronary Guide Catheter

a. The guide catheter is inserted into the body to engage the target coronary artery and the length of the guide that is outside the femoral sheath is marked.

b. The guide is removed from the body and the marked segment is cut and removed (Fig. 2.2A).

c. A sheath (1 Fr size smaller than the guide catheter; i.e., 6 Fr sheath for a 7 Fr guide catheter, etc.) is cut to create a 3–4 cm connecting segment for the two guide pieces (Fig. 2.2B and C). Both ends of this connecting segment are flared with a dilator (of equal size to the guide) to facilitate insertion (Fig. 2.2D).

d. This connecting sheath segment is used to reconnect the proximal and distal guide catheter pieces minus the portion that was removed to shorten the guide (Fig. 2.2E and F; final result in Fig. 2.2G). Placing a Tegaderm (3M) over the connection site may help prevent accidental disconnection.

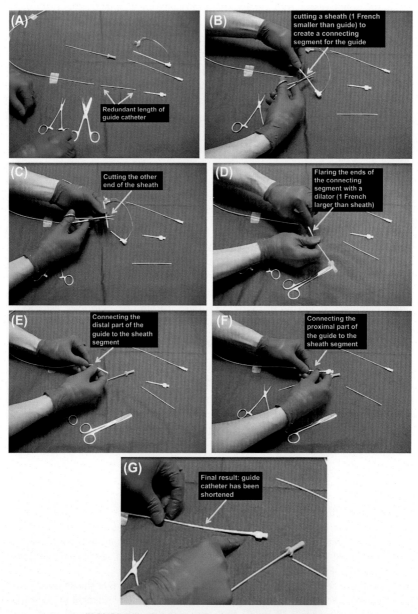

FIGURE 2.2 Overview of the guide shortening technique.

A limitation of shortened guide catheters is that they have poor torque trans-mission during vessel engagement and guide manipulations (especially during long procedures).

2.3.3 Side Holes

For right coronary artery CTOs guide catheters with side holes (Fig. 2.3) are commonly used (especially for proximal occlusions), because they can prevent pressure dampening, may allow antegrade flow into the vessel, and may decrease the risk for hydraulic dissection during antegrade contrast injection. In contrast, unprotected left main coronary arteries should not be engaged with side-hole guide catheters (with the exception of ostial left main CTOs), because subopti-mal guide catheter position may not be recognized and may lead to decrease in antegrade left main flow, global ischemia, and hemodynamic collapse. Side-hole guides, may provide a false sense of security, as hydraulic dissections can still occur upon injection. Moreover, dampening of the pressure ensures that minimal antegrade flow is allowed, when antegrade dissection/reentry techniques are used, minimizing expansion of subintimal hematoma. Side holes also increase contrast volume[3] and potentially degrade angiographic image quality. A strategy of thoughtful active guide manipulation is often chosen over use of side-hole guide catheters, depending on the preference and comfort of the operator.

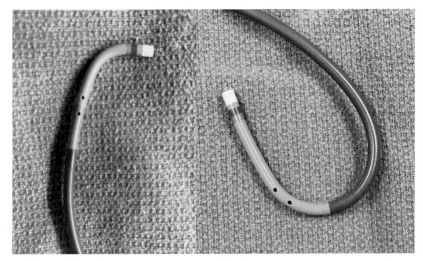

FIGURE 2.3   Example of guide catheter with side holes. *Courtesy of Dr. William Nicholson.*

If no side-hole guides are available, an 18G needle or a scalpel can be used to create side holes in the guide catheter, followed by careful flushing with saline before use. However, side holes made by hand may prevent advancement of a guide catheter extension within the modified guide and can also weaken the guide and lead to kinking.

2.3.4 Y-Connectors With Hemostatic Valves

CTO PCIs can be lengthy procedures and result in significant blood loss from the Y-connector. Using a Y-connector with a hemostatic valve (such as the Co-pilot, Abbott Vascular or Guardian, Vascular Solutions; Fig. 2.4) can help minimize blood loss from back bleeding (which is particularly important for larger guide catheters, such as 8 Fr) and is easier to use compared with standard rotating hemostatic valves.

Co-pilot **Guardian**

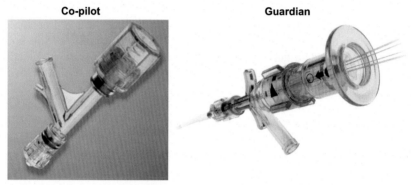

FIGURE 2.4   Types of Y-connectors with hemostatic valves. *Image of Co-pilot courtesy of Abbott Vascular. ©2016/2017 Abbott. All rights reserved.*

2.3.5 Guide Catheter Extensions

Three guide catheter extensions are currently available in the United States: the GuideLiner V3 catheter (Vascular Solutions, Fig. 2.5), the Trapliner (Vascular Solutions, Fig. 2.6) and the Guidezilla II (Boston Scientific, Fig. 2.7). The Trapliner is a rapid exchange guide catheter extension with a guidewire trapping balloon. Another guide catheter extension, the Guidion (Interventional Medical Device Solutions, Fig. 2.8) is available in Europe (Table 2.2).

GuideLiner V3 Dimensions

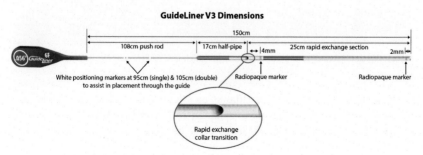

FIGURE 2.5   Illustration of the GuideLiner V3 catheter.

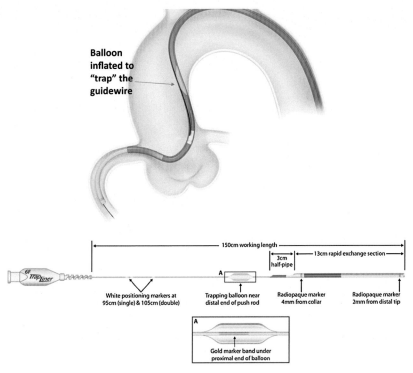

FIGURE 2.6   Illustration of the Trapliner catheter.

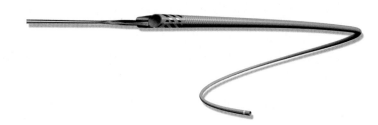

FIGURE 2.7   Illustration of the Guidezilla II catheter. *Image provided courtesy of Boston Scientific. ©2017 Boston Scientific Corporation or its affiliates. All rights reserved.*

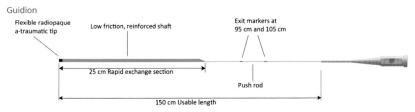

FIGURE 2.8   Illustration of the Guidion catheter. *Reproduced with permission from IMDS.*

TABLE 2.2 Overview of Guide Catheter Extensions

Name	Sizes (Fr)	Internal Diameter	Total Length (cm)	Distal Cylinder Length
GuideLiner V3	5	0.046″ (1.17 mm)	150	25 cm
	5.5	0.051″ (1.30 mm)		XL: 40 cm
	6	0.056″ (1.42 mm)		
	7	0.062″ (1.57 mm)		
	8	0.071″ (1.80 mm)		
Trapliner	6	0.056″ (1.42 mm)	150	13 cm
	7	0.062″ (1.57 mm)		
	8	0.071″ (1.80 mm)		
Guidezilla II	6	0.057″ (1.45 mm)	145	25 cm
	7	0.063″ (1.60 mm)		XL: 40 cm
	8	0.072″ (1.83 mm)		
Guidion	5	0.041″ (1.04 mm)	150	25 cm
	6	0.056″ (1.42 mm)		
	7	0.062″ (1.57 mm)		
	8	0.071″ (1.80 mm)		

All guide catheter extensions consist of a push rod and a distal cylinder (25 cm in the GuideLiner V3 and the Guidezilla II, and 13 cm in the Trapliner) that are advanced into the coronary vessel. They are manufactured in various sizes (Table 2.2) to fit various guide catheters, resulting in an inner diameter that is approximately 2 Fr smaller than that of the guide catheter. In addition to the rod and cylinder, the Trapliner also has a balloon proximal to the proximal collar that allows trapping of equipment (Fig. 2.6).

Guide Catheter Extensions Tips and Tricks

1. To minimize the risk of the guidewire wrapping around the guide catheter extension shaft, after inserting the guide catheter extension the external push rod should be placed into a towel at the side of the Y-connector (Fig. 2.9).
2. Advancing the guide catheter extension may be easier to achieve by inflating a balloon halfway inside the distal part of guide catheter extension cylinder

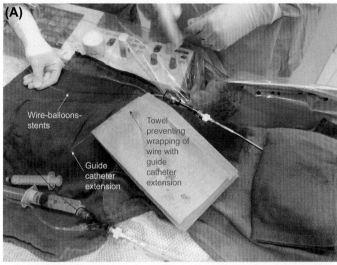

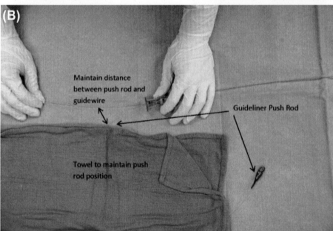

FIGURE 2.9 (A) Guide catheter extension manipulation to minimize the risk for guidewire wrap-around. (B) *Photo from Chad Kugler* showing how to angle the guideliner away from the wire in a stationary position to avoid wire wrap inhibiting balloon/stent entry into the back of the guideliner.

and the vessel (Fig. 2.10). The guide catheter extension is then advanced upon balloon deflation. Advancement over a balloon catheter or microcatheter is preferred to advancement over a 0.014″ coronary wire to minimize the risk of catching a plaque edge and causing a dissection (see Online Case 44).

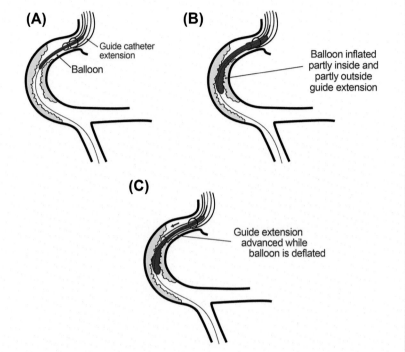

FIGURE 2.10 Delivery of a guide catheter extension to the target coronary segment using a balloon.

3. Guide catheter extension advancement may also be facilitated by use of a dedicated dilator (GuideLiner Navigation catheter, Vascular Solutions, Figs. 2.11 and 2.12).

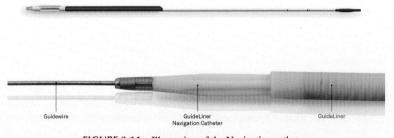

FIGURE 2.11 Illustration of the Navigation catheter.

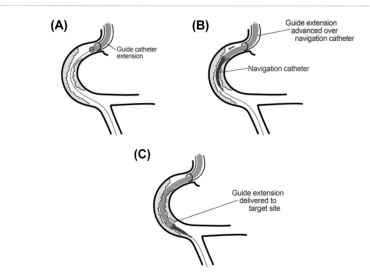

FIGURE 2.12   Delivery of a guide catheter extension to the target coronary segment using the Navigation catheter (Vascular Solutions).

4. Deformation of guidewires, stents, or other equipment can occur during advancement through the guide catheter extension collar (Fig. 2.13A).[7] To avoid this, it may be necessary in some cases to advance the stent or other equipment into the guide extension catheter outside the body and then introduce everything as a single unit into the guide catheter, or use fluoroscopic guidance to visualize the stent as it enters the guide catheter extension collar.

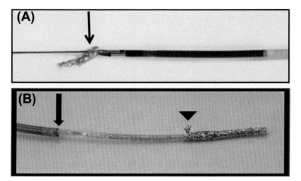

FIGURE 2.13   Complications of stent delivery through guide catheter extensions. (A) Stent deformation while attempting to deliver it through a GuideLiner catheter. (B) Stent deformation while attempting to retrieve an undeployed stent into the distal tip of a Guidezilla catheter. The tip of the Guidezilla prolapsed on itself (*arrow*) during attempted stent retrieval, resulting in catching the proximal edge of the stent and causing deformation (*arrowhead*). *(A) Reproduced with permission from Papayannis AC, Michael TT, Brilakis ES. Challenges associated with use of the GuideLiner catheter in percutaneous coronary interventions. J Invasive Cardiol 2012;**24**:370–71; (B) Courtesy of Dr. William Nicholson.*

5. Deformation of stents or other equipment can also occur while withdrawing the equipment back into the distal tip of the catheter after a failed attempt to advance to the target lesion, as shown in Fig. 2.13B.
6. Attempts to advance guidewires through a guide that contains a guide catheter extension smaller than the size of the guide (e.g., a 6 Fr extension within an 8 Fr guide) should be avoided, as the wire is likely to advance between the guide catheter extension cylinder and the guide catheter wall.
7. Deep advancement of the guide catheter extension may cause coronary dissection.[8]
8. During retrograde interventions, a guide catheter extension can be advanced through the antegrade guide catheter to facilitate the reverse controlled antegrade and retrograde tracking and dissection (CART) technique ("GuideLiner Reverse CART," as described in Chapter 6, Fig. 6.27). Guide catheter extensions can also be used from the retrograde side to increase support for retrograde gear delivery.
9. Every effort should be undertaken to minimize pressure dampening, but if dampening occurs, it is important to verify that adequate antegrade flow is preserved and no vessel injury has occurred before proceeding with the intervention.[8] Injection through a guide catheter extension with dampened pressure waveform may cause dissection that can propagate either antegrade or retrograde (Fig. 2.14). It is important to hold the guide catheter extension shaft during forceful contrast injection to minimize the risk of ejecting the guide catheter extension![9]

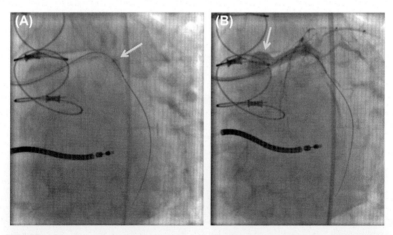

FIGURE 2.14 Retrograde dissection caused by contrast injection through a guide catheter extension with dampened waveform (also video is provided, click here). (A) Guide catheter extension with its tip (*arrow*) deep-seated into the circumflex artery. (B) Retrograde dissection into aortic root (*arrow*) from contrast injection through the deep seated guide catheter extension.

10. Although distal to proximal stenting is preferred, proximal to distal stenting can be done if needed, followed by insertion of the guide catheter extension through the proximal stent to allow distal stent delivery.[10]
11. Although coaxial alignment of the guide catheter is ideal, the guide catheter extension may be particularly effective in facilitating vessel engagement and equipment delivery when guide coaxial alignment is not possible, for example in anomalous coronary arteries (Fig. 2.15).[11] Distal anchoring (or use of the balloon technique shown in Fig. 2.10 or the Navigation catheter shown in Fig. 2.12) may be needed to deliver the guide catheter extension in such cases.[8]
12. The proximal collar of the guide catheter extension should not be advanced outside the guide catheter.

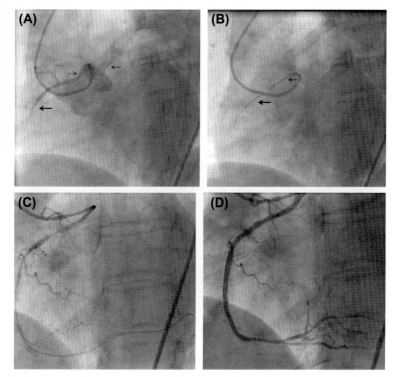

FIGURE 2.15   Use of the GuideLiner catheter for treating a CTO of an anomalous right coronary artery. (A) Chronic total occlusion (*arrow*) of an anomalous right coronary artery arising from the left sinus of Valsalva. (B) A GuideLiner catheter was placed over a Fielder guidewire with the support of an uninflated balloon kept in the proximal right coronary artery for better support. (C) The CTO was crossed with Confianza Pro 9 wire with the support of the GuideLiner. (D) Successful recanalization of the right coronary artery with TIMI 3 flow. *Reproduced with permission from Senguttuvan NB, Sharma SK, Kini A. Percutaneous intervention of chronic total occlusion of anomalous right coronary artery originating from left sinus – use of mother and child technique using GuideLiner. Indian Heart J 2015;67(Suppl. 3):S41–42, Elsevier Publication.*

13. Because the guide catheter extension decreases the original guide size by 2 Fr, special attention to pressure dampening and to activated clotting time (ACT) is needed to decrease the risk of thrombus formation.

14. Trapping of equipment using a guide extension is difficult: (a) the trap balloon needs to be placed proximal to the proximal collar and (b) the equipment needs to be retracted to this location in order for successful trapping to occur. The Trapliner provides a clever solution to this problem by incorporating the trapping balloon into its push rod.

15. Guide catheter extensions are very flexible and can advance even through highly tortuous lesions.[12]

16. Two guide catheter extensions can be used simultaneously in a mother–daughter–granddaughter configuration (i.e., a 6 Fr extension through an 8 Fr extension) when multiple extreme bends need to be navigated, for example when performing PCI through angulated saphenous vein grafts (Fig. 2.16) (Online Case 87).[13]

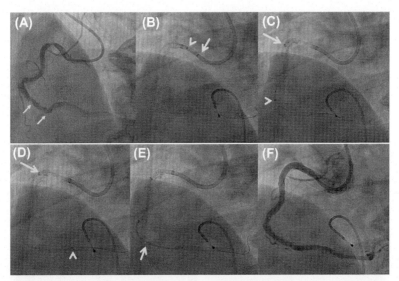

FIGURE 2.16 Illustration of the "mother–daughter–granddaughter" technique. (A) Diagnostic angiography demonstrating lesions in the distal right coronary artery (arrows), which was very tortuous. (B) A 6 Fr GuideLiner (arrowhead) is advanced inside an 8 Fr GuideLiner (arrow) into the guide catheter. (C) The 6 Fr GuideLiner (arrowhead) and the 8 Fr GuideLiner (arrow) are advanced into the right coronary artery. (D) The 6 Fr GuideLiner (arrowhead) is advanced past the distal right coronary artery lesion. (E) A stent (arrow) is delivered through the "mother–daughter–granddaughter" system to the distal right coronary artery. (F) Excellent final result after stent implantation. *Courtesy of Dr. William Nicholson.*

2.4 MICROCATHETERS AND SUPPORT CATHETERS

Antegrade CTO crossing should always be attempted using an over-the-wire system, i.e., a microcatheter or an over-the-wire balloon, because such a system:
a. Provides better support and increases wire tip stiffness, enhancing its penetration capacity (Fig. 2.17).
b. Allows reshaping of the guidewire tip.
c. Allows easy guidewire exchanges.
d. Protects the proximal part of the vessel from guidewire-induced injuries.

Microcatheters are preferred to over-the-wire balloons as they allow accurate visualization of their tip location (because the marker is located at the tip, whereas in small balloons (1.20–1.50 mm in diameter) the marker is located in mid-shaft and the tip is not angiographically visible (Fig. 2.18).

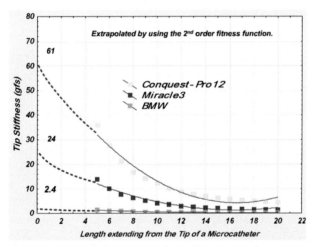

FIGURE 2.17   Change in guidewire tip stiffness with various guidewire lengths extending past a microcatheter tip. *Reproduced with permission from Waksman, Saito. Chronic total occlusions: a guide to revascularization. Wiley-Blackwell; 2013.*

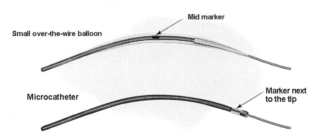

FIGURE 2.18   Comparison of over-the-wire balloons and microcatheters used for chronic total occlusion percutaneous coronary intervention.

Several microcatheters and support catheters are commercially available (Table 2.3). Some of the more commonly used microcatheters in CTO PCI are the following: Corsair, Corsair Pro, and Caravel (Asahi Intecc); Turnpike,

TABLE 2.3 Overview of Commercially Available Microcatheters and Support Catheters

Manufacturer	Catheter	Length	Distal Shaft Outer Diameter
Microcatheters			
Asahi Intecc	**Tornus**	**135 cm**	**2.1 and 2.6 Fr**
	Corsair and Corsair Pro[a]	**135 cm, 150 cm**	**2.6 Fr**
	Caravel[a]	**135 cm, 150 cm**	**1.9 Fr**
Boston Scientific	Renegade 18	105 cm, 115 cm, 135 cm	2.5 Fr
	Mamba	135 cm	2.3 Fr
	Mamba Flex	135 cm, 150 cm	2.1 Fr
Cordis	Transit	135 cm	2.5 Fr
	Prowler	150 cm	1.9 Fr
IMDS	NHancer Rx[b] (dual lumen)	135 cm	
	NHancer ProX[b]	135 cm, 155 cm	2.3 Fr
Kaneka	Crusade[b] (dual lumen)	140 cm	1.3 Fr distal tip 3.1 Fr crossing profile
	Mizuki[b]	135 cm, 150 cm	1.8 Fr distal tip 2.5 Fr shaft
	Mizuki FX[b]	135 cm, 150 cm	1.7 Fr distal tip 2.5 Fr shaft
Roxwood	MicroCross 14 and MicroCross 14 es	155 cm	1.6 Fr
Spectranetics	Quick Cross	135 cm, 150 cm	2.0 Fr
Terumo	Progreat	110 cm, 130 cm	2.4 and 2.7 Fr
	Finecross MG[a]	**130 cm, 150 cm**	**1.8 Fr**
	FineDuo[b] (dual lumen)	140 cm	

TABLE 2.3 Overview of Commercially Available Microcatheters and Support Catheters—cont'd

Manufacturer	Catheter	Length	Distal Shaft Outer Diameter
Microcatheters			
Vascular Solutions	Minnie	90 cm, 135 cm, 150 cm	2.2 Fr
	SuperCross	130 cm, 150 cm With preformed tip angle options of straight, 45, 90, or 120 degrees	2.1 Fr
	Venture	145 cm (rapid exchange) 140 cm (over-the-wire)	2.2 Fr
	Twin Pass and Twin Pass Torque (dual lumen)	140 cm	1.9 Fr distal tip 3 Fr crossing profile
	TurnPike[a]	**135 cm 150 cm**	**2.6 Fr**
	TurnPike LP[a]	**135 cm 150 cm**	**2.2 Fr**
	Turnpike Spiral[a]	**135 cm 150 cm**	**3.1 Fr**
	Turnpike Gold	135 cm	3.2 Fr
Volcano	Valet	135 cm 150 cm	1.8 Fr shapeable distal tip
Support Catheters			
Roxwood Medical	MultiCross	135 cm	
Roxwood Medical	CenterCross	135 cm	
Radius Medical	Prodigy	125 cm	
Nitiloop	NovaCross	135 cm	

[a]Most commonly used.
[b]Not available in the United States as of 2017.

Turnpike LP, Turnpike Spiral, and Turnpike Gold (Vascular Solutions); Finecross (Terumo); and Micro 14 (Roxwood Medical). Support catheters, such as the MultiCross and CenterCross (Roxwood Medical) are also available to increase the support of the guidewire and/or microcatheter during antegrade crossing attempts.

2.4.1 Over-the-Wire Balloons

Either a microcatheter or an over-the-wire balloon can be used to support antegrade CTO PCI. In general microcatheters are preferred because:
a. They allow better understanding of distal tip position (a marker is placed at the microcatheter tip, whereas in small balloons the marker is located in the middle of the balloon) (Fig. 2.18).
b. Are more flexible and track better than over-the-wire balloons.
c. Have less tendency to kink than over-the-wire balloons (kinking of balloon shaft prohibits future wire exchanges and often necessitates balloon catheter and wire removal and replacement with new gear, losing the crossing progress achieved). Over-the-wire balloons, however, may provide better support than many microcatheters.

2.4.2 Corsair and Corsair Pro

The Corsair microcatheter (Asahi Intecc, Fig. 2.19) was developed as a septal channel dilator to facilitate retrograde CTO PCI.[14] The Corsair proprietary Shinka shaft is constructed with eight thin wires wound with two larger wires, which facilitate torque transmission (Fig. 2.19). The inner lumen is lined with a polymer that enables contrast injection and facilitates wire advancement. The distal 60 cm of the catheter are coated with a hydrophilic polymer to enhance crossability. The tip is tapered and soft and is loaded with tungsten powder to enhance visibility. A platinum marker coil is placed 5 mm from the tip.

In the Corsair Pro (Fig. 2.20) the distal radiopaque marker band was removed, the tip flexibility was increased, and the hub was redesigned to encompass the proximal section of the catheter, reducing the likelihood for wire kinking and entrapment.

Corsair and Corsair Pro Tips and Tricks
1. Two Corsair lengths are currently available (135 cm long with light blue proximal hub and 150 cm long with dark blue proximal hub).
2. The Corsair can be used in the antegrade direction for wire support and exchange (usually the 135 cm length).
3. The Corsair is initially advanced by pushing; if difficulty in advancement is encountered then rotation of the catheter starts. Rotation of the catheter should be avoided when not needed, to reduce the risk for Corsair fatigue.
4. The Corsair catheter can be advanced by rotating in either direction, although it is braided to have better torque transmission with counterclockwise rotation. If resistance is encountered, a counterclockwise rotation combined with

forward push is the most powerful maneuver. However, the Corsair should not be overrotated (>10 consecutive turns without release), as overrotation could cause catheter deformation and entrapment, fracture proximal to the catheter tip, or result in the wire binding to the microcatheter (Corsair fatigue) (Fig. 2.21).

5. Rotation of the catheter with both hands and gentle antegrade pressure while the guidewire is kept in stable position allows for displacement of friction and tracking along the guidewire. This maneuver should be performed with pinning of the guidewire with the little finger to avoid inadvertent advancement. Advancing a Corsair retrogradely across a septal collateral may take several minutes (see Chapter 6, step 6 on options when a microcatheter does not advance through a collateral channel).

6. Contrast can be injected through the Corsair for distal vessel visualization, but the catheter should subsequently be flushed to minimize the risk for guidewire stickiness. Rarely the wire may get stuck, requiring removal of both Corsair and guidewire.

7. If difficulty is encountered while attempting to advance the Corsair catheter after prolonged use, the cause may be Corsair fatigue, and you should consider exchanging for a new Corsair. Also the Corsair tip may become flared and advance poorly, also requiring exchange for a new catheter (Chapter 6, step 6).

8. After wire externalization and during antegrade gear delivery, the tip of antegrade equipment (such as balloons and stents) should never come in contact with the tip of the retrograde Corsair catheter over the same guidewire to avoid interlocking and equipment entrapment (as described in Chapter 6 Step 9 and in Chapter 12).

2.4.3 Caravel

The Caravel microcatheter (Fig. 2.22) was developed to advance through small and tortuous collaterals. It has a very low distal tip profile (1.4 Fr) and low distal shaft profile (1.9 Fr) with a hydrophilic coating. It also has a braided shaft. It was designed to advance with forward push, but could also be rotated to cross challenging collaterals. However, the Caravel was not designed to withstand aggressive rotation and advancement. Such an approach can strain the distal tip connection to the shaft of the microcatheter and result in fracturing off its tip (Fig. 2.23) (Online Case 87).

2.4.4 Finecross

The Finecross (Terumo) microcatheter is very flexible and has a low crossing profile (1.8 Fr distal tip). The Finecross catheter has a stainless steel braid (to enhance torquability) and a distal marker located 0.7 mm from the tip (Fig. 2.24).

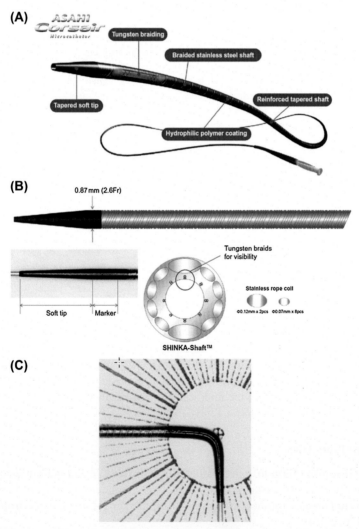

FIGURE 2.19 Illustration of the Corsair microcatheter. Overview (A) and construction (B) of the Corsair microcatheter. (C) The flexibility of the Corsair catheter distal tip. *Reproduced with permission from Asahi Intecc.*

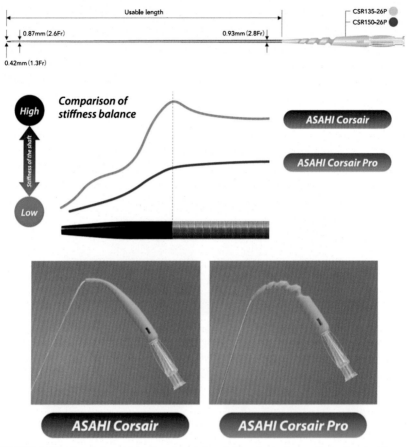

FIGURE 2.20   Comparison of the Corsair and Corsair Pro microcatheters. *Reproduced with permission from Asahi Intecc.*

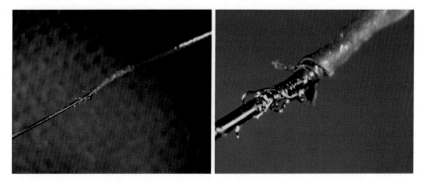

FIGURE 2.21   Images of the tip of a Corsair catheter permanently bound to a Pilot 200 guidewire with destruction of the wire's polymer jacket and entanglement of the wire coil by the tip of the Corsair catheter. *Courtesy of Dr. William Nicholson.*

FIGURE 2.22 Illustration of the Caravel microcatheter. *Reproduced with permission from Asahi Intecc.*

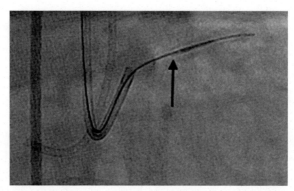

FIGURE 2.23 The tip of the Caravel microcatheter broke off from the remainder of the body of the catheter while attempting torqueing through a calcified stenotic segment of the left anterior descending artery. *Courtesy of Dr. William Nicholson.*

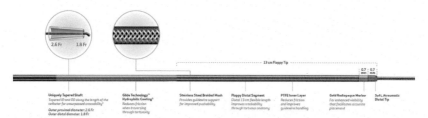

FIGURE 2.24 Construction of the Finecross MG microcatheter. *Image provided by Terumo Medical Corporation.*

2.4.5 Micro 14

The Micro 14 (Roxwood Medical) microcatheter is the longest (155 cm long) and one of the lowest crossing profile microcatheters currently available (1.6 Fr distal tip) (Fig. 2.25). It has variable pitch braid, and a hydrophilic coating. It is available in two versions, Micro 14 and Micro 14 es (extra support). Micro 14 is more flexible for advancing through tortuosity or for retrograde crossing, whereas the Micro 14 es is designed for enhanced antegrade crossing.

FIGURE 2.25 Construction of the Micro 14 microcatheter. *Reproduced with permission from Roxwood Medical.*

Micro 14 Catheter Tips and Tricks

1. The longer length of this microcatheter can be particularly useful in retrograde cases through bypass grafts or through very long or distal collaterals, especially when 100-cm long guide catheters are being used.
2. Rotation of the catheter is possible but the catheter is designed to advance mainly by pushing.

2.4.6 Turnpike, Turnpike LP, Turnpike Spiral, Turnpike Gold

The Turnpike (Vascular Solutions) has a dual layer bidirectional coil (Fig. 2.26) that facilitates torque transmission while allowing flexibility and preventing kinking. It also has a soft, tapered tip facilitating collateral branch crossing.

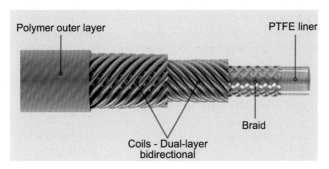

FIGURE 2.26 Construction of the Turnpike microcatheter.

The Turnpike catheter is produced in four versions (Fig. 2.27): Turnpike, Turnpike LP, Turnpike Spiral, and Turnpike Gold.

Turnpike is the standard catheter with a 1.6 Fr outside diameter at the distal tip and 2.6 Fr outside distal shaft diameter.

Turnpike LP is a lower profile version with 1.6 Fr outside diameter at the distal tip and 2.2 Fr outside distal shaft diameter.

The Turnpike Spiral has a distal nylon coil and the Turnpike Gold has a gold-plated, threaded metallic tip and distal nylon coil to increase trackability.

Similar to the Tornus catheter, the Turnpike Spiral and Gold are well suited for crossing balloon uncrossable lesions (Section 2.8 and Chapter 8).

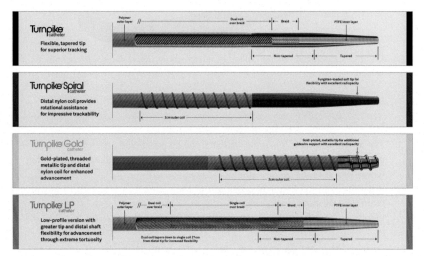

FIGURE 2.27 Illustration of the Turnpike microcatheters.

Turnpike Catheter Tips and Tricks

1. The Turnpike and Turnpike LP catheters can be rotated in either direction. In contrast the Turnpike Spiral and Gold are rotated clockwise to advance and counterclockwise for withdrawal (opposite direction compared with the Tornus catheter).
2. The Turnpike LP catheter is the go-to catheter for many operators for retrograde crossing of tortuous epicardial collaterals.
3. Some operators routinely use the Turnpike Spiral for antegrade crossing attempts, as it provides strong support.

2.4.7 Mamba and Mamba Flex

The Mamba and Mamba Flex catheters (Boston Scientific, Fig. 2.28) have a flexible, tapered shaft braid (11 wires are tightly wound on the proximal end to provide stiffness, torque, and pushability and then taper to allow for a lower profile and optimized flexibility at the distal end). They both have a durable, lubricious hydrophilic coating on the distal 60 cm.

The Mamba is only available in 135 cm length, has a higher distal crossing profile (0.032″, 0.81 mm), and is designed to provide strong guidewire support for antegrade crossing. The Mamba Flex is available on both 135 and 150 cm length, has a lower distal crossing profile (0.028″, 0.71 mm), and is designed to advance through tortuosity and for retrograde crossing.

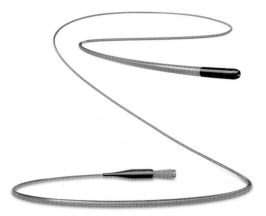

FIGURE 2.28   Illustration of the Mamba microcatheter.

2.4.8 Mizuki

The Mizuki catheter (Kaneka, Fig. 2.29) is available in two versions, one with standard tip stiffness (Mizuki) and one with a more flexible tip (Mizuki FX). It has hydrophilic coating and fluoro-resin inner surface for lubricity.

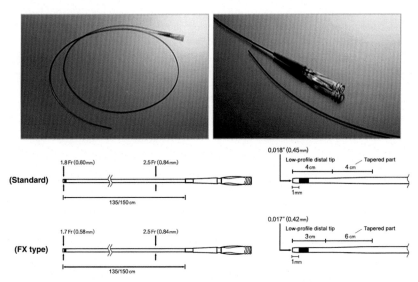

FIGURE 2.29   Illustration of the Mizuki and Mizuki FX microcatheters. *Provided by Kaneka.*

2.4.9 NHancer ProX

The NHancer ProX catheter (IMDS, Fig. 2.30) is available in 135 and 155 cm lengths, and has a soft, tapered tip and tip-to-hub variable braid.

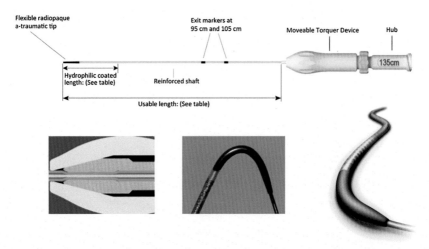

FIGURE 2.30   Illustration of the NHancer Pro X microcatheter. *Courtesy of IMDS.*

2.4.10 Venture (Online Cases 48, 96, 97)

The Venture catheter (Vascular Solutions, Fig. 2.31) has an 8 mm radiopaque tor-quable distal tip that has a bend radius of 2.5 mm.[15–20] The tip can be deflected up to 90 degrees by clockwise rotation of a thumb wheel on the external handle. With rotation of the entire catheter, steering in all planes is possible. It is compatible with 6 Fr guiding catheters and with 0.014″ guidewires. Both a rapid exchange and an over-the-wire catheter are available, but only the over-the-wire Venture catheter should be used for CTO PCI, as it allows for wire exchanges.

FIGURE 2.31   Illustration of the over-the-wire Venture catheter. *Reproduced with permission from Vascular Solutions.*

Venture Tips and Tricks

1. The Venture catheter has a deflectable tip, which can be utilized to assist with accessing difficult side branch vessels. As shown in Fig. 2.32, the catheter design allows the operator to rotate the tip deflector twist knob to transmit increasing tip deflection to the distal tip of the microcatheter.

2. Usually the Venture catheter is delivered to the target vessel in a straight configuration over a workhorse guidewire (Fig. 2.33A and B). Once it reaches the target coronary segment the workhorse guidewire is withdrawn inside the Venture catheter (Fig. 2.33C) and the tip deflector twist knob is clockwise

rotated to deflect the catheter tip. The deflected catheter is rotated and with-drawn until it points to the target vessel and lesion (Fig. 2.33D), followed by CTO guidewire advancement into the occlusion (Fig. 2.33E). The Venture catheter can then be removed leaving the guidewire in place (Fig. 2.33F).

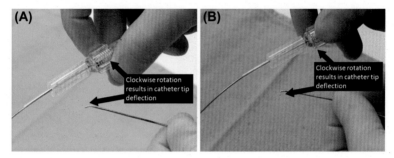

FIGURE 2.32 Illustration of Venture catheter manipulation. *Courtesy of Dr. William Nicholson.*

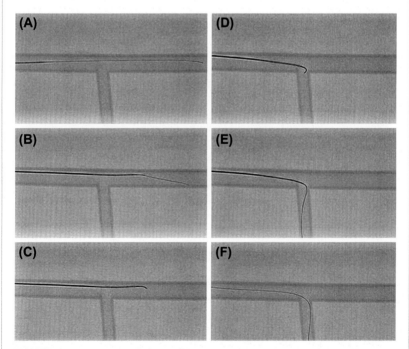

FIGURE 2.33 Illustration of the use of the Venture catheter. *Courtesy of Dr. William Nicholson.*

3. The classic example for use of the Venture catheter is to treat ostial circumflex CTOs (Fig. 2.34).

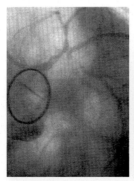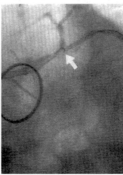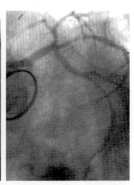

FIGURE 2.34 Example of Venture catheter use to cross an ostial circumflex chronic total occlusion. *Reproduced with permission from McNulty E, Cohen J, Chou T, Shunk K. A "grapple hook" technique using a deflectable tip catheter to facilitate complex proximal circumflex interventions.* Catheter Cardiovasc Interv *2006;67:46–8.*

4. The Venture catheter can also prevent the guidewire from prolapsing into a side branch when CTO penetration is challenging.[21]
5. During retrograde CTO PCI, the Venture catheter can be used to enable wiring of collateral branches with a difficult, acutely angulated takeoff.
6. Removal of the Venture catheter using a trapping balloon technique requires the use of 8 Fr guide catheters, given its larger profile as compared with an over-the-wire balloon or other microcatheters.[20] For the same reason an 8 Fr guide catheter is needed to perform the parallel-wire technique when one of the wires is inserted through the Venture catheter.
7. The Venture catheter is stiff, which can be both an advantage and disadvantage, as it can provide extra support, but can also predispose to vessel injury (Online Case 97). Since the bend radius is 2.5 mm, special care must be exercised when deflecting the tip in <2.5 mm diameter arteries.
8. The Venture catheter bend should be released and the tip straightened during advancement or removal to prevent vessel damage.

2.4.11 Dual Lumen Microcatheters

Only the Twin Pass (Vascular Solutions, Fig. 2.35A) and Twin Pass Torque (Vascular Solutions, Fig. 2.35B) dual lumen microcatheters are available in the United States in 2017. The Twin Pass Torque has a braided shaft that facilitates torqueing and positioning of the over-the-wire port, which has a 10 degrees exit angle and is located 1.5 mm distal to the proximal marker. Three other dual lumen microcatheters are available outside the United States: Crusade (Kaneka, Fig. 2.35C), FineDuo (Terumo, Fig. 2.35D), and NHancer Rx (IMDS, Fig. 2.35E).

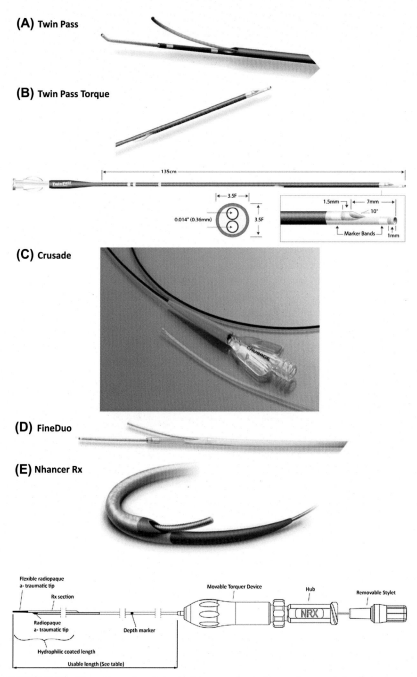

(A) Twin Pass

(B) Twin Pass Torque

(C) Crusade

(D) FineDuo

(E) Nhancer Rx

FIGURE 2.35 Illustration of various dual lumen microcatheters. *Image of the Crusade was provided courtesy of Kaneka. Image of the FineDuo was provided courtesy of Terumo Medical Corporation. Image of the Nhancer Rx was provided courtesy of IMDS.*

All dual lumen microcatheters consist of a rapid exchange delivery system in the distal segment together with an over-the-wire lumen that runs the length of the catheter. A radiopaque marker band identifies the distal tip of each lumen; the distal band corresponds to the exit point of the rapid exchange segment and the proximal band marks the exit point of the over-the-wire lumen.

Dual lumen microcatheters have multiple uses in CTO and non-CTO PCIs (Fig. 2.36):

1. Parallel wiring (Chapter 4, step 7b).
2. Wiring CTOs with a side branch adjacent to the proximal cap.
3. Wiring the distal true lumen if guidewire crossing is achieved into a side branch near the distal cap: instead of pulling back and redirecting the guidewire (risking inability to recross the occlusion), the dual lumen microcatheter enables wiring the main distal vessel without losing access to the side branch.
4. Wiring the side branch of a bifurcation, including wiring through jailed side branches during bifurcation stenting.
5. Facilitating the reversed guidewire (otherwise called hairpin guidewire) technique.
6. Wiring various septal branches during retrograde crossing attempts.
7. Antegrade wiring of the distal true lumen if the externalized retrograde guidewire crossed a collateral in close proximity to the distal cap, precluding safe antegrade dilation of the CTO over the retrograde wire.

Dual Lumen Microcatheters Tips and Tricks

1. Dual lumen microcatheters sometimes may be challenging to deliver. Some of them come with a stiffening mandrel (NHancer Rx) that, when loaded into the OTW lumen, can provide support and pushability during catheter insertion.
2. Controlling the direction of the side port can be challenging and may require removing the microcatheter and reinserting it. Controlling the direction may be easier with the Twin Pass Torque microcatheter.
3. Once successful wiring is achieved, maintaining distal wire position while withdrawing the dual lumen microcatheter can be very difficult and is best achieved with the assistance of a trapping balloon.

2.4.12 SuperCross

The SuperCross (Fig. 2.37) (Vascular Solutions) is a microcatheter that is manufactured with either a straight tip or with various distal tip angulations (45, 90, and 120 degrees) and can facilitate guidewire advancement through areas of tortuosity. The SuperCross microcatheters can be helpful in scenarios similar to those in which the Venture Catheter is employed. The 120 degrees bend can be

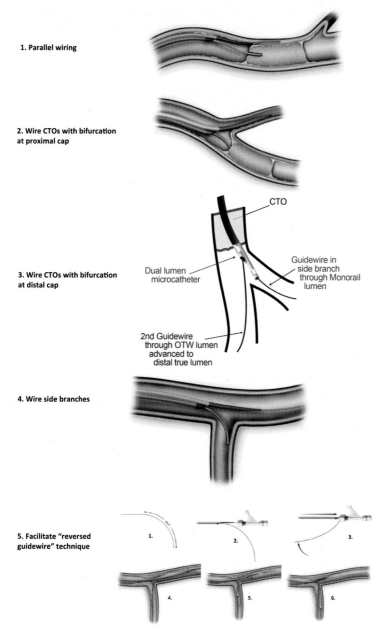

1. Parallel wiring

2. Wire CTOs with bifurcation at proximal cap

3. Wire CTOs with bifurcation at distal cap

Dual lumen microcatheter

CTO

Guidewire in side branch through Monorail lumen

2nd Guidewire through OTW lumen advanced to distal true lumen

4. Wire side branches

5. Facilitate "reversed guidewire" technique

FIGURE 2.36   Various uses of the dual lumen microcatheters.

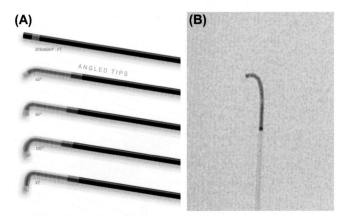

FIGURE 2.37   **Illustration of the SuperCross catheter.** (A) Various types of SuperCross angulation. (B) Angiographic appearance of the SuperCross microcatheter.

very helpful when trying to make the turn antegrade up the target vessel after reaching the anastomotic site of reverse angled septal and epicardial collaterals. When using a saphenous vein graft as the retrograde conduit, making the turn antegrade up the target vessel when reaching the anastomotic site can also be facilitated by using the SuperCross 120 degrees bend microcatheter.

2.4.13 MultiCross and CenterCross

The MultiCross (Fig. 2.38) and CenterCross (Fig. 2.39) (Roxwood Medical) are two support catheters that provide anchoring to the vessel wall. Both have a stabilizing self-expanding scaffold (expands up to 4.5 mm in diameter) that is deployed proximal to the target lesion. The Multicross contains three separate lumens within the scaffold, each located 120 degrees apart.[22] The CenterCross has a single, large central lumen that can accommodate a microcatheter. These catheters increase the backup support and the penetration force to cross the proximal cap and traverse through the occlusion.

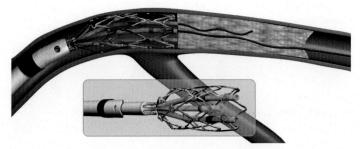

FIGURE 2.38   Illustration of the MultiCross catheter. *Reproduced with permission from Roxwood Medical.*

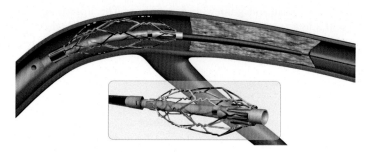

FIGURE 2.39   Illustration of the CenterCross catheter. *Reproduced with permission from Roxwood Medical.*

Multicross and Centercross Tips and Tricks

1. The CenterCross inner lumen can accommodate all ≥150 cm long single lumen microcatheters discussed earlier (Corsair, Caravel, Finecross, Turnpike, Turnpike LP, Micro14, etc.).
2. Due to their profile, neither the CenterCross nor MultiCross can be exchanged using the trapping technique, hence long (300 cm) guidewires should be used to remove them from the guide catheter.
3. Both catheters require at least 10 mm proximal landing zone in the target coronary vessel.

2.4.14 Prodigy

Similar to MultiCross and CenterCross, the Prodigy catheter (Fig. 2.40, Radius Medical) was designed to provide strong guidewire support.[23] It consists of a 5 Fr catheter with a soft, atraumatic elastomeric balloon at its distal tip that can be expanded up to 6 mm in diameter. The inflation lumen has a pressure relief valve that limits inflation pressure to 1 mmHg, anchoring the catheter in place while minimizing the risk for proximal vessel injury and allowing enhanced guidewire pushability.

2.4.15 NovaCross

The NovaCross catheter (Nitiloop) has a 10 mm long flexible Nitinol element, which upon axial compression, deforms by curving outward several helical struts (Fig. 2.41).[24] The extent of compression is controlled by the operator and correlates with the curvature of the nitinol struts so that the greater the compression the greater the radial protrusion of the struts. The fine nitinol loops are highly elastic, allowing gentle abutment of the distal end of the microcatheter to the vessel lumen. Once penetration of the guidewire is achieved, a 0.6 mm in diameter inner portion of the NovaCross can be extended distally up to 4 cm, to assist in interocclusion guidewire penetration. The upcoming NovaCross Xtreme will allow for the nitinol element to remain deployed while advancing the microcatheter inner portion through the lesion, thus improving interocclusion crossing.

Prodigy catheter

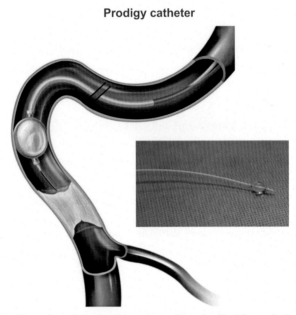

FIGURE 2.40   Illustration of the Prodigy catheter. *Reproduced with permission from Radius Medical.*

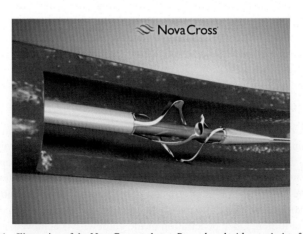

FIGURE 2.41   Illustration of the NovaCross catheter. *Reproduced with permission from Nitiloop.*

2.5 GUIDEWIRES

This area has the largest number of options, as well as personal preferences (Table 2.4).

However, many high-volume operators advocate limiting the options down to a few must-have wires (listed in no particular order) (Fig. 2.42):

Must-have Wires for CTO PCI

1. **Wires for crossing microchannels and knuckle formation**
    The Fielder XT (Asahi Intecc) or Fighter (Boston Scientific) are soft, polymer-jacketed, tapered wires for initial antegrade crossing via microchannels and for knuckle formation during antegrade or retrograde dissection and reentry.
2. **Wires for penetration**
    - The Gaia guidewires have a stiff, tapered tip with moderate penetrating power and excellent torquability and are used for CTO crossing attempts if the course of the target vessel is well understood.
    - The Confianza Pro 12 (Asahi Intecc) is a stiff, tapered-tip, penetration wire for crossing resistant caps.
    - The Pilot 200 (Abbott Vascular) is a polymer-jacketed and moderately stiff, nontapered tip wire, used for crossing when the course of the target lesion and vessel is uncertain. The Pilot 200 can also be used for knuckle formation, but it makes larger knuckles than the Fielder XT wire.
3. **Wires for crossing collateral channels**
    The Sion (hydrophilic, highly torquable soft guidewire with excellent shape retention, Asahi Intecc), Fielder FC, XT-R, and Sion black (polymer-jacketed soft wire, Asahi Intecc) and the Suoh 03 (very soft tip for epicardial collaterals) are used for wiring collaterals during retrograde crossing.
4. **Wires for externalization**
    The RG3 (330 cm long, Asahi Intecc) and R350 (350 cm long, Vascular Solutions) wires are used for externalization.

Except for the retrograde and externalization guidewires, many operators currently use only short (180–190 cm) guidewires, and use the trapping technique to remove or exchange the microcatheters or over-the-wire balloons (Section 3.7). However, for operators performing transradial CTO PCI availability of 300 cm long guidewires and guidewire extensions is important, as the trapping technique for exchanging over-the-wire to rapid exchange equipment may not always be feasible through a 6 Fr guide catheter, especially if a guide catheter extension is utilized. The 6 Fr Trapliner can be helpful in these cases as it allows easy trapping in 6 Fr guides.

2.5.1 Fielder, Gladius, and Fighter Guidewires

The Fielder family of guidewires (Fig. 2.43, Asahi Intecc) are polymer-jacketed, soft-tip guidewires, with either tapered (Fielder XT, 0.009″ taper) or nontapered (Fielder FC) distal tip.

TABLE 2.4 Description of Coronary Guidewires Commonly Utilized in Chronic Total Occlusion Percutaneous Coronary Intervention

Wire Category	Tip Style	Commercial Name	Tip Stiffness (g)	Manufacturer	Properties
Polymer Jacket					
	Tapered	**Fielder XT**[a] **Fielder XT-A** **Fielder XT-R**	**0.8** **1.0** **0.6**	**Asahi Intecc**	Front-line wire for antegrade crossing. Can also be used for knuckle wire formation and for retrograde crossing. Fielder XT-R is designed for retrograde collateral crossing.
		Fighter[a]	**1.2**	**Boston Scientific**	
	Straight (nontapered), low tip stiffness	**Fielder FC**[a]	**0.8**	**Asahi Intecc**	Used to cross through collateral vessels during the retrograde approach.
		Sion Black	**0.8**	**Asahi Intecc**	
		Whisper LS, MS, ES	0.8, 1.0, 1.2	Abbott Vascular	
		Pilot 50	1.5	Abbott Vascular	
		Choice PT Floppy	2.1	Boston Scientific	
	Straight (nontapered), high tip stiffness	**Pilot 150|200**[a]	**2.7|4.1**	**Abbott Vascular**	Antegrade crossing, especially when the course of the occluded vessel is unclear. Also useful for knuckle wire formation and for reentry into true lumen during LAST technique.
		Gladius	3	Asahi Intecc	
		Crosswire NT	7.7	Terumo	
		PT Graphix Intermediate	1.7	Boston Scientific	
		PT2 Moderate support	**2.9**	**Boston Scientific**	
		Shinobi	7.0	Cordis	
		Shinobi Plus	6.8	Cordis	
Open Coil (No Polymer Jacket)					
	Straight, ultralow tip stiffness	**Suoh 03**[a]	0.3	**Asahi Intecc**	Designed for crossing tortuous epicardial collaterals.

	Product		Manufacturer	Comments
Straight, low tip stiffness	SION (hydrophilic)[a]	0.7	**Asahi Intecc**	Probably the most commonly used guidewire for collateral crossing.
	SION blue[a]	0.5	**Asahi Intecc**	Excellent workhorse guidewire.
Tapered, low tip stiffness	Samurai RC	1.2	Boston Scientific	
	Cross-it 100XT (0.010")	1.7	Abbott Vascular	
	Runthrough NS tapered (0.008")	1.0	Terumo	
Tapered, high tip stiffness, hydrophilic coating	**Gaia 1st (0.010″), 2nd (0.011″), and 3rd (0.012″)[a]**	**1.7, 3.5, 4.5**	**Asahi Intecc**	Antegrade crossing when vessel course is known. Can also be used for retrograde crossing.
	Confianza Pro 9, 12[a] (0.009″)	9.3, 12.4	**Asahi Intecc**	
	Astato 20 (0.008″)	20	Asahi Intecc	
	PROGRESS 140T, 200T (0.0105″, 0.009″)	12.5, 13.3	Abbott Vascular	
	Persuader 9 (0.011″)	9.1	Medtronic	
	ProVia 9, 12 (0.009″)	11.8, 13.5	Medtronic	
	Hornet 10, 14 (0.008″)	10, 14	Boston Scientific	
Straight tip, high tip stiffness	MiracleBros 3, 4.5, 6	3.9, 4.4, 8.8	Asahi Intecc	Antegrade crossing when vessel course is known.
	MiracleBros 12	13.0	Asahi Intecc	
	Ultimate 3	3	Abbott Vascular	
	PROGRESS 40, 80, 120	5.5, 9.7, 13.9	Abbott Vascular	
	Persuader 3, 6 (-philic and -phobic)	5.1, 8.0	Medtronic	
	Provia 3, 6 (-philic and -phobic)	8.3, 9.1	Medtronic	

Continued

TABLE 2.4 Description of Coronary Guidewires Commonly Utilized in Chronic Total Occlusion Percutaneous Coronary Intervention—cont'd

Wire Category	Tip Style	Commercial Name	Tip Stiffness (g)	Manufacturer	Properties
	Tapered, high tip stiffness, hydrophobic coating	Confianza 9 and 12 (hydrophobic)	8.6, 12	Asahi Intecc	Antegrade crossing when vessel course is known.
		Persuader 9 (hydrophobic)	9.1	Medtronic	
		ProVia 9, 12 (hydrophobic)	11.8, 13.5	Medtronic	
Externalization wires		RG3[a]		Asahi Intecc	330 cm in length.
		R350[a]		Vascular Solutions	350 cm in length.
Wiggle wire	Curved distal portion	Wiggle	1.0	Abbott Vascular	190 cm and 300 cm long.
Extra support guidewires	Soft, nontapered	Iron Man BHW	1.0	Abbott Vascular	190 cm and 300 cm long.
		Grand Slam	0.7	Asahi Intecc	180 and 300 cm long.
		Mailman		Boston Scientific	182 and 300 cm long.

[a]Most commonly utilized guidewires.

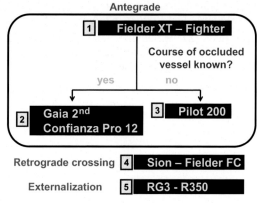

FIGURE 2.42   Simplified algorithm for guidewire selection.

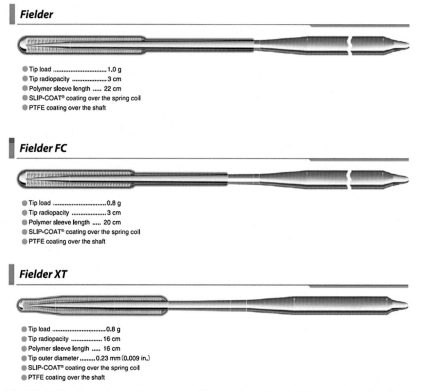

FIGURE 2.43   Illustration of the Fielder guidewires. *Reproduced with permission from Asahi Intecc.*

The Fielder XT-A and XT-R (Fig. 2.44) are composite core versions of the Fielder XT guidewire. The Fielder XT-A has slightly stiffer tip (1.0 g) than the Fielder XT (0.8 g) to enhance crossability, whereas the Fielder XT-R has softer tip (0.6 g) to enhance trackability. The Gladius guidewire (Asahi Intecc) is a stiffer (3 g tip) polymer-jacketed guidewire with composite core technology.

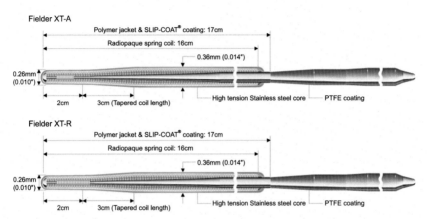

FIGURE 2.44 Illustration of the Fielder XT-A and XT-R guidewires. *Reproduced with permission from Asahi Intecc.*

The Fighter guidewire (Fig. 2.45, Boston Scientific) is a polymer-jacketed, soft-tip guidewire with a 0.009″ tapered tip.

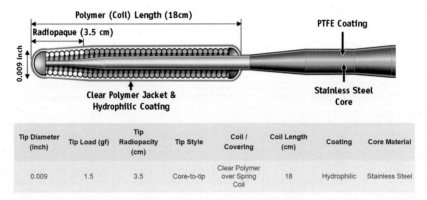

Tip Diameter (inch)	Tip Load (gf)	Tip Radiopacity (cm)	Tip Style	Coil / Covering	Coil Length (cm)	Coating	Core Material
0.009	1.5	3.5	Core-to-tip	Clear Polymer over Spring Coil	18	Hydrophilic	Stainless Steel

FIGURE 2.45 Illustration of the Fighter guidewire. *Image provided courtesy of Boston Scientific.* ©2017 Boston Scientific Corporation or its affiliates. All rights reserved.

Fielder and Fighter Guidewires: Tips and Tricks

1. The Fielder XT and XT-A and the Fighter guidewire are among the most commonly used initial guidewires for antegrade wire escalation.[25]
2. The Fielder XT and Fighter guidewires are also very useful for forming tight knuckles, both in the antegrade and the retrograde approach. Although knuckle formation is usually safe, there are reported cases of fracture of the knuckled guidewire.[26] When performing the knuckling technique, the wire should be pushed and not rotated to avoid wire knotting and entrapment.
3. The Fielder FC guidewire, that has a nontapered tip, may be useful in antegrade wiring through a visible microchannel, as it may be less likely to enter the subintimal space and cause dissection as compared with the Fielder XT. It is also useful for retrograde crossing via collateral channels, although many operators prefer the Sion or Sion Black as the wire of choice for this purpose.

2.5.2 MiracleBros, Ultimate, Confianza, Astato, Gaia, and Hornet Family of Wires

The MiracleBros wires (Asahi Intecc) are stiff, nontapered wires (up to 12 g distal tip stiffness) with high penetrating power (Fig. 2.46). These wires are nonhydrophilic (hydrophobic) and are favored for delivering gear to CTO segments, caps, or spaces, as they are less likely to slip out of place. The Ultimate 3 wire (Asahi Intecc) is a 3 g nontapered tip wire with composite core technology (Fig. 2.47).

The Confianza guidewires (Asahi Intecc) are stiff but also have a tapered tip (Fig. 2.48).

The Astato (Asahi Intecc) is a peripheral guidewire with tapered tip and 20 g distal tip stiffness that is occasionally used in hard-to-penetrate coronary lesions (Fig. 2.49).

The Gaia guidewires (Fig. 2.50, Asahi Intecc) are low-to-stiff, microcone tip guidewires that have a composite core designed to enhance torquability, maneuverability, and ability to cross long tortuous occlusions, but require slower, more meticulous and precise manipulation compared with other currently available CTO guidewires.[27–29] The distal 1 mm tip is preshaped to a 45 degrees angle. The Gaia wires are available in three types (Gaia First, Gaia Second, and Gaia Third) with increasing tip stiffness and tapered tip diameter.

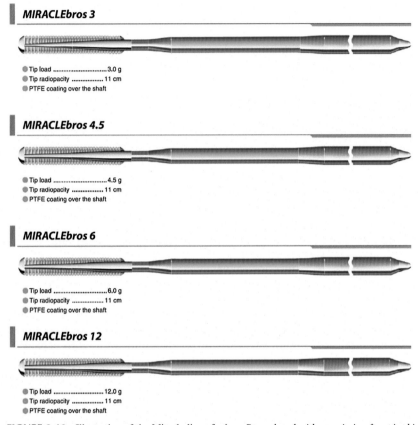

MIRACLEbros 3

- Tip load ....3.0 g
- Tip radiopacity .... 11 cm
- PTFE coating over the shaft

MIRACLEbros 4.5

- Tip load ....4.5 g
- Tip radiopacity .... 11 cm
- PTFE coating over the shaft

MIRACLEbros 6

- Tip load ....6.0 g
- Tip radiopacity .... 11 cm
- PTFE coating over the shaft

MIRACLEbros 12

- Tip load ....12.0 g
- Tip radiopacity .... 11 cm
- PTFE coating over the shaft

FIGURE 2.46 Illustration of the Miracle line of wires. *Reproduced with permission from Asahi Intecc.*

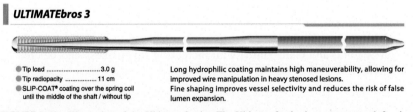

ULTIMATEbros 3

- Tip load ....3.0 g
- Tip radiopacity .... 11 cm
- SLIP-COAT® coating over the spring coil until the middle of the shaft / without tip

Long hydrophilic coating maintains high maneuverability, allowing for improved wire manipulation in heavy stenosed lesions.
Fine shaping improves vessel selectivity and reduces the risk of false lumen expansion.

FIGURE 2.47 Illustration of the Ultimate 3 wire. The Ultimate 3 wire has a nontapered, 3 g tip guidewire with composite core technology. *Reproduced with permission from Asahi Intecc.*

CONFIANZA

- Tip load ....9.0 g
- Tip radiopacity ....20 cm
- Tip outer diameter ....0.23 mm (0.009 inch)
- Silicone coating over the spring coil

Silicone coated and tapered to 0.23mm (0.009inch) at the tip. The 9g stiff, tapered tip helps to penetrate highly stenosed lesions.

CONFIANZA PRO

- Tip load ....9.0 g
- Tip radiopacity ....20 cm
- Tip outer diameter ....0.23 mm (0.009 inch)
- SLIP-COAT® coating over the spring coil, excluding the tip

Similar structure and tip stiffness as Conquest with SLIP-COAT® coating for lubricity. The distal tip is not coated to allow it to catch on the entry point of the lesions.

CONFIANZA PRO 12

- Tip load ....12.0 g
- Tip radiopacity ....20 cm
- Tip outer diameter ....0.23 mm (0.009 inch)
- SLIP-COAT® coating over the spring coil, excluding the tip

A tapered tip with 12g tip load. For penetration of calcification and proximal or distal thick, fibrous caps.

CONFIANZA PRO 8-20

- Tip load ....20.0 g
- Tip radiopacity ....17 cm
- Tip outer diameter ....0.20 mm (0.008 inch)
- SLIP-COAT® coating over the spring coil, excluding the tip

Designed for crossing complex lesions with heavy calcifications and tough fibrous tissues. The wire has a tip load of 20g and is tapered to 0.20mm (0.008inch). It is the finest and stiffest guidewire in the current Asahi series.

FIGURE 2.48 Illustration of the Confianza line of wires. The Confianza wires have a tapered tip and the Pro wires have hydrophilic coating. *Reproduced with permission from Asahi Intecc.*

Astato XS 20   0.014"

- Tip load .... 20.0 g
- Tip radiopacity .... 17 cm
- Total length 180cm, 300cm
- Total length 0.014inch
- SLIP COAT® hydrophilic coating over the spring coil
- PTFE coating over the shaft

FIGURE 2.49 Illustration of the Astato 20 guidewire. *Reproduced with permission from Asahi Intecc.*

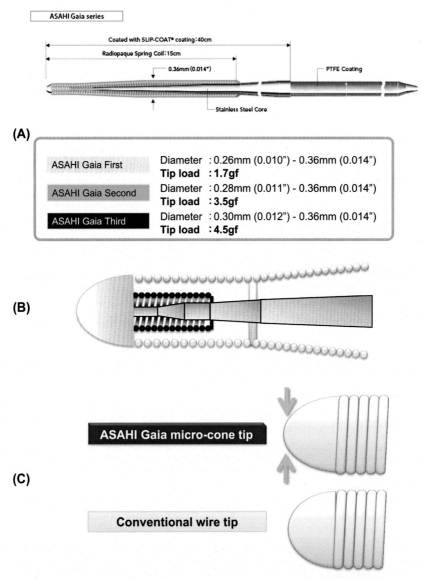

FIGURE 2.50 **Illustration of the Gaia guidewires.** (A) Construction and types of the Gaia guidewires. (B) Illustration of the composite core, dual-coil construction. (C) Illustration of the microcone tip of the guidewire. *Reproduced with permission from Asahi Intecc.*

Finally, the Hornet guidewires (Fig. 2.51, Boston Scientific) are stiff, tapered-tip (0.008″) guidewires with various degrees of penetration force depending on distal stiffness.

Miracle/Ultimate/Confianza/Gaia/Hornet Wires: Tips and Tricks

1. The Confianza Pro 12 guidewire can be very useful in puncturing a calcified proximal or distal CTO cap, given its high penetration power. Following puncture, step-down wiring should be considered by exchanging the Confianza Pro 12 for a less aggressive guidewire, such as the Ultimate 3 guidewire (escalation/deescalation concept as described in Chapter 4, step 5)
2. Given better trackability (due to composite core technology) the Ultimate 3 and Gaia wires are preferred over the Confianza Pro 12 for antegrade wire escalation, except in cases of very hard to penetrate proximal cap. The Gaia 2nd is the most commonly used Gaia guidewire.
3. The Confianza Pro 12 guidewire can be very useful when the proximal cap is located adjacent to a large side branch and initial wires keep deflecting into the side branch. Its stiff, tapered, uncoated tip allows it to be directed away from the side branch to grab tissue and engage and puncture the cap.
4. The Confianza Pro 12 guidewire can also be very useful for reentering into the true lumen (wire-based reentry, as described in Chapter 5, Section 5.6.2).
5. Because of its high penetration power, the Confianza Pro 12 wire should not be used in cases where the course of the CTO vessel is not well understood, as it may easily exit the vessel architecture and cause perforation and/or dissection.
6. The MiracleBros wires are very supportive and hydrophobic, with good tactile feedback and are the preferred wires (especially the Miracle Bros 12) for delivering the Stingray LP balloon to the reentry zone, as described in Chapter 5, Section 5.5, step 2).
7. Overrotation of these wires, especially in highly calcified lesions, can result in tip fracture and separation.

2.5.3 Pilot Family of Guidewires

Similar to the Fielder guidewires (Asahi Intecc), the Pilot guidewires (Abbott Vascular, Fig. 2.52) are polymer-jacketed, but have higher tip load (with the exception of the Pilot 50). The Pilot 200 guidewire is a 4.1 g tip guidewire and is the next preferred wire for wire escalation if initial CTO crossing attempts with a soft, tapered polymer-jacketed guidewire fail and the course of the target vessel is not well understood (Figure 2.42). The Pilot 200 can engage and traverse microchannels, occasionally resulting in true-to-true lumen crossing. It can also be used for knuckle formation both in the antegrade and the retrograde approach.

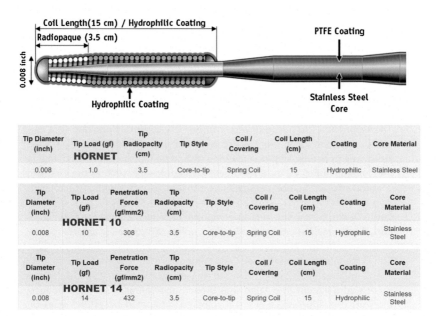

Tip Diameter (inch)	Tip Load (gf)	Tip Radiopacity (cm)	Tip Style	Coil / Covering	Coil Length (cm)	Coating	Core Material
HORNET							
0.008	1.0	3.5	Core-to-tip	Spring Coil	15	Hydrophilic	Stainless Steel

Tip Diameter (inch)	Tip Load (gf)	Penetration Force (gf/mm2)	Tip Radiopacity (cm)	Tip Style	Coil / Covering	Coil Length (cm)	Coating	Core Material
HORNET 10								
0.008	10	308	3.5	Core-to-tip	Spring Coil	15	Hydrophilic	Stainless Steel

Tip Diameter (inch)	Tip Load (gf)	Penetration Force (gf/mm2)	Tip Radiopacity (cm)	Tip Style	Coil / Covering	Coil Length (cm)	Coating	Core Material
HORNET 14								
0.008	14	432	3.5	Core-to-tip	Spring Coil	15	Hydrophilic	Stainless Steel

FIGURE 2.51 Illustration of the Hornet guidewires. *Image provided courtesy of Boston Scientific. ©2017 Boston Scientific Corporation or its affiliates. All rights reserved.*

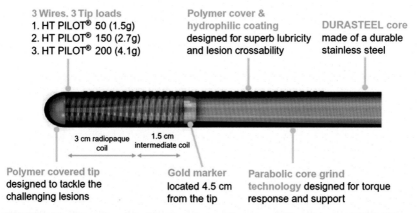

FIGURE 2.52 Illustration of the Pilot line of wires. *Courtesy of Abbott Vascular. ©2017 Abbott. All rights reserved.*

Pilot Wires: Tips and Tricks

1. The Pilot 200 guidewire is useful for knuckle formation, but because it is stiffer than Fielder XT, it tends to form a larger, wider knuckle that could cause larger subintimal hematomas.
2. Because of its transitionless core, the Pilot 200 is less likely to prolapse into side branches in its knuckled conformation. As such, it is often used as a knuckle wire when the initial knuckled wire continues to track a side branch pathway.

2.5.4 Sion, Sion Black, and Samurai RC

The Sion guidewire (Asahi Intecc) has a composite core and double coil technology (Fig. 2.53). These features provide excellent torque transmission and resistance to deformation.

The Sion Black (Fig. 2.54, Asahi Intecc) has the composite core technology and a 20 cm polymer jacket that greatly enhances its lubricity.

The Samurai RC (Fig. 2.55, Boston Scientific) has a tip-to-core construction. The Sion Black and the Samurai RC wires are useful for wiring both septal and epicardial collaterals.

2.5.5 Suoh 03 Guidewire

The Suoh 03 guidewire (Fig. 2.56, Asahi Intecc) is an ultrasoft tip (0.3 g) wire, specifically designed for crossing tortuous collaterals. It has a 19 cm long coil and a hydrophilic coating in its distal 52 cm.

2.5.6 Externalization Guidewires

There are two guidewires specifically designed for externalization. The RG3 wire (Fig. 2.57, Asahi Intecc) is 330 cm in length, has a 0.010 inch shaft with hydrophilic coating for more than half of its length, and has excellent resistance

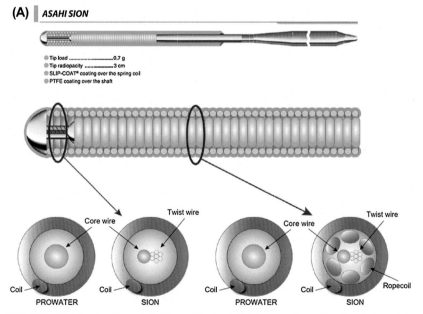

FIGURE 2.53 Construction of the Sion guidewire (A) and comparison with other Asahi guidewires (B). *Reproduced with permission from Asahi Intecc.*

(B)

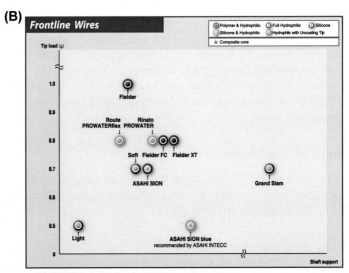

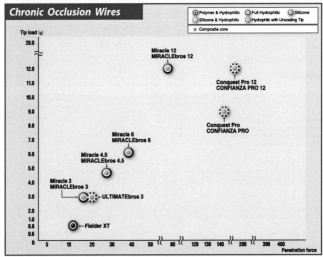

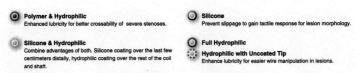

FIGURE 2.53 cont'd.

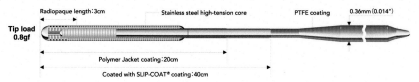

FIGURE 2.54 Illustration of the Sion Black guidewire. *Reproduced with permission from Asahi Intecc.*

SAMURAI RC Guidewire

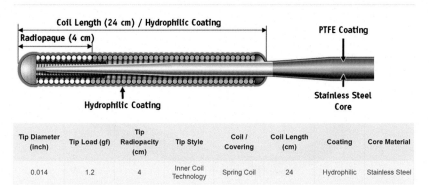

Tip Diameter (inch)	Tip Load (gf)	Tip Radiopacity (cm)	Tip Style	Coil / Covering	Coil Length (cm)	Coating	Core Material
0.014	1.2	4	Inner Coil Technology	Spring Coil	24	Hydrophilic	Stainless Steel

FIGURE 2.55 Illustration of the Samurai RC guidewire. *Image provided courtesy of Boston Scientific. ©2017 Boston Scientific Corporation or its affiliates. All rights reserved.*

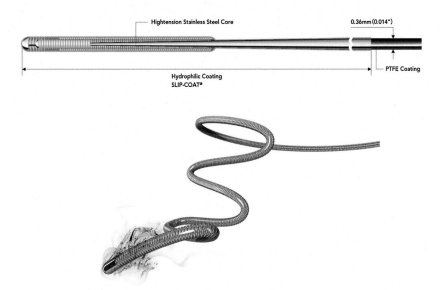

FIGURE 2.56 Illustration of the Suoh 03 guidewire. *Reproduced with permission from Asahi Intecc.*

Spring Coil : 8cm

Radiopaque Spring Coil : 3cm

Silicone Coating : 160cm

0.26mm (0.010″)

Coated with SLIP-COAT ⁺ Coating : 170cm

Usable Length : 330cm

FIGURE 2.57 Illustration of the RG3 guidewire. *Reproduced with permission from Asahi Intecc.*

to kinking. The R350 wire (Fig. 2.58, Vascular Solutions) is 350 cm long, has a 0.013 inch shaft and a 5 cm gold-plated tungsten coil on the distal end that is visible under fluoroscopy. The proximal 150 cm has a PTFE coating, whereas the distal 200 cm has a hydrophilic coating.

FIGURE 2.58 Illustration of the R350 guidewire.

Externalization Wires: Tips and Tricks

1. Externalization is easier when the tip of the retrograde microcatheter enters into the antegrade guide catheter or an antegrade guide catheter extension. Trapping the retrograde wire with a balloon in the antegrade guide facilitates advancement of the microcatheter into the antegrade guide.
2. Avoidance of wire kinking is very important to facilitate delivery of the externalized guidewire.
3. The wire introducer should be inserted through the hemostatic valve of the antegrade guide catheter to thread the externalization wire, followed by reattachment of the Y-connector to the guide catheter (Chapter 6, step 8).
4. The externalized guidewire may cause injury to the collateral circulation if the ends are pulled and the wire segment within the collateral is not protected by a microcatheter or over-the-wire balloon.
5. Managing the loop (between the guides) and actively maintaining the guide catheter positions is also critical. Avoid "choking" the heart, as this may cause global ischemia and hemodynamic collapse.

2.5.7 Wiggle Wire

The Wiggle wire (Fig. 2.59, Abbott Vascular) has a curved distal shaft, designed to facilitate equipment delivery through tortuosity and calcification.

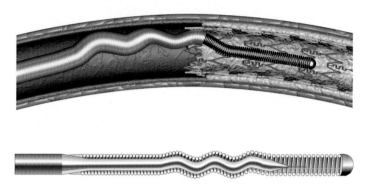

FIGURE 2.59 Illustration of the Wiggle wire. *Courtesy of Abbott Vascular. ©2016/2017 Abbott. All rights reserved.*

2.5.8 Extra Support Wires

Several guidewires with strong, supportive body and soft tip are available for facilitating equipment delivery (and guide catheter exchanges), such as the Iron Man, BHW (Abbott Vascular), Grand Slam (Asahi Intecc), and Mailman (Boston Scientific).

2.6 DISSECTION/REENTRY EQUIPMENT

Even though dissection/reentry can be accomplished using guidewires, the CrossBoss catheter, Stingray LP balloon, and Stingray guidewire (Boston Scientific) are preferred to improve success rates and procedural efficiency.

2.6.1 CrossBoss Catheter

The CrossBoss catheter (Fig. 2.60) is a stiff, metallic, over-the-wire catheter with a 1 mm blunt, rounded, hydrophilic-coated distal tip that can advance through the occlusion when the catheter is rotated rapidly using a proximal torque device (fast-spin technique) (Chapter 5 section 3). If the catheter enters the subintimal space, it creates a limited dissection plane making reentry into the distal true lumen easier. The risk of perforation is low provided that the CrossBoss catheter is not advanced into side branches. If the CTO is crossed subintimally, the Stingray LP balloon and guidewire can be used to assist with reentry into the distal true lumen, as described next.[21,30,31]

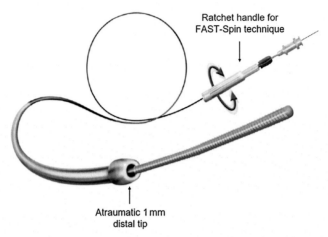

Ratchet handle for
FAST-Spin technique

Atraumatic 1 mm
distal tip

FIGURE 2.60 Illustration of the CrossBoss catheter. *Image provided courtesy of Boston Scientific. ©2017 Boston Scientific Corporation or its affiliates. All rights reserved.*

2.6.2 Stingray LP Balloon and Guidewire

The Stingray LP (low-profile) balloon is 2.5 mm wide when inflated and 10 mm in length and has a flat shape with two side exit ports. Upon low-pressure (2–4 atm) inflation it orients one exit port automatically toward the true lumen, especially when the space created by dissection is small (Fig. 2.61).[32] The Stingray guidewire is a stiff guidewire with a 20 cm distal radiopaque segment and a 0.009″ tapered tip with a 0.0035″ distal prong. The Stingray guidewire can be directed toward one of the two side ports of the Stingray LP balloon under fluoroscopic guidance to reenter the distal true lumen.[21,30,31]

CrossBoss and Stingray Tips and Tricks

1. It is best to avoid advancing stiff guidewires through the CrossBoss catheter, since the CrossBoss greatly enhances the guidewire penetration power and increases the risk for perforation.
2. Frequently, a knuckled wire is required to redirect the CrossBoss catheter away from side branches.
3. The CrossBoss catheter may cross from true to true lumen in approximately one-third of the cases.[33]
4. A guidewire should be left within the CrossBoss catheter during advancement to prevent retrograde blood entry into the CrossBoss lumen.
5. The CrossBoss may be particularly effective for crossing in-stent restenotic CTOs, as the stent strut may act as a barrier preventing advancement of the CrossBoss catheter behind the restenosed stent.[34,34a]

6. Inserting the Stingray guidewire into the hub of a balloon or microcatheter should be done using an introducer to prevent deformation of the distal guidewire tip.

7. The Gaia guidewires (that are preshaped with a similar bend to that of the Stingray guidewire) are excellent alternatives to the Stingray guidewire.

8. Extending a knuckled wire dissection plane for the final few centimeters with the CrossBoss will help manage the subintimal space ("finish with the Boss," Chapter 5, Section 4, Step 4.2). This minimizes subintimal hematoma formation and helps maintain close proximity to the true lumen at the target reentry zone, enhancing the likelihood of successful reentry.

9. In the right coronary artery CTO reentry should be attempted in a nontortuous, noncalcified segment closest to the reconstitution zone and proximal to bifurcations if possible.

10. The stick-and-swap technique (Chapter 5, Section 5, step 3.2) is frequently used for reentry: a stiff wire (such as the Stingray wire) is used to create a channel toward the distal true lumen followed by a polymer-jacketed guidewire (usually the Pilot 200) to track through the channel into the distal true lumen.

11. The double-blind stick-and-swap technique (Chapter 5, Section 5, step 3.3) is increasingly being performed instead of doing multiple angiographic projections for reentry into the distal true lumen.[35]

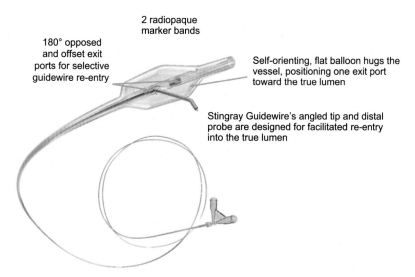

2 radiopaque marker bands

180° opposed and offset exit ports for selective guidewire re-entry

Self-orienting, flat balloon hugs the vessel, positioning one exit port toward the true lumen

Stingray Guidewire's angled tip and distal probe are designed for facilitated re-entry into the true lumen

Stingray Balloon
  • 6F (2.0mm) guide catheter compatible
  • 0.014" (0.36mm) & 0.018" (0.46mm) guidewire compatible
Stingray Guidewire
  • 0.014" (0.36mm) hydrophilic coated

FIGURE 2.61 Illustration of the Stingray LP catheter and Stingray guidewire. *Image provided courtesy of Boston Scientific©.*

2.7 SNARES

Snares are often needed to capture the retrograde guidewire into the antegrade guide. Three-loop (tulip) snares, such as the Ensnare (Merit Medical) and the Atrieve (Angiotech) are more effective in capturing the guidewire compared with the single-loop snares, such as the Amplatz Gooseneck snare (Covidien) (Fig. 2.62).

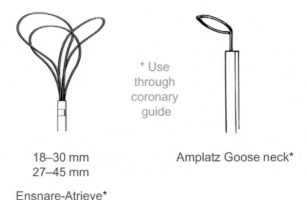

* Use through coronary guide

18–30 mm
27–45 mm

Amplatz Goose neck*

Ensnare-Atrieve*

FIGURE 2.62 Illustration of the three-loop and single-loop snares that are currently commercially available.

2.7.1 Snaring Technique (Fig. 2.63)

The snare is withdrawn and collapsed into the introducer tool (Fig. 2.63A–C). It is then introduced into the antegrade guide catheter (Fig. 2.63D and E) and advanced until it exits from the distal guide tip (Fig. 2.63F).

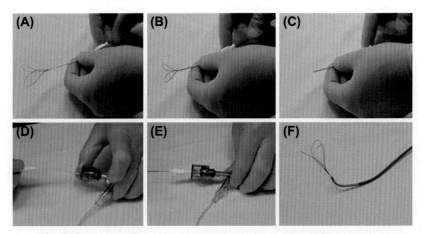

FIGURE 2.63 Illustration of snare preparation. *Courtesy of Dr. William Nicholson.*

The retrograde guidewire is advanced through the snare (Fig. 2.64A). The snare is withdrawn into the guide catheter along with the retrograde guidewire (Fig. 2.64B–D). The guidewire is pushed through the retrograde guide until it exits from the antegrade guide hub allowing the antegrade guide to be reseated into the antegrade coronary artery (Fig. 2.64E and F).

Snaring Tips and Tricks

1. Large snares (for example 27–45 or 18–30 mm Ensnare or Atrieve snares) are preferred to maximize the likelihood of capturing the retrograde guidewire.
2. Each snare comes with a delivery sheath, which is discarded as the antegrade guide catheter is used for snaring the retrograde guidewire.
3. Do not discard the snare collapsing tool, as it is necessary for reintroducing the snare into the guide catheter, if needed.
4. Ideally the retrograde microcatheter should be advanced into the aorta before attempting to snare the retrograde guidewire to minimize the risk of the guidewire being retracted back into the CTO requiring recrossing of the occlusion. Furthermore, this precaution will protect the proximal coronary artery from traction-induced trauma while snaring the retrograde wire creating tension.
5. It may be easier to snare the retrograde guidewire if it is positioned in the brachiocephalic artery.
6. The floppy radiopaque part of the guidewire is the safest to snare, followed by careful sweeping into the antegrade guide.
7. Alternatively, if a short retrograde guide has been used, some operators prefer to snare 300 cm long Pilot 200 guidewires.
8. Short 180 cm wires should never be snared. If done inadvertently, the wire should be pulled out from the antegrade guide, without attempting to retrieve it from the donor artery, as the snared segment of the wire may become kinked precluding removal via the retrograde guide without also removing the support catheter.

2.8 UNCROSSABLE AND UNDILATABLE LESION EQUIPMENT

While failing to cross the CTO segment is the most common cause of CTO PCI failure, the second most common reason for failure is inability to cross or dilate the lesion after wire crossing. This can be facilitated by increasing guide catheter support (e.g., using a guide catheter extension and various anchor balloon techniques) or by modifying the lesion with the Tornus catheter (Asahi Intecc) or various other modalities, such as rotational or orbital atherectomy and laser, as described in detail in Chapter 8.

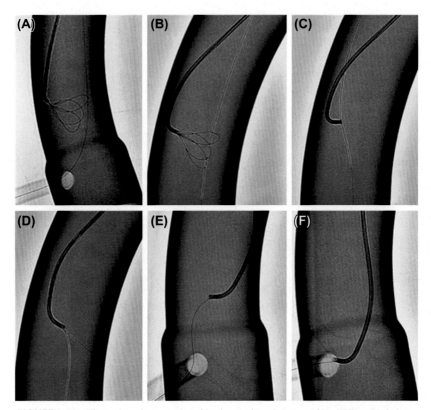

FIGURE 2.64 Illustration of retrograde guidewire snaring. *Courtesy of Dr. William Nicholson.*

2.8.1 Balloons: Threader

The first step in crossing a balloon-uncrossable lesion is advancing a small caliber balloon. Balloons that are 1.20–1.50 mm in diameter are typically used, as they have a single marker in mid-shaft, reducing the distal crossing profile.

Another first line catheter for balloon-uncrossable lesions is the Threader microdilation catheter (Fig. 2.65, Boston Scientific), which combines a 1.20 × 12 mm balloon at its tip with a kink-resistant shaft that enhances deliverability. The Threader is available both as rapid exchange and over-the-wire system, although the rapid exchange system provides more support and is preferred by most operators.

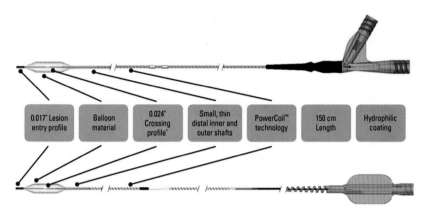

*Crossing profile is defined as the maximum diameter found between the proximal end of the balloon and the distal tip of the catheter.

Threader PowerCoil™ Technology

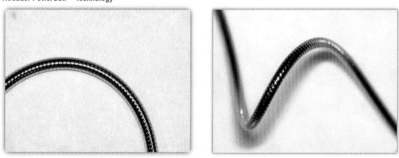

FIGURE 2.65 Illustration of the Threader microdilation catheter. *Image provided courtesy of Boston Scientific. ©2017 Boston Scientific Corporation or its affiliates. All rights reserved.*

2.8.2 Microcatheters for Balloon-Uncrossable Lesions: Tornus, Turnpike Spiral, and Turnpike Gold

The Tornus catheter (Fig. 2.66) consists of eight stainless steel wires stranded in a coil.[36] It comes in two sizes (2.1 and 2.6 Fr), with the latter providing more guidewire support. It has a platinum marker located 1 mm from the tip. Unlike other microcatheters, the Tornus does not have an inner polymer and as a result injection of contrast cannot be done through it, but it provides strong support. The Tornus is advanced by counterclockwise rotation and withdrawn by clockwise rotation. To avoid catheter kinking and unraveling of the stranded steel wires, no more than 10 rotations should be done in either direction.

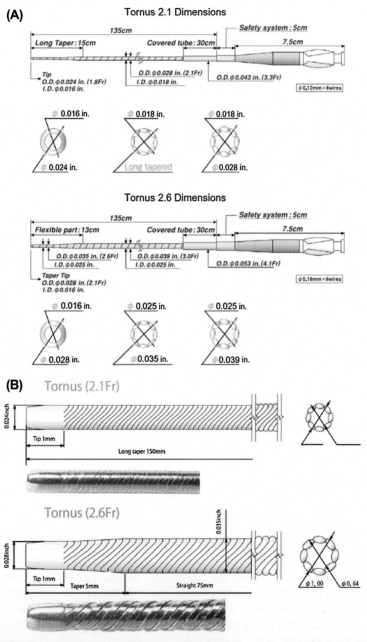

FIGURE 2.66 Illustration of the dimensions (A) and the distal segment (B) of the Tornus 2.1 and 2.6 catheters. *Reproduced with permission from Asahi Intecc.*

Similar to the Tornus, the Turnpike Spiral and Turnpike Gold (see Section 2.4.6) can be used to modify a resistant lesion. In contrast to Tornus, the Turnpike Spiral and Gold are advanced using clockwise rotation and withdrawn using counter-clockwise rotation.

Microcatheter use in balloon-uncrossable lesions

1. Any microcatheter (such as Corsair, Caravel, Turnpike, Turnpike LP, Micro 14, etc.) can be used in balloon-uncrossable lesions. If the microcatheter crosses the lesion, it may sufficiently modify it to enable subsequent crossing of balloons; alternatively microcatheter crossing allows exchange of the guidewire for a stiffer, more supportive guidewire, or an atherectomy guidewire.

2.8.3 Laser

(Online Cases 1, 5, 18, 27, 47, 52, 73, 86) Excimer laser coronary atherectomy (ELCA, Fig. 2.67, Spectranetics) uses ultraviolet energy (wavelength: 308 nm) delivered by a xenon-chlorine pulsed laser catheter with pulse frequency of 25–80 Hz and fluence of 30–80 mJ/mm. ELCA uses energy for disruption and disintegration of the molecular bonds within the atherosclerotic plaque in a highly controlled manner through ablation rather than burning. Since laser can be used over any 0.014″ guidewire, it is highly valuable for the treatment of balloon uncrossable- (and also balloon-undilatable) lesions (Chapter 8).[37] In some instances, the laser catheter can facilitate lesion crossing by modifying the proximal cap, without actually crossing the lesion.

FIGURE 2.67 Illustration of coronary laser catheter. *Reproduced with permission from Spectranetics.*

The laser catheters currently available for coronary use range in diameter from 0.9 to 2.0 mm and are available as rapid exchange (0.9, 1.4, 1.7, and 2.0 mm) or over-the-wire (0.9 mm). The catheter used in coronary CTOs is nearly always the 0.9 mm laser that is compatible with 6 Fr guide catheters. Since the coronary laser catheters allow lasing for only 5 s before the laser is switched off for 10 s, many operators prefer to use off-label the 0.9-mm Turbo-Elite catheter (Spectranetics), which has no lasing duration limits and allows repetition up to 80 Hz. Laser is usually performed during saline inflation but in rare cases laser can be used with simultaneous contrast injection to expand an underexpanded stent (see online video "Impact of contrast on laser activation").[38]

2.8.4 Atherectomy

Both rotational (Fig. 2.68, Boston Scientific) and orbital (Fig. 2.69, CSI) atherectomy (Online Case 39) can successfully modify balloon-uncrossable lesions, however they both require specialized guidewires, which may not be able to be delivered across the lesion if a balloon or a microcatheter fail to cross.

Atherectomy Tips and Tricks

1. Temporary pacemaker insertion is typically not required for atherectomy of lesions in the left coronary system, but can be useful for atherectomy of right coronary artery lesions.
2. Administration of aminophylline (adenosine antagonist, usually as a dose of 250 mg infused over 10 min) during atherectomy may obviate the need for temporary pacemaker insertion among right coronary artery or dominant circumflex lesions.
3. In heavily calcified lesions, early atherectomy should be considered to facilitate equipment delivery and optimal lesion expansion.
4. Rotational atherectomy has been performed in the subintimal space for balloon-uncrossable lesions (see Online Case 53), however this should be done with extreme caution to minimize the risk for perforation.

2.8.5 Cutting and Scoring Balloons

The cutting balloon (Boston Scientific, Fig. 2.70) can be used for treating balloon undilatable lesions and also for releasing intramural hematomas that develop distal to the stents, after reentry. Cutting balloon inflation should be done very slowly (1 atm every 5 s).

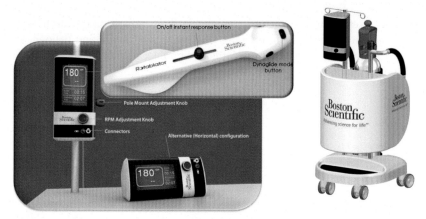

FIGURE 2.68 Illustration of the rotational atherectomy system. *Image provided courtesy of Boston Scientific. ©2017 Boston Scientific Corporation or its affiliates. All rights reserved.*

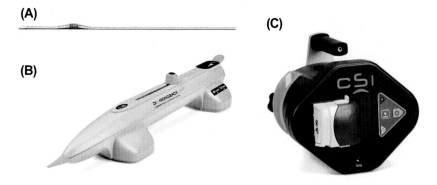

FIGURE 2.69 Illustration of the orbital atherectomy system. (A) 1.25 mm coronary crown. (B) Handle. (C) Pump. *Reproduced with permission from CSI.*

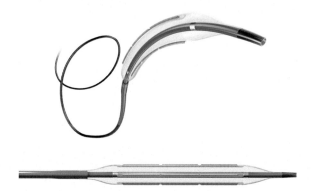

FIGURE 2.70 Illustration of the Wolverine cutting balloon. *Image provided courtesy of Boston Scientific. ©2017 Boston Scientific Corporation or its affiliates. All rights reserved.*

The AngioSculpt scoring balloon catheter (Spectranetics) (Fig. 2.71) is composed of a semicompliant balloon encircled by three nitinol spiral struts to score the target lesion on balloon inflation. Compared with the cutting balloon, the AngioSculpt balloon has a lower profile and produces more scoring marks per millimeter of plaque.

(A)

(B)

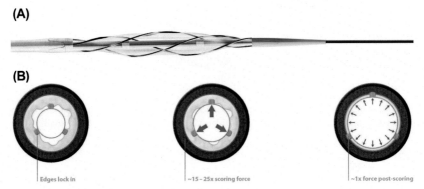

Edges lock in   ~15–25x scoring force   ~1x force post-scoring

FIGURE 2.71 Illustration of the Angiosculpt balloon (A) and its mechanism of action (B). *Reproduced with permission from Spectranetics.*

The Blimp scoring balloon catheter (IMDS) (Fig. 2.72) has a short distal monorail segment with the balloon being adjacent to the coronary guidewire.

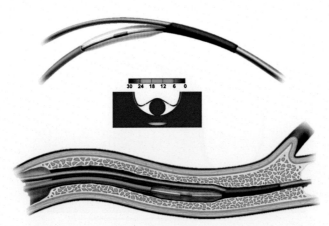

FIGURE 2.72 Illustration of the Blimp scoring balloon and its mechanism of action. *Courtesy of IMDS.*

2.9 IMAGING

Intravascular ultrasound (IVUS) is the most commonly used intravascular imaging modality in CTO PCI.[39]

Utility of IVUS in CTO PCI

IVUS can help to:
a. Identify the proximal cap location in cases with proximal cap ambiguity.
b. Confirm that the antegrade guidewire has engaged the CTO lesion (Fig. 2.73).
c. Facilitate reentry into the true lumen during both antegrade and retrograde crossing.
d. Confirm that the retrograde guidewire has entered the proximal true lumen before externalization.
e. Determine the appropriate balloon size for the CART and reverse CART techniques.
f. Evaluate stent expansion and stent strut apposition.

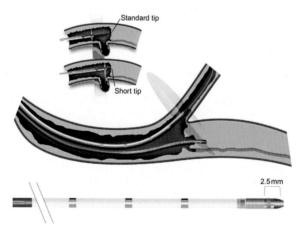

FIGURE 2.73 Illustration of use of intravascular ultrasound for guiding the entry of the crossing guidewire into the chronic total occlusion. *Reproduced with permission from Philips Volcano.*

IVUS Tips and Tricks

1. Solid-state, phased array systems (Eagle Eye, Philips Volcano) are preferred over rotational IVUS systems, such as the Revolution (Philips Volcano), Altantis SR Pro (Boston Scientific), and TVC (InfraRedx) because the imaging transducer is closer to the IVUS catheter tip.
2. A short-tip solid-state IVUS catheter (Eagle Eye Short Tip, Philips Volcano; Fig. 2.73) is preferred for imaging during CTO PCI, as it minimizes the extent of distal dissection required for distal imaging and is more deliverable.

Optical coherence tomography can be used for stent optimization (see Online Case 47), but its use to guidewire crossing is limited, because (1) it has low penetration depth and (2) it requires injection of contrast or dextran, which can cause or hydraulically enlarge a dissection.

Preprocedural **computed tomography coronary angiography** can be used to clarify proximal cap ambiguity, the course of the occluded segment, and the size and morphology of the distal target segment, but is limited in its ability to characterize collateral vessels (Section 3.3.6).

2.10 COMPLICATION MANAGEMENT EQUIPMENT

Although this equipment is rarely needed, it must be available not only for CTO PCI, but also for any other PCI.

2.10.1 Covered Stents

As of 2017, only one coronary covered stent is commercially available in the United States, the Graftmaster Rx (Abbott Vascular). It is approved through a humanitarian device exemption[40] for use in large vessel perforations (Fig. 2.74). In other countries there are more deliverable covered stents, such as the Papyrus stent (Biotronic).

Covered Stents Tips and Tricks[41]

a. The Graftmaster Rx consists of two stainless steel stents with a middle layer of ePTFE.

b. It is bulky and difficult to deliver, hence excellent guide catheter support is important.

c. It is available in diameters of 2.8–4.8 mm and lengths between 16 and 26 mm.

d. It requires a 6 Fr guide catheter for the 2.8–4.0 mm stents and a 7 Fr guide catheter for the 4.5 and 4.8 mm stents.

e. The Graftmaster Rx may be difficult to advance through previously deployed stents, necessitating techniques such as distal anchor or use of 8 Fr guide catheter extensions.

f. Minimum inflation pressure is 15 atm, but even higher pressures for up to 60 seconds (and use of intravascular ultrasound) are preferred to ensure adequate stent expansion.

g. After expansion the stent may shorten up to 1.6 mm on each side (for a total of 3.2 mm at nominal pressure—15 atm). Hence, adequate overlap of stents is important to cover long areas of perforation.

h. Use of a **dual catheter** (ping-pong guide) **technique** (see Chapter 12, Section 12.1.1.2.3) is often required to minimize bleeding into the pericardium while preparing for covered stent delivery and deployment,[42] although delivery of both a balloon and a covered stent is feasible through a single 8 Fr guide catheter (block and deliver technique, see Chapter 12, section 12.1.1.2.3).

i. Postdilatation of the shoulders of the stent may be necessary to fully oppose the stent to the vessel wall if extravasation persists behind the stent despite covering the perforation.

FIGURE 2.74   Illustration of the Graftmaster covered stent. *Courtesy of Abbott Vascular.*

2.10.2 Coils

Coils should be available for use in case of distal branch or collateral vessel perforation. They can also be used to stop a main vessel perforation in a CTO vessel by reoccluding the vessel. Coils are permanent embolic agents that can be deployed either through 0.014″ microcatheters (neurovascular coils, such as Axium, Medtronic; Fig. 2.75) or through larger 0.018″ microcatheters (standard coils, such as Interlock, Boston Scientific; Fig. 2.76).[43] Coils are usually made of stainless steel or platinum alloys and some of them have polymers or synthetic wool or dacron fibers attached along the length of the wire to increase thrombogenicity. Once advanced into the target vessel, the coils assume a preformed shape, sealing the perforation. Particular attention is needed when coiling branches to prevent the coil from prolapsing in the main vessel.

See Online Video: "How to deliver and deploy an Axium coil".

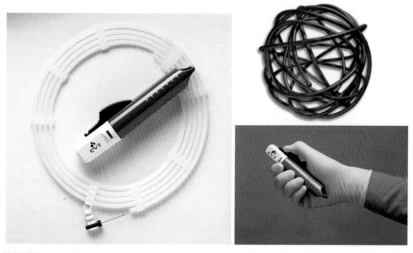

FIGURE 2.75   Illustration of the Axium detachable coil system (Medtronic), which is compatible with 0.014 inch wire compatible microcatheters. *Reproduced with permission from Medtronic.*

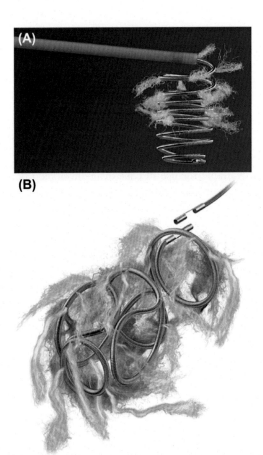

FIGURE 2.76 **Example of a detachable coil that can be used for embolization in case of distal coronary perforation (Interlock, Boston Scientific).** (A) Deployment of the coil. (B) The coil configuration after delivery. *Image provided courtesy of Boston Scientific. ©2017 Boston Scientific Corporation or its affiliates. All rights reserved.*

Coils Tips and Tricks[41]

a. Given that coils are used very infrequently in cardiac catheterization laboratories, it is important for each operator to be familiar with the principles underlying their use and with 1–2 specific coil types, so that he/she can deliver them rapidly in case of perforation.

b. There are two broad categories of coils according to **mechanism of release**: **pushable** and **detachable**. **Pushable** coils are inserted into a microcatheter and pushed with a coil pusher or the front end of a guidewire until they exit into the vessel, hence deployment can be unpredictable and is irreversible.

Detachable coils are released using a dedicated release device once their position into the target vessel is confirmed; conversely if their position is not satisfactory, they can be retrieved. Detachable coils are preferred for treating distal coronary perforations, as they allow optimal and predictable positioning.

c. There are also two broad categories of coils according to the **size of the delivery microcatheter. Coils compatible with 0.014″ microcatheters (such as Axium, Medtronic) are preferred**, as they can be delivered through the standard microcatheter used for CTO PCI (such as the Corsair, Caravel, Turnpike, and Finecross) without requiring change to a larger microcatheter. **Coils compatible with 0.018 inch microcatheters** (such as Interlock [Boston Scientific], Azur [Terumo], and Micronester [Cook]) cannot be delivered through the standard microcatheters used during CTO PCI, and require change to a larger microcatheter, such as the Progreat (Terumo), Renegade (Boston Scientific) or Transit (Cordis).

d. Simultaneous balloon inflation (to stop bleeding into the pericardium) and coil delivery (through a microcatheter) can be achieved through a single 7 or 8 Fr guide catheter ("block-and-deliver" technique see Chapter 12, section 12.1.1.2.4).[44,45]

2.10.3 Pericardiocentesis Tray

Pericardiocentesis can be performed using a standard 0.018 needle, a J-tip 0.035″ guidewire, and a standard pigtail catheter; however, having all equipment assembled in a premanufactured pericardiocentesis tray can facilitate and speed up the procedure.

2.11 RADIATION PROTECTION

Given that CTO procedures can be lengthy, it is ideal to minimize radiation exposure for both the patient and the operator, as described in detail in Chapter 10. There are several radiation protection pads that decrease scatter radiation from the patient, such as the RadPad and No Brainer radiation protection hat (Worldwide Innovations & Technologies, Inc) (Fig. 2.77), and the Zero Gravity ceiling suspended radiation protection system (Biotronic).

2.12 BALLOONS AND STENTS

Small balloons and the Threader microdilation catheter (Section 2.8.1) are needed for crossing tight lesions. Also long (30 mm or longer) noncompliant balloons are helpful to minimize the number of balloon inflations needed for pre- and postdilation of long target coronary segments.

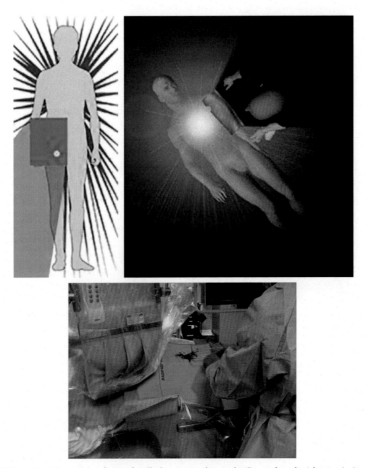

FIGURE 2.77 Illustration of use of radiation protection pads. *Reproduced with permission from Worldwide Innovations & Technologies, Inc.*

Aorto-ostial CTOs could benefit from use of the Ostial Flash balloon (Fig. 2.78, Cardinal Health), which flares the proximal stent struts against the aortic wall and facilitates expansion and reengagement with a catheter. Ostial Flash is a dual balloon angioplasty catheter with a larger proximal low pressure balloon and a higher pressure distal balloon.[46,47] There are three markers that can be visualized under fluoroscopy: a proximal marker that marks the proximal end of the anchoring balloon and should be located in the aorta, a mid-marker that marks the proximal end of the angioplasty balloon and should be placed at the vessel ostium, and a distal marker that marks the distal end of the angioplasty balloon. The distal balloon length is 8 mm and is available in diameters from 3.0 to 4.5 mm. The proximal anchoring balloon can expand up to 14 mm. The distal balloon is inflated with an inflating device whereas the proximal balloon is inflated with a 1 cc syringe.

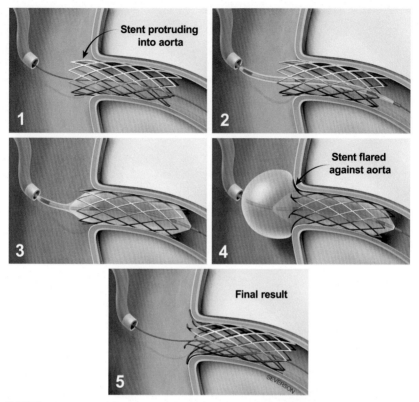

FIGURE 2.78 Illustration of the Ostial Flash balloon. *Reproduced with permission from Nguyen-Trong PJ, Martinez Parachini JR, Resendes E, et al. Procedural outcomes with use of the flash ostial system in aorto-coronary ostial lesions.* Catheter Cardiovasc Interv *2016;88:1067–74.*

Trapping (Section 3.7) can be performed with standard balloons, however a dedicated trapping balloon (that does not have a wire lumen and is, therefore, resistant to kinking and deformation) is also available (Trapper, Boston Scientific, Fig. 2.79; and Trap it, IMDS, Fig. 2.80). Trapping can also be performed with the Trapliner guide catheter extension (Section 2.3.5).

Drug-eluting stents (DESs) should be used in all CTO PCI (unless the patient has a contraindication to prolonged antiplatelet therapy), as they significantly reduce the risk for restenosis and reocclusion.[48] Second-generation DES appear to be more efficacious compared with first-generation DES.[49] A full discussion of CTO lesion stenting follows in Chapter 11.

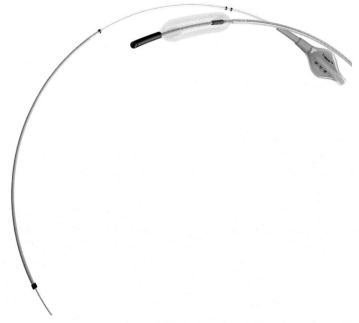

FIGURE 2.79 Illustration of the Trapper balloon. *Image provided courtesy of Boston Scientific.* *©2017 Boston Scientific Corporation or its affiliates. All rights reserved.*

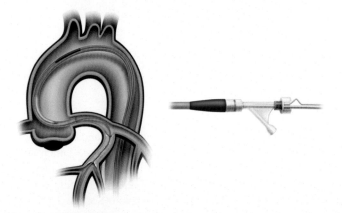

FIGURE 2.80 Illustration of the Trap it balloon. *Reproduced with permission from IMDS.*

2.13 HEMODYNAMIC SUPPORT

Online cases 11, 14, 20, 29, 31, 46, 51, 67.

Hemodynamic support can be used either prophylactically or after occurrence of a complication during CTO PCI. Given the potential for serious, even life-threatening complications during CTO PCI (such as donor vessel injury), availability of hemodynamic support devices is important as part of a comprehensive higher risk and complex PCI program.[50] A comprehensive review of hemodynamic support options is beyond the scope of this book, but basic principles of hemodynamic support are discussed herein.

Four devices are currently available in the United States for providing percutaneous left ventricular hemodynamic support: the intraaortic balloon pump (IABP), the Impella (2.5, CP, and 5.0, Abiomed Inc, Danvers, Massachusetts), the Tandem Heart (Cardiac Assist Inc., Pittsburgh, PA), and venoarterial extracorporeal membrane oxygenator (VA-ECMO) (Fig. 2.81; Table 2.5). Moreover, the PHP (Abbott Vascular) is available in Europe.

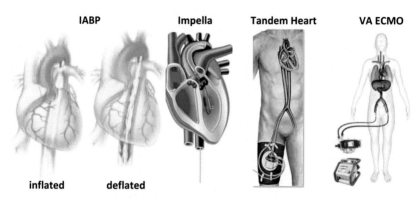

IABP   **Impella**   **Tandem Heart**   **VA ECMO**

inflated   **deflated**

FIGURE 2.81   Overview of hemodynamic support devices.

The IABP is the smallest device but also provides the least hemodynamic support. The IABP is usually inserted percutaneously through a 7–8 Fr femoral arterial sheath. The IABP mechanism of action is inflation of a balloon with helium in the aorta during diastole, displacing blood peripherally and increasing cardiac output, while reducing left ventricular end diastolic pressure.

The Impella is a nonpulsatile axial flow pump that is advanced through the aortic valve and moves blood from the left ventricle into the aorta. Use of the Impella results in left ventricular unloading with reduction in end-diastolic pressure and volume. The Impella is available in two percutaneous types (2.5 and CP) and one device that requires surgical cutdown (5.0). The Impella RP device can provide right ventricular support by pumping blood from the inferior vena cava to the pulmonary artery and consists of a 22 Fr motor mounted on an 11 Fr catheter.

The Tandem Heart is a centrifugal pump that propels blood from the left atrium into the femoral artery. It uses a 21 Fr cannula placed through a transeptal puncture and a 15–19 Fr arterial cannula. With appropriately sized arterial cannulae, full perfusion can be achieved. The Tandem Heart requires continuous monitoring to prevent displacement of the transeptal cannula into the right atrium. In isolation, the Tandem Heart does not oxygenate the blood, but an oxygenator can be placed into the circuit to allow full cardiopulmonary support.

TABLE 2.5 Hemodynamic Support Devices

	IABP	Impella	Tandem Heart	VA ECMO
Feasibility				
Availability	+++	++	+	+
Arterial access size required	7–8 Fr	12 Fr (Impella 2.5) 14 Fr (Impella CP) 21 Fr (Impella 5.0)	15–17 Fr arterial 21 Fr venous	14–17 Fr arterial 18–21 Fr venous
Contraindications	• High bleeding risk • Severe aortic regurgitation • Thoracic or abdominal aortic aneurysm	• High bleeding risk • Severe aortic regurgitation • Severe PAD[b] • Left ventricular thrombus • Mechanical aortic valve • Ventricular septal defect	• High bleeding risk • Severe aortic regurgitation • Severe PAD[b]	• High bleeding risk • Severe aortic regurgitation • Severe PAD[b]
Efficacy				
Cardiac output increase (L/min)	0.3–0.5	≈2.5 (Impella 2.5) ≈4.0 (Impella CP) ≈5.0 (Impella 5.0)	4–5[a]	4–5[a]
Affected by arrhythmias	Yes	No	No	No
Requires adequate right ventricular function	Yes	Yes	Yes	No
Can correct respiratory failure	No	No	Yes[c]	Yes

Complications

Risk for lower limb ischemia	+	++	+++	+++
Transeptal puncture required	No	No	Yes	No
Risk for bleeding	+	++	++	++
Risk for hemolysis	+	++	++	++

[a]Depending on arterial cannula size.
[b]Subclavian or transcaval access can be used for placing the arterial cannula in case of severe peripheral arterial disease.
[c]Adding an oxygenator to the TandemHeart circuit.
PAD, peripheral arterial disease.

VA ECMO consists of a centrifugal, nonpulsatile pump that circulates the blood and has a membrane oxygenator. Venous blood is aspirated through a venous cannula, advanced through the oxygenator, and returned to the patient through an arterial cannula. Similar to Tandem Heart, VA ECMO requires large size cannulae and a perfusionist to manage the system. VA ECMO can generate adequate flows to maintain systemic perfusion, but also increases myocardial oxygen demand due to increase of left ventricular end diastolic pressure and volume, which could adversely impact myocardial recovery.[51]

Determining the need for prophylactic or urgent insertion of a hemodynamic support device depends on the patient's clinical condition (hemodynamic status, left ventricular systolic function, and end-diastolic pressure), procedural risk (e.g., retrograde CTO PCI through the last remaining vessel), and local device availability and expertise.

2.14 THE CTO CART

Having a dedicated CTO cart (Fig. 2.82) with all commonly used CTO PCI equipment (including equipment for managing complications, such as covered stents and coils) can facilitate and make CTO PCI more efficient. The CTO cart should be readily available during the procedure, restocked regularly, and known by not only the CTO operators but also the technicians and nurses.

FIGURE 2.82 Example of a CTO cart.

In summary, there are many pieces of equipment that can be used in CTO PCI. While there is no universal CTO PCI equipment checklist and there is a great component of personal preference, experience, and availability in constructing a CTO checklist, Table 2.1 could serve as a starting point. Having the right tool for the right job can significantly simplify the procedure and enhance success rates in CTO PCI.

REFERENCES

1. Brilakis ES. The essential equipment for CTO interventions. *Cardiol Today's Interv* May June 2013.
2. Joyal D, Thompson CA, Grantham JA, Buller CEH, Rinfret S. The retrograde technique for recanalization of chronic total occlusions: a step-by-step approach. *JACC Cardiovasc Interv* 2012;**5**:1–11.
3. Brilakis ES, Grantham JA, Thompson CA, et al. The retrograde approach to coronary artery chronic total occlusions: a practical approach. *Catheter Cardiovasc Interv* 2012;**79**:3–19.
4. Brilakis ES, Grantham JA, Rinfret S, et al. A percutaneous treatment algorithm for crossing coronary chronic total occlusions. *JACC Cardiovasc Interv* 2012;**5**:367–79.
5. Agelaki M, Koutouzis M. Balloon-assisted tracking for challenging transradial percutaneous coronary intervention. *Anatol J Cardiol* 2017;**17**:E1.
6. Wu EB, Chan WW, Yu CM. Retrograde chronic total occlusion intervention: tips and tricks. *Catheter Cardiovasc Interv* 2008;**72**:806–14.
7. Papayannis AC, Michael TT, Brilakis ES. Challenges associated with use of the GuideLiner catheter in percutaneous coronary interventions. *J Invasive Cardiol* 2012;**24**:370–1.
8. Luna M, Papayannis A, Holper EM, Banerjee S, Brilakis ES. Transfemoral use of the GuideLiner catheter in complex coronary and bypass graft interventions. *Catheter Cardiovasc Interv* 2012;**80**:437–46.
9. Chang YC, Fang HY, Chen TH, Wu CJ. Left main coronary artery bidirectional dissection caused by ejection of guideliner catheter from the guiding catheter. *Catheter Cardiovasc Interv* 2013;**82**:E215–20.
10. Mamas MA, Fath-Ordoubadi F, Fraser DG. Distal stent delivery with Guideliner catheter: first in man experience. *Catheter Cardiovasc Interv* 2010;**76**:102–11.
11. Senguttuvan NB, Sharma SK, Kini A. Percutaneous intervention of chronic total occlusion of anomalous right coronary artery originating from left sinus – use of mother and child technique using guideliner. *Indian Heart J* 2015;**67**(Suppl. 3):S41–2.
12. Repanas TI, Christopoulos G, Brilakis ES. "Candy Cane" guide catheter extension for stent delivery. *J Invasive Cardiol* 2015;**27**:E169–70.
13. Finn MT, Green P, Nicholson W, et al. Mother-daughter-granddaughter double GuideLiner technique for delivering stents past multiple extreme angulations. *Circ Cardiovasc Interv* 2016;**9**.
14. Tsuchikane E, Katoh O, Kimura M, Nasu K, Kinoshita Y, Suzuki T. The first clinical experience with a novel catheter for collateral channel tracking in retrograde approach for chronic coronary total occlusions. *JACC Cardiovasc Interv* 2010;**3**:165–71.
15. McClure SJ, Wahr DW, Webb JG. Venture wire control catheter. *Catheter Cardiovasc Interv* 2005;**66**:346–50.
16. Naidu SS, Wong SC. Novel intracoronary steerable support catheter for complex coronary intervention. *J Invasive Cardiol* 2006;**18**:80–1.

17. McNulty E, Cohen J, Chou T, Shunk K. A "grapple hook" technique using a deflectable tip catheter to facilitate complex proximal circumflex interventions. *Catheter Cardiovasc Interv* 2006;**67**:46–8.
18. Aranzulla TC, Colombo A, Sangiorgi GM. Successful endovascular renal artery aneurysm exclusion using the Venture catheter and covered stent implantation: a case report and review of the literature. *J Invasive Cardiol* 2007;**19**:E246–53.
19. Aranzulla TC, Sangiorgi GM, Bartorelli A, et al. Use of the Venture wire control catheter to access complex coronary lesions: how to turn procedural failure into success. *EuroIntervention* 2008;**4**:277–84.
20. Iturbe JM, Abdel-Karim AR, Raja VN, Rangan BV, Banerjee S, Brilakis ES. Use of the venture wire control catheter for the treatment of coronary artery chronic total occlusions. *Catheter Cardiovasc Interv* 2010;**76**:936–41.
21. Brilakis ES, Lombardi WL, Banerjee S. Use of the Stingray® guidewire and the Venture® catheter for crossing flush coronary chronic total occlusions due to in-stent restenosis. *Catheter Cardiovasc Interv* 2010;**76**:391–4.
22. Mitsutake Y, Ebner A, Yeung AC, Taber MD, Davidson CJ, Ikeno F. Efficacy and safety of novel multi-lumen catheter for chronic total occlusions: from preclinical study to first-in-man experience. *Catheter Cardiovasc Interv* 2015;**85**:E70–5.
23. Moualla SK, Khan S, Heuser RR. Anchoring improved: introduction of a new over-the-wire support balloon. *J Invasive Cardiol* 2014;**26**:E130–2.
24. Walsh S, Dudek D, Bryniarski L, et al. Efficacy and Safety of Novel NovaCross Microcatheter for Chronic Total Occlusions: First-in-human Study. *J Invasive Cardiol* 2016;**8**: 88–91.
25. Karatasakis A, Tarar MN, Karmpaliotis D, et al. Guidewire and microcatheter utilization patterns during antegrade wire escalation in chronic total occlusion percutaneous coronary intervention: insights from a contemporary multicenter registry. *Catheter Cardiovasc Interv* 2017;**80**: E90–8.
26. Danek BA, Karatasakis A, Brilakis ES. Consequences and treatment of guidewire entrapment and fracture during percutaneous coronary intervention. *Cardiovasc Revasc Med* 2016;**17**:129–33.
27. Tomasello SD, Giudice P, Attisano T, Boukhris M, Galassi AR. The innovation of composite core dual coil coronary guide-wire technology: a didactic coronary chronic total occlusion revascularization case report. *J Saudi Heart Assoc* 2014;**26**:222–5.
28. Galassi AR, Ganyukov V, Tomasello SD, Haes B, Leonid B. Successful antegrade revascularization by the innovation of composite core dual coil in a three-vessel total occlusive disease for cardiac arrest patient using extracorporeal membrane oxygenation. *Eur Heart J* 2014;**35**:2009.
29. Khalili H, Vo MN, Brilakis ES. Initial experience with the Gaia composite core guidewires in coronary chronic total occlusion crossing. *J Invasive Cardiol* 2016;**28**:E22–5.
30. Werner GS. The BridgePoint devices to facilitate recanalization of chronic total coronary occlusions through controlled subintimal reentry. *Expert Rev Med Dev* 2011;**8**:23–9.
31. Brilakis ES, Badhey N, Banerjee S. "Bilateral knuckle" technique and Stingray re-entry system for retrograde chronic total occlusion intervention. *J Invasive Cardiol* 2011;**23**:E37–9.
32. Michael TT, Papayannis AC, Banerjee S, Brilakis ES. Subintimal dissection/reentry strategies in coronary chronic total occlusion interventions. *Circ Cardiovasc Interv* 2012;**5**:729–38.
33. Whitlow PL, Burke MN, Lombardi WL, et al. Use of a novel crossing and re-entry system in coronary chronic total occlusions that have failed standard crossing techniques: results of the FAST-CTOs (Facilitated Antegrade Steering Technique in Chronic Total Occlusions) trial. *JACC Cardiovasc Interv* 2012;**5**:393–401.
34. Papayannis A, Banerjee S, Brilakis ES. Use of the crossboss catheter in coronary chronic total occlusion due to in-stent restenosis. *Catheter Cardiovasc Interv* 2012;**80**:E30–6.

34a. Wilson WM, Walsh S, Hanratty C, Strange J, Hill J, Sapontis J, Spratt JC. A novel approach to the management of occlusive in-stent restenosis (ISR). EuroIntervention 2014;**9**:1285–93.

35. Christopoulos G, Kotsia AP, Brilakis ES. The double-blind stick-and-swap technique for true lumen reentry after subintimal crossing of coronary chronic total occlusions. *J Invasive Cardiol* 2015;**27**:E199–202.

36. Fang HY, Lee CH, Fang CY, et al. Application of penetration device (Tornus) for percutaneous coronary intervention in balloon uncrossable chronic total occlusion-procedure outcomes, complications, and predictors of device success. *Catheter Cardiovasc Interv* 2011;**78**: 356–62.

37. Karacsonyi J, Karatasakis A, Danek BA, Banerjee S, Brilakis ES. Laser applications in the coronaries, In: *Textbook of Atherectomy.* HMP Communications; 2016.

38. Karacsonyi J, Danek BA, Karatasakis A, Ungi I, Banerjee S, Brilakis ES. Laser coronary atherectomy during contrast injection for treating an underexpanded stent. *JACC Cardiovascular Interventions* 2016;**9**:e147–8.

39. Galassi AR, Sumitsuji S, Boukhris M, et al. Utility of intravascular ultrasound in percutaneous revascularization of chronic total occlusions: an overview. *JACC Cardiovasc Interv* 2016;**9**:1979–91.

40. Romaguera R, Waksman R. Covered stents for coronary perforations: is there enough evidence? *Catheter Cardiovasc Interv* 2011;**78**:246–53.

41. Brilakis ES, Karmpaliotis D, Patel V, Banerjee S. Complications of chronic total occlusion angioplasty. *Interv Cardiol Clin* 2012;**1**:373–89.

42. Ben-Gal Y, Weisz G, Collins MB, et al. Dual catheter technique for the treatment of severe coronary artery perforations. *Catheter Cardiovasc Interv* 2010;**75**:708–12.

43. Pershad A, Yarkoni A, Biglari D. Management of distal coronary perforations. *J Invasive Cardiol* 2008;**20**:E187–91.

44. Tarar MN, Christakopoulos GE, Brilakis ES. Successful management of a distal vessel perforation through a single 8-French guide catheter: combining balloon inflation for bleeding control with coil embolization. *Catheter Cardiovasc Interv* 2015;**86**:412–6.

45. Garbo R, Oreglia JA, Gasparini GL. The Balloon-Microcatheter technique for treatment of coronary artery perforations. *Catheter Cardiovasc Interv* 2017;**89**:E75–83.

46. CardinalHealth. FLASH™ and FLASH™MINI Dual Balloon. Angioplasty Catheters; 2015. https://www.cordis.com/content/dam/cordis/web/documents/brochure/Cordis-FLASHOstial SystemAnimationBrochure.pdf.

47. Nguyen-Trong PJ, Martinez Parachini JR, Resendes E, et al. Procedural outcomes with use of the flash ostial system in aorto-coronary ostial lesions. *Catheter Cardiovasc Interv* 2016;**88**:1067–74.

48. Brilakis ES, Kotsia A, Luna M, Garcia S, Abdullah SM, Banerjee S. The role of drug-eluting stents for the treatment of coronary chronic total occlusions. *Expert Rev Cardiovasc Ther* 2013;**11**:1349–58.

49. Lanka V, Patel VG, Saeed B, et al. Outcomes with first- versus second-generation drug-eluting stents in coronary chronic total occlusions (CTOs): a systematic review and meta-analysis. *J Invasive Cardiol* 2014;**26**:304–10.

50. Kirtane AJ, Doshi D, Leon MB, et al. Treatment of higher-risk patients with an indication for revascularization: evolution within the field of contemporary percutaneous coronary intervention. *Circulation* 2016;**134**:422–31.

51. Tomasello SD, Boukhris M, Ganyukov V, et al. Outcome of extracorporeal membrane oxygenation support for complex high-risk elective percutaneous coronary interventions: a single-center experience. *Heart Lung* 2015;**44**:309–13.

Chapter 3

The Basics: Timing, Dual Injection, Studying the Lesion, Access, Anticoagulation, Guide Support, Trapping, Pressure and Electrocardiogram Monitoring

3.1 TIMING

In general, chronic total occlusion (CTO) percutaneous coronary interventions (PCIs) **should not be performed ad hoc in order to**[1]:

a. Allow time for thorough procedural planning and preparation for both the operator and the cardiac catheterization laboratory staff, which is essential for success.
b. Minimize the amount of contrast administered and radiation dose.
c. Minimize patient and operator fatigue.
d. Allow time to collect sufficient information on the viability and/or the extent of ischemia of the territory supplied by the occluded vessel.
e. Allow for a detailed discussion with the patient and family about the indications, goals, risks, and alternatives (such as medical therapy and coronary artery bypass graft surgery) to the procedure. Risks that may be increased in CTO PCI compared with non-CTO PCI include radiation injury and perforation.

In some cases, however, ad hoc PCI may be the best option, such as in patients who present with an acute coronary syndrome due to failure of a highly diseased saphenous vein graft, in whom treatment of the native coronary artery CTO is considered to be the preferred treatment strategy (Online Case 87 and 103).[2] Also in patients with acute coronary occlusions resulting in cardiogenic shock, successful use of CTO PCI techniques can achieve urgent complete revascularization and help stabilize the patient's hemodynamics.[3–5]

Manual of Chronic Total Occlusion Interventions. http://dx.doi.org/10.1016/B978-0-12-809929-2.00003-X

101

3.2 DUAL INJECTION

Dual injection angiography is of critical importance in CTO PCI. This is the simplest and most effective technique for increasing CTO PCI success rates and decreasing complications and should be performed in all patients with contralateral collaterals.[6]

Even in antegrade-only cases, placing a safety workhorse guidewire in the donor vessel may stabilize the catheter, preventing disengagement of the guide catheter and allowing prompt treatment in case of a complication.

3.2.1 Why Is Dual Injection Important?

Benefits of dual injection

Dual injection provides the following benefits[7]:

Before PCI

a. Nonsimultaneous, single-catheter injection often provides suboptimal visualization of the CTO segment and limited ability to assess both the proximal and distal cap and the distal vessel beyond the CTO due to collateral competitive flow (Figs. 3.1 and 3.2). Occasionally, dual injection will reveal that the CTO is not a total occlusion, but rather a functional occlusion with a central patent channel (Fig. 3.1). In other cases there may be more than one tandem CTO (see Online Case 6). Dual injections also provide a more accurate assessment of the true length of the CTO.

During PCI

a. Contralateral injection during CTO PCI allows **visualization of the guidewire position** during antegrade crossing attempts. If the guidewire is outside the vessel or in a side branch it can be repositioned before advancing equipment, significantly reducing the risk of perforation or other complications (Fig. 3.3).

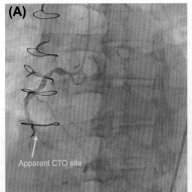 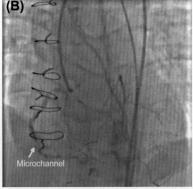

(A) Apparent CTO site

(B) Microchannel

FIGURE 3.1   Example of dual injection revealing a "microchannel" at the assumed occlusion site. Injection of the right coronary artery revealed a distal right coronary artery occlusion (A), but the length of the occlusion and the quality of the distal vessel could not be determined. Dual injection (via the right coronary artery and the left internal mammary graft that supplied collaterals to the right posterior descending artery) demonstrated a microchannel (B) at the occlusion site, very short occlusion length, and diffusely diseased distal vessel. Crossing of the chronic total occlusion was easily achieved with a Fielder XT guidewire.

b. Even if there are ipsilateral collaterals at baseline, during CTO PCI the collateral flow direction and strength of flow can shift from one source to another (Online Case 95) due to ipsilateral collateral damage, which can occur frequently in antegrade attempts, not allowing determination of distal guidewire position.

c. When using dissection/reentry techniques antegrade contrast injections may result in hydraulic enlargement of the subintimal space and reduce the likelihood of successful reentry (Fig. 3.4). We recommend removing the injecting syringe from the antegrade guide manifold during dissection/reentry attempts to prevent inadvertent contrast injection and expansion of the subintimal space, as demonstrated in Chapter 5, Fig. 5.15.

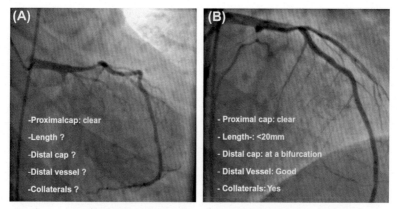

FIGURE 3.2   **Example of how dual injection can significantly improve the understanding of the chronic total occlusion (CTO) anatomy and CTO crossing options.** Injection of the left main coronary artery demonstrates a proximal circumflex CTO (A), but the characteristics of the lesion remained unknown. Using dual injection (B) the characteristics of the CTO (proximal cap ambiguity, lesion length, bifurcation at distal cap, quality of distal vessel, and presence of collaterals) were clarified. *Courtesy of Dr. Santiago Garcia.*

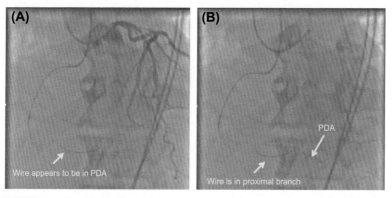

FIGURE 3.3   **Dual injection to determine distal guidewire position after crossing.** The guidewire appeared to be in the right posterior descending artery (PDA; A), but was actually in a proximal side branch (B). Dual injection allowed correction of guidewire position before balloon inflation and stent deployment.

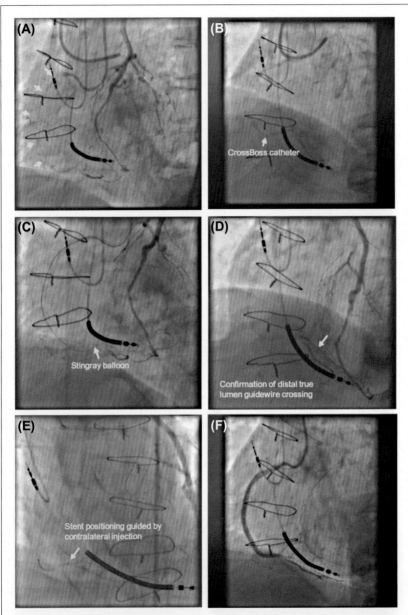

FIGURE 3.4 Contralateral injection for guiding stent placement after subintimal chronic total occlusion (CTO) crossing. A right coronary artery CTO (A) was subintimally crossed with the CrossBoss catheter (*arrow*, B), followed by reentry using the Stingray balloon and a Pilot 200 wire (stick-and-swap technique; C, D). Contralateral injection was used to guide stent placement at the posterior descending artery/right posterolateral branch bifurcation (*arrow*, E) with an excellent final result (F).

3.2.2 Dual Injection Technique

How to Perform Dual Injection

a. **Introduce the right coronary artery guide catheter first** (before inserting the left coronary artery guide catheter) to allow for unimpeded torqueing necessary to engage the right coronary artery ostium, as well as to prevent guide–guide interaction with a left main guiding catheter during the procedure. In case of difficult cannulation of the left main, introducing a stabilizing guidewire into the right coronary artery can prevent disengagement of the right guide catheter.

b. Administer sublingual or intracoronary **nitrates** before injections to maximally dilate the vessels and help show collateral flow.

c. Use **low magnification** (13-inch instead of 8-inch) to enable visualization of the entire coronary circulation.

d. **Do not pan** the table to facilitate recognition of collaterals.

e. Obtain **long cine acquisition** to allow for the contrast to travel through the collateral vessels and fill the distal vessel.

f. **Inject the donor vessel** (vessel that supplies the territory distal to the CTO) **first,** followed by injection of the occluded vessel after collaterals have filled the distal vessel. To reduce radiation exposure the donor vessel can be injected before cine is recorded.

g. Avoid using a side-hole guide catheter on the donor side to achieve better distal opacification and decrease contrast volume.

h. Use of **8 French** (Fr) guide catheters for angiography provides better filling of the vessel, improving visualization, especially of small collateral vessels. Too little volume and low rate of contrast injection can mask important information about the lesion and result in image artifacts.

i. In complex cases several contralateral injections might be necessary. The amount of contrast required may be decreased by inserting a **microcatheter selectively into the donor branch** and administering only 1–2 cc contrast for each image shot. This method cannot be applied in case of several collaterals due to the competitive blood flow from other branches.

j. As in all coronary angiography, looking at the electrocardiogram and pressure waveform **before** and **after** each contrast injection is a must. Electrocardiographic changes can provide early warning of an impending complication, such as ischemia during collateral vessel crossing. Injection in the setting of severe pressure dampening can cause severe coronary and/or aortocoronary artery dissection.

k. If collateral visualization is still suboptimal, a cine recording done at 30 frames per second (fps) can enhance visualization. This technique uses more radiation than 15 fps and should be used sparingly, remembering to reset the acquisition back to 15 fps.

l. Most interventionalists will use the **right femoral artery to cannulate the right coronary artery and the left femoral artery to cannulate the left main**. This is done to avoid confusion. If the right coronary artery is not being used the right femoral artery is often chosen for the antegrade guide and the left femoral artery for the retrograde guide. There is no magic about this choice but it is easy to remember and strongly recommended that it be done the same way every time. Advancing or pulling on the wrong wire or catheter can easily negate any progress that has been made.

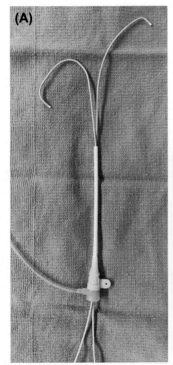
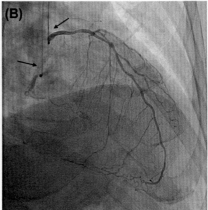

FIGURE 3.5 Illustration of using two 4 Fr diagnostic catheters through an 8 Fr sheath (A) and a dual angiogram obtained using this technique (B). *Courtesy of Dr. William Nicholson.*

One technique for simultaneous injection of both coronaries **during the diagnostic angiogram without requiring a second point of access** is to upsize the femoral arterial access point to an 8 Fr sheath. Two 4 Fr catheters can then be passed through a single 8 Fr sheath to allow simultaneous injection of both coronaries (Figs. 3.5 and 3.6).[8] However, coronary visualization may be poor with a 4 Fr catheter even with use of automatic injectors and bleeding may occur through the sheath's hemostatic valve.

3.3 STUDYING THE LESION

3.3.1 How to Evaluate the Lesion

Spending enough time to study the CTO angiographic parameters will make the procedure easier and will increase the likelihood of success. Any other previous angiograms should also be located and reviewed carefully.

When? Before the case begins.

By whom? Ideally the films should be reviewed by the entire CTO team, including the physicians, technicians, and fellows.

How long? Usually 15–30 min per patient. This much time is necessary to fully understand the CTO anatomy and determine the best action plan. With each repeat viewing of the images, new anatomic information becomes evident; for

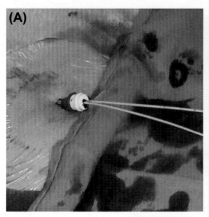 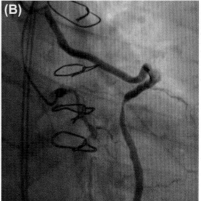

FIGURE 3.6   Illustration of dual angiography (B) using two 4 Fr diagnostic catheters through an 8 Fr sheath (A). *Courtesy of Dr. Gabriele Gasparini.*

example, the course of collaterals or the location of the proximal cap is better appreciated.

Which parameters should be assessed?

There are four key parameters that should be evaluated during the angiographic review (Fig. 3.7)[7]:

Four Key Angiographic Parameters to Guide CTO PCI

1. Proximal cap and proximal vessel
2. Lesion length and quality
3. Quality of distal target vessel
4. Collateral circulation

These parameters can help us understand: (1) where the CTO starts and what the vessel proximal to the CTO looks like; (2) the course and quality of the CTO segment; (3) the ending point of the CTO and the quality of the vessel distally; and (4) potential retrograde pathways for getting to the distal cap.

Assessment Tips and Tricks

- Using **slow replay** and **magnified views** may help clarify the CTO vessel course and the collateral connections. In case of multiple collaterals a frame-by-frame replay can help determine the direction of flow in the branches of the distal vessel and identify the dominant collateral.
- Occasionally, **tracing the collaterals backward** may help identify their origin and course.
- Some **image postprocessing techniques**, such as color inversion and increasing the contrast, may help to discover usable collaterals.
- In some patients, **preprocedural coronary computed tomography** may help decipher the course of the occluded vessel and evaluate the presence and extent of calcification and tortuosity (Section 3.3.6). This is especially helpful in patients with very long occluded segments. However, coronary computed tomography is not useful for assessing collaterals.

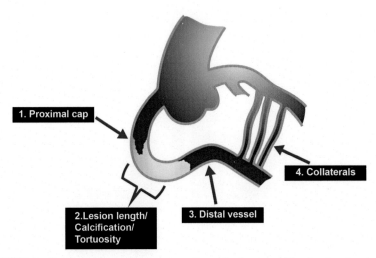

FIGURE 3.7 Four key angiographic parameters that need to be assessed for planning chronic total occlusion percutaneous coronary intervention.

Why? To understand the CTO anatomy and collateral circulation, which enables the operator to map out all possible options for crossing the occlusion and create a strategic plan (Fig. 3.8). Such procedural plans are often provided by proctors before planned cases.

3.3.2 Proximal Cap (Fig. 3.9)

3.3.2.1 Proximal Vessel

Diffusely diseased proximal vessels can cause pressure dampening upon guide catheter engagement. Pressure dampening may not cause ischemia in proximally occluded vessels supplying a small distal territory. Conversely, dampening in large vessels giving multiple side branches proximal to the occlusion could cause ischemia and/or hypotension, and may require intermittent guide catheter disengagement, use of a smaller guide catheter, or use of a guide catheter with side holes. Injection through a guide catheter with dampened pressure waveform could cause coronary and/or aortocoronary dissection (Section 12.1.1.1.2) and should be avoided, often by removing the injection syringe (Fig. 5.15). Pressure dampening can also predispose to air embolization and to thrombus formation within the guide catheter.

Careful wiring of diffusely diseased proximal vessels (usually using workhorse guidewires with standard tip bends) is critical to minimize the risk for proximal vessel dissection. A microcatheter is then advanced and

EXAMPLE OF PROCEDURAL PLAN FOR RIGHT CORONARY ARTERY CTO

Proximal cap: ambiguous with side branch

Occlusion length: around 30–40mm

Distal vessel: diffusely diseased

Collaterals: multiple septal collaterals - would need to assess proximal and mid LAD to make sure there are np significant lesions before going retrograde.

There is also an occluded SVG-PDS that could be used for retrograde approach if needed.

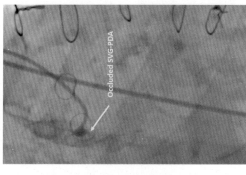

Occluded SVG-PDA

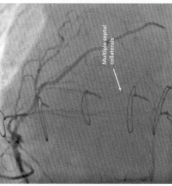

Multiple septal collaterals

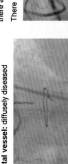

Distal cap in mid RCA

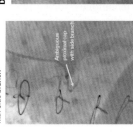

Ambiguous proximal cap with side branch

Plan

1. Would do lateral view to see if a stump is present in proximal RCA (unlikely)

2. Primary retrograde (AFTER CHECKING THAT PROXIMAL/MID LAD ARE OK) via septal collaterals

3. BASE (balloon-assisted subintimal entry) (balloon to dissect the proximal RCA then go subintimal with knuckle wire)

4. Retrograde via occluded SVG

FIGURE 3.8 Example of a strategic plan for chronic total occlusion percutaneous coronary intervention.

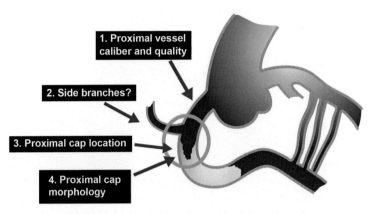

FIGURE 3.9 Assessment of the proximal cap and vessel.

the workhorse wire is exchanged for specific CTO wires with much smaller, CTO-specific bends, which can be more difficult to advance through larger vessels.

3.3.2.2 Side Branches

Side branches near or at the proximal cap may hinder antegrade wiring, since guidewires (especially polymer-jacketed guidewires) may preferentially enter those branches instead of engaging the CTO. Some of those branches could potentially be used for side branch anchoring to increase guide catheter support (Fig. 3.10, Section 3.6.5).

3.3.2.3 Proximal Cap Location

Understanding the start of the CTO (proximal cap) is critical for both success and safety. If the proximal cap is clearly defined and unambiguous, up front use of an antegrade approach is favored. CTOs with poorly defined, ambiguous proximal caps (Fig. 3.11) may be best approached with a primary retrograde approach, or using advanced subintimal techniques, such as the "move the cap" techniques (Section 9.1.4).

Multiple angiographic projections, including some unconventional ones, may be needed to clarify the location of the proximal cap (Section 9.1.1); for example, the straight lateral projection for right coronary artery occlusions. In some cases, angiography alone may be inconclusive, but intravascular ultrasound can help elucidate the proximal part of the CTO (Section 9.1.3). A full discussion of how to approach a CTO with an ambiguous proximal cap is presented in Section 9.1.

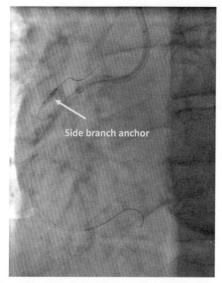

FIGURE 3.10   Example of side branch anchor technique.

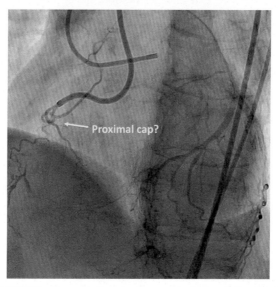

FIGURE 3.11   Example of proximal cap ambiguity (see Online Case 32).

3.3.2.4 Proximal Cap Morphology

Tapered proximal caps are more favorable than blunt caps, as they facilitate guidewire entry into the occlusion.[9,10] The proximal cap is usually more resistant than the distal cap, likely due to exposure to arterial pressure. Severe calcification can make wire penetration challenging, requiring highly penetrating guidewires, such as the Confianza Pro 12, Hornet 14, and Astato 20.

3.3.3 Occlusion Length and Quality (Fig. 3.12)

3.3.3.1 Length

Lesion length is almost always overestimated with single injections, hence dual injections (Section 3.2) are important for accurate estimation of the lesion length. Lesion length and quality assessment can also be performed using multiple detector computed tomography. Longer lesion length is usually associated with higher difficulty in crossing and lower success rates. A cutoff point frequently used is 20 mm[9]: according to the hybrid algorithm ≥20 mm long lesions may be best approached with a primary dissection/reentry strategy (because subintimal guidewire entry is highly likely to occur), whereas <20 mm long lesions are usually first approached with antegrade wire escalation (Chapter 7).[7]

3.3.3.2 Quality

Intraocclusion calcification and tortuosity increase the difficulty of CTO crossing. Calcification (Section 9.9) increases the likelihood of subintimal guidewire entry and may cause difficulty crossing with a balloon or microcatheter in case of successful guidewire crossing. Tortuosity increases the likelihood of guidewire exit and perforation, hence highly tortuous occlusions may be best crossed subintimally with a knuckled guidewire (Section 5.4). The guidewire knuckle is less traumatic than the tip of a guidewire and more likely to remain within the vessel architecture, which is reflected in the motto, "trust the knuckle." Occasionally, small contrast-filled islands can be discovered in the body of long occlusions. These islands are supplied by side

Lesion Length and Quality

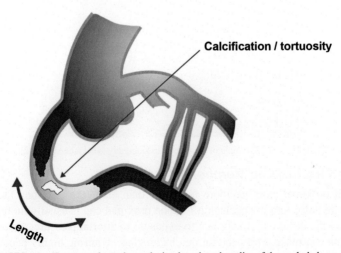

FIGURE 3.12 How to evaluate the occlusion length and quality of the occluded segment.

branches communicating with collaterals and can be helpful for wire-based strategies to follow the track of the vessel and avoid wire exit from the vessel architecture.

3.3.4 Quality of the Distal Vessel (Fig. 3.13)

Evaluating the **size** and quality of the vessel distal to the occlusion is important for deciding on a procedural plan and estimating the likelihood of success. Large vessels distal to the CTO are associated with easier crossing and higher likelihood of success. Conversely, small, diffusely diseased distal vessels can be much harder to recanalize and are associated with lower procedural success rates, in part due to difficulty reentering in the true lumen if the antegrade guidewire crosses into the subintimal space. The size of the distal vessel may be small due to chronic hypoperfusion, but it can increase significantly both acutely and within a few months after successful recanalization. In some patients, the size of the distal vessel may be underestimated due to partial filling caused by competitive flow via ipsilateral and contralateral collaterals. Having access to prior films could greatly assist with determining the true size of the distal vessel.

Distal vessel **calcification** may hinder reentry attempts in the case of subintimal guidewire entry and/or predispose to perforation during or after balloon inflation or stent deployment. Very high-pressure postdilations should be avoided in heavily calcified vessels, especially with oversized balloons.

The morphology of the distal cap often cannot be visualized as clearly as the proximal cap because of lower perfusion pressure through the collaterals (Fig. 3.14). Having a **bifurcation** at the distal cap makes CTO crossing harder, in part

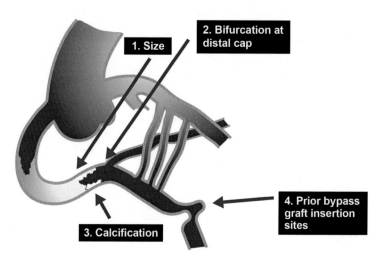

FIGURE 3.13   How to evaluate the quality of the vessel distal to the chronic total occlusion.

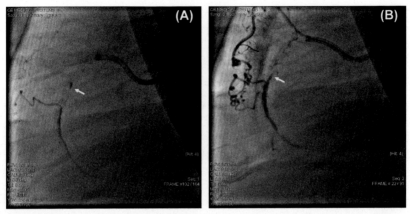

FIGURE 3.14 Visualization of the chronic total occlusion distal cap (*arrow*) by contralateral (A) versus ipsilateral (B) injection. The distal cap is best visualized via a bridging collateral. *Courtesy of Dr. Imre Ungi.*

because if subintimal guidewire crossing occurs, reentry into the distal true lumen may result in loss of one of the branches.[11] Such lesions may be best approached by a primary retrograde approach, or alternatively by performing double reentry into both branches of the bifurcation (Section 9.6, Online cases 5, 7, 8, 12, 25, 34, 45, 49, 64, 80). Another option is to combine an antegrade approach for one of the branches with the retrograde approach for salvaging the other branch.

In patients with prior coronary artery bypass graft surgery the distal vessel may be distorted at the **distal anastomotic site**, as many times the bypass graft causes tenting of the native vessel at the anastomosis.

3.3.5 Collaterals

3.3.5.1 Source and Course of the Collaterals (Fig. 3.15)

a. **Source**

Collaterals may arise from the CTO artery itself, proximal to the occlusion (ipsilateral collaterals), or from another coronary artery (contralateral collaterals). These would include right to left collaterals or left to right collaterals. Aortocoronary bypass grafts (patent or occluded) are not true collaterals, but may still serve as retrograde conduits. For example, in patients with severely degenerated saphenous vein grafts (SVGs), recanalization of the native coronary artery provides superior short- and long-term outcomes as compared with treating the SVG.[12]

b. **Course (septal, epicardial, bypass grafts)**

As their name implies, septal collaterals course through the septum, whereas epicardial collaterals course at the heart's surface. Septal collaterals are preferred over epicardial collaterals, because they are safer to cross: septal collateral perforation rarely results in a complication, whereas epicardial collateral perforation can cause tamponade or localized chamber compression in patients

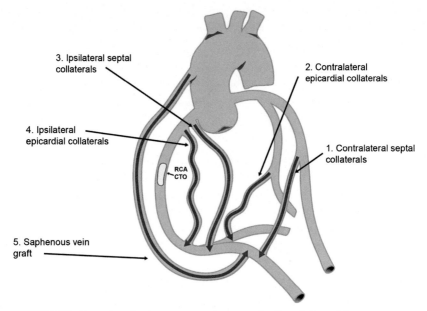

FIGURE 3.15   Examples of various sources and courses of collaterals for a right coronary artery chronic total occlusion.

with prior coronary bypass graft surgery resulting in a very difficult to manage scenario referred to as dry tamponade (Online Case 43, Section 12.1.1.2.5). Treatment of epicardial collateral perforations requires occlusion from both sides of the perforation for successful sealing. Both patent and occluded SVGs can be used for retrograde crossing.[2,13] Retrograde crossing through internal mammary artery grafts is feasible (Online Cases 29, 37, 46, 50, and 57), but may result in significant ischemia and hemodynamic compromise (Online Case 46).

Not all epicardial collaterals are equal. For example, Mashayekhi et al. demonstrated that ipsilateral epicardial collaterals from acute marginal to acute marginal branch (Type B, Fig. 3.16) have high risk for perforation (Fig. 3.17) and should not be used for retrograde crossing.[14]

3.3.5.2 Collateral Evaluation (Figs. 3.18 and 3.19)

Collateral vessels can be assessed for:

1. Quality of the vessel proximal to the collateral
2. Entry angle
3. Size (CC classification)
4. Number
5. Tortuosity
6. Bifurcations
7. Exit angle to distal vessel
8. Location of entry into distal vessel

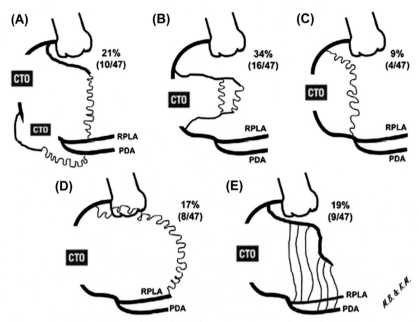

FIGURE 3.16 The varying courses and frequency of ipsilateral collateral connections (CCs) of the right coronary artery (RCA). (A) Type A: CCs originating from a high right marginal (RM) branch inserting to the right posterolateral artery (RPLA) or from a lower RM to the posterior descending artery (PDA). (B) Type B: CCs linking distal ends of higher and lower RMs, thereby bridging the chronic total occlusion of the RCA. (C) Type C: CCs originating directly from the proximal RCA and inserting close to the crux cordis. (D) Type D: CCs with a longer epimyocardial course inserting at the distal part of the RPLA. (E) Type E: septal intramyocardial, ipsilateral CCs, also called right superior descending artery. (*Curved lines* correspond to an epimyocardial course, *straight lines* correspond to an intramyocardial course of CCs.) *Reproduced with permission from Mashayekhi K, Behnes M, Akin I, Kaiser T, Neuser H. Novel retrograde approach for percutaneous treatment of chronic total occlusions of the right coronary artery using ipsilateral collateral connections: a European centre experience.* EuroIntervention *2016;11:e1231–6.*

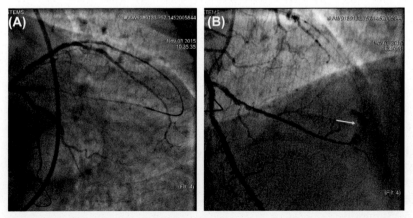

FIGURE 3.17 Example of epicardial collateral that was successfully crossed with a Corsair microcatheter (A), but developed a perforation upon microcatheter withdrawal (*arrow*, B). *Courtesy of Dr. Imre Ungi.*

Collateral assessment

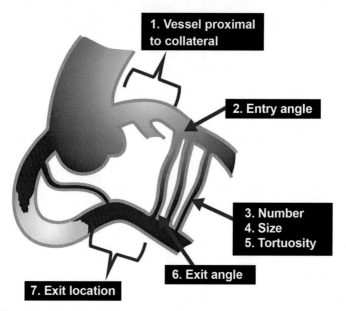

FIGURE 3.18   How to evaluate the suitability of a collateral vessel for retrograde crossing.

3.3.5.2.1 Quality of the Donor Vessel Proximal to the Collateral

When the collaterals originate distal to a severe lesion in the donor vessel, advancement of the microcatheter through the lesion can cause ischemia, proportionate to the size of the affected myocardium. For example in
patients with severe left main disease, advancing microcatheters through the
stenosis could cause severe ischemia and possibly hemodynamic collapse.
In such cases, it is preferable to first treat the vessel proximal to the collateral takeoff before attempting retrograde CTO PCI. It is also advisable
to allow enough time for the proximal vessel stent to heal or endothelialize
prior to bringing the patient back for CTO PCI. All efforts should be made
to avoid occluding or jeopardizing the collaterals during PCI of the proximal
diseased segment.

Access to the septal collateral through previously implanted stents can
be difficult in certain cases.[15] Even if the septal branch wiring is successful,
catheter passage through the stent cells might be impossible or it may
damage the tip or the hydrophilic coverage of the microcatheter. However,
predilation with a low profile monorail balloon usually can solve this
problem.

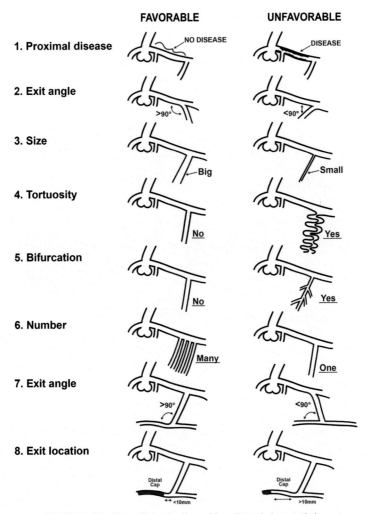

FIGURE 3.19   Favorable and unfavorable collateral characteristics.

Sometimes, extensive disease in the donor vessel may lead to selection of an alternative revascularization strategy, such as coronary artery bypass graft surgery (see Online Case 75).

3.3.5.2.2 Entry Angle

Obtuse angle (>90 degrees) facilitates entry into the collateral–acute angle (<90 degrees) makes entry more difficult and may require use of angled microcatheters (such as Venture and Supercross), a dual lumen microcatheter, or the hairpin wire technique (see Section 9.6.2).

TABLE 3.1 Werner Classification of Coronary Collateral Circulation

Werner Collateral Connection Grade[16]	
CC0	No continuous connection
CC1	Thread-like continuous connection
CC2	Side branch-like connection (≥0.4 mm)
CC3	>1 mm diameter of direct connection (not included in the original description)

3.3.5.2.3 Size (CC Classification)

Larger collaterals are easier to cross and less likely to lose flow during wire and microcatheter advancement. The most commonly used classification of collateral size is the Werner classification (Table 3.1; Fig. 3.20).

3.3.5.2.4 Number

Multiple collaterals not only offer more retrograde crossing options, but also minimize the risk of ischemia in case collateral flow is compromised during retrograde crossing attempts. Compromise of flow of a solitary collateral vessel supplying a large area of myocardium could result in profound ischemia, chest pain, and even hemodynamic collapse. Sometimes, however, if a dominant collateral gets compromised, other nonvisible collaterals may be recruited to supply the distal vessel. That is another reason why dual injection is important, even for CTOs without apparent contralateral collaterals.

On the other hand, a great number of septal collaterals can be confusing and make it difficult to select the best path for crossing with the wire. Epicardial collateral perforation carries high risk, because even if the antegrade route is successfully closed, there can be continuous retrograde bleeding through the other collaterals resulting in pericardial tamponade.

3.3.5.2.5 Tortuosity

Less tortuous collaterals are preferred, especially when they are small. In a single center analysis of 157 retrograde CTO recanalizations collateral tortuosity was one of the strongest predictors of failure.[17]

Extremely tortuous collaterals pose challenges in crossing with both the guidewire and microcatheter and are at higher risk of perforation. The Suoh 03 guidewire can be very useful in navigating such collaterals. Z-curve tortuosity should generally be avoided. Evaluation of the degree of tortuosity of a

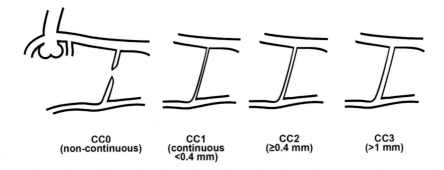

CC0
(non-continuous)

CC1
(continuous
<0.4 mm)

CC2
(≥0.4 mm)

CC3
(>1 mm)

Werner Classification

FIGURE 3.20   Werner classification of collateral size.

collateral during the cardiac cycle can provide useful information as to whether the collateral will be navigable with wires and microcatheters.

3.3.5.2.6 Bifurcations

The presence of a bifurcation within a collateral makes distal crossing of the collateral more challenging. Bifurcations pose a particular challenge during septal crossing if they are located in the first 3 mm of the septal branch.

3.3.5.2.7 Exit Angle

Similar to entry angle, an obtuse (>90 degrees) exit angle facilitates guidewire advancement into the distal vessel in the direction of the distal cap. Acute angles can be difficult to navigate and/or the guidewire may preferentially advance toward the distal vessel instead of advancing toward the distal cap. This is commonly seen when using SVGs for retrograde access.

3.3.5.2.8 Location of Entry Into the Distal Vessel

Ideally collaterals used for the retrograde approach should connect to the distal vessel, distal (>10 mm) to the distal cap, in a straight coronary segment. Collateral insertion at the distal cap (Fig. 3.21) makes CTO crossing challenging or impossible as the guidewire may preferentially enter the distal vessel or the proximal subintimal space instead of engaging the CTO distal cap.

Examples of favorable and unfavorable collateral branches are shown in Figs. 3.22 and 3.23.

> **Note**
> The presence of collaterals does not necessarily signify viability of the collateralized myocardial territory, as collaterals can also develop to nonviable territories.[18]

3.3.5.3 Optimal Views for Visualizing Collaterals

1. Septal collaterals
   a. Right anterior oblique (RAO) cranial is best for determining the origin of the collateral.
   b. Straight RAO or RAO caudal is best for visualizing the part of the collateral closer to the posterior descending artery (PDA), which is usually more tortuous than the part that is closer to the left anterior descending artery (LAD).
   c. Left anterior oblique (LAO) view may be helpful during wiring attempts.

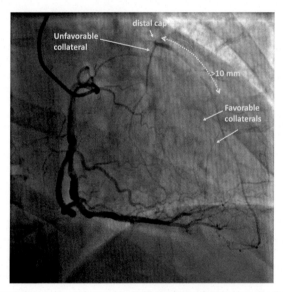

FIGURE 3.21   Example of the impact of collateral entry location into the distal vessel.

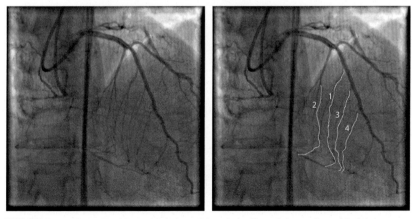

FIGURE 3.22   Example of the multiple favorable septal collaterals in a patient with right coronary artery chronic total occlusion. In such patients collaterals can be graded according to suitability, with crossing of more favorable collaterals attempted first (1>2>3>4).

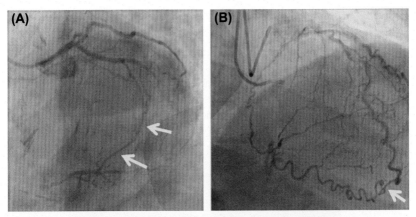

FIGURE 3.23 **Examples of favorable and unfavorable epicardial collateral vessels.** Favorable epicardial collateral from the circumflex to the right posterolateral branch with minimal tortuosity and adequate size (A). Unfavorable epicardial collateral from the left anterior descending artery to the right coronary artery because of extreme tortuosity and because it enters the right coronary artery very close to the distal cap (B). *Modified with permission from Joyal D, Thompson CA, Grantham JA, Buller CEH, Rinfret S. The retrograde technique for recanalization of chronic total occlusions: a step-by-step approach.* JACC Cardiovasc Interv *2012;5:1–11, Elsevier.*

2. Epicardial collaterals in the lateral wall (diagonal–obtuse marginal vessels)
   a. LAO cranial
   b. RAO cranial
3. Epicardial collaterals between the proximal circumflex and right coronary artery (RCA)
   a. Anteroposterior caudal
   b. RAO

3.3.5.4 Rentrop Classification

The Rentrop classification is commonly used to describe the filling of the distal vessel (Table 3.2).[19]

3.3.6 Use of Computed Tomography Angiography

Although infrequently used at present, preprocedural coronary computed tomography (CT) angiography may provide assessment of the proximal cap location, calcification, tortuosity, vessel course, and length of the occluded segment. It can also help identify the best angiographic projection for CTO

TABLE 3.2 The Rentrop Classification of Distal Vessel Filling[19]

Rentrop Classification[19] (Developed for Occluded and Nonoccluded Arteries)	
0	No filling of collateral vessels
1	Filling of collateral vessels without any epicardial filling of the target artery
2	Partial epicardial filling by collateral vessels of the target artery
3	Complete epicardial filling by collateral vessels of the target artery (in chronic total occlusions, Rentrop 3 is prevalent in 85% of lesions)

crossing.[20] In aorto-ostial right coronary occlusion with no visible calcification, CT angiography may help identify the location of the ostium or an aberrant origin of the vessel. CT angiography can be very helpful in cases with poor visualization of the distal target vessel and identifying the origin of anomalous coronary arteries (Fig. 3.24). CT angiography is, however, limited in identifying collateral circulation.

Potential future applications of preoperative CT angiography include the prediction of a successful antegrade technique[21,22] and possible improvements in the selection of patients for referral to more experienced CTO operators. In addition, improvements in coregistration of coronary CT angiography with real-time fluoroscopy (Fig. 3.25)[23] might help the operator better identify distal wire position with intraprocedural use of coronary CT angiography.

3.4 VASCULAR ACCESS

3.4.1 Dual Arterial Access

This is the most important step for increasing CTO PCI success rates (see benefits of dual injection in Section 3.2). It should be used in nearly all cases, except possibly those with no contralateral collaterals (however, even in those cases contralateral collateral recruitment can occur during the procedure if ipsilateral collaterals are compromised).

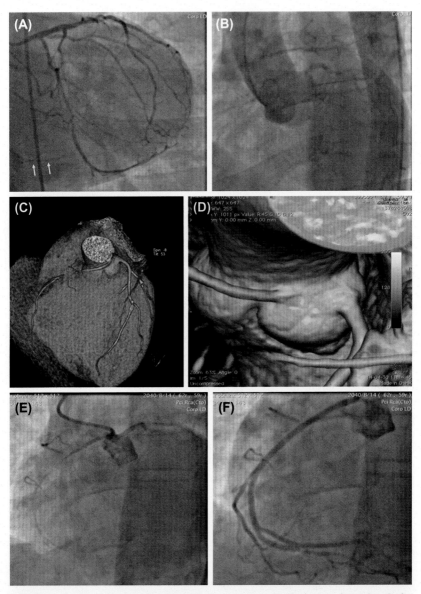

FIGURE 3.24 **Example of using computed tomography to facilitate chronic total occlusion (CTO) percutaneous coronary intervention.** A patient presented with angina and inferior ischemia. Although collaterals to the right posterior descending artery were visualized (*arrows*, A), it was impossible to find the ostium of the right coronary artery (B). Computed tomography demonstrated an anomalous right coronary artery originating from the left sinus of Valsalva (C, D). The right coronary artery had a short CTO (C) but did not have any calcification. The right coronary artery was engaged with a 3D right coronary artery guide (E) and successfully recanalized (F) using antegrade wire escalation. *Courtesy of Dr. Leszek Bryniarski.*

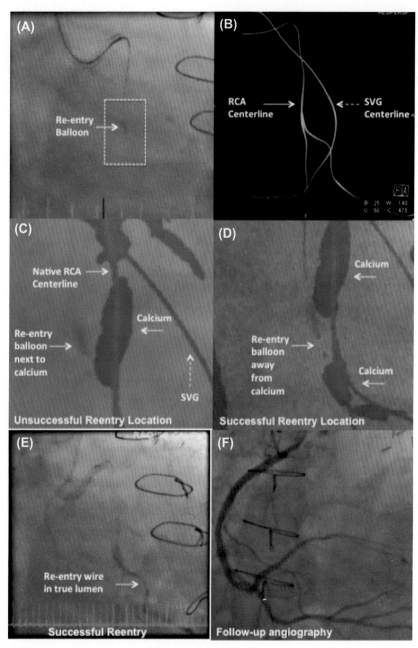

FIGURE 3.25   Analysis of unsuccessful then successful antegrade dissection reentry during CTO PCI. In this example, coronary CTA fusion data display was chosen to show both the centerline and the arterial calcium. (A) Fluoroscopic image showing a reentry balloon

---

Arterial Access Selection for CTO PCI

- Bifemoral access is preferred by most operators, especially early in the learning curve, as it may allow easier vessel engagement and enhanced guide catheter support (with use of large 8 Fr guide catheters); it carries, however, increased risk of vascular access complications compared with the radial approach. Bifemoral access may be particularly important for more complex lesions: in a Japanese study of complex CTO PCI cases (J-CTO score ≥3) the transradial group had a significantly lower success rate than the transfemoral group (35.7% vs. 58.2%; $P = .04$).[24]
- Femoral–radial access may be used, especially when the retrograde approach is unlikely to be utilized. In most men, a 7 Fr guide catheter can be used without any injury of the radial artery using a 7 Fr slender sheath (Terumo). From the right radial artery an excellent support can be achieved in the left main with a 7 Fr EBU 4.0 guide catheter, therefore even retrograde RCA recanalization can be accomplished with this hybrid vascular access.
- Biradial access can be successfully used in centers with significant radial access expertise (Section 2.1).[25,26] However many centers will use radial sheaths larger than the traditional 6 Fr for CTO cases (such as 7 Fr Slender sheaths) or use sheathless guides.

Long sheaths (usually 45 cm) enhance support and facilitate guide catheter manipulations.

3.4.2 Advanced: Triple Arterial Access

Occasionally triple arterial access may be needed for CTO PCI. An example is an RCA CTO with distal filling via the LAD, which is supplied by a left internal mammary artery graft (Fig. 3.26), when a retrograde approach is planned via a septal collateral that originates proximal to the LAD occlusion.

←────────────────────────────────────

placed distal to an RCA CTO (as confirmed by contralateral contrast injection; image not shown). The initial attempt at reentry from this location was unsuccessful. (B) Centerlines from CTA segmentation showing the RCA and SVG. (C) Magnified CTA/fluoroscopy fusion of the initial unsuccessful reentry site showing that the reentry balloon resided next to a large zone of calcium (*yellow arrow*). The reentry balloon was then advanced more distally, approximately 1 cm, and successful reentry was performed using the stick-and-swap technique. (D) Magnified CTA/fluoroscopy fusion revealed that the successful reentry zone was between two areas of calcification. (E) Fluoroscopic image during contralateral contrast injection shows the reentry wire (*white arrow*) in the distal RCA true lumen. (F) Follow-up angiogram showing successful CTO PCI revascularization of the RCA. *CTA*, computed tomography angiography; *CTO*, chronic total occlusion; *PCI*, percutaneous coronary intervention; *RCA*, right coronary artery; *SVG*, saphenous vein graft. *Reproduced with permission from Ghoshhajra BB, Takx RA, Stone LL, et al. Real-time fusion of coronary CT angiography with x-ray fluoroscopy during chronic total occlusion PCI. Eur Radiol 2016.*

3.5 ANTICOAGULATION

- Unfractionated heparin is the preferred anticoagulation agent for CTO PCI, because it can be reversed in case of severe perforation. The recommended activated clotting times (ACTs) are:
  - >300 s for antegrade CTO PCI.
  - >350 s for retrograde CTO PCI (some operators use >300 s but check ACT very frequently if it is in the low 300 s range).
- The ACT should be checked every 20–30 min, depending on how high above the target ACT the most recent measurement was. Any contamination of the blood specimen with water, contrast, and drugs must be strictly avoided because it may strongly influence the result.
- Some operators administer a heparin drip in addition to the initial bolus to minimize changes in anticoagulation levels.
- A small (4 Fr) femoral venous sheath can be inserted to facilitate ACT checking by the cath lab staff to minimize interruptions of the physician tasks. Some operators use an 8 Fr sheath with a 7 Fr guide catheter, which allows them to draw blood from the side arm of the sheath.
- Bivalirudin is best avoided because its anticoagulant effect cannot be reversed. Moreover, there are unpublished cases in which guide thrombosis occurred during long procedures.
- Glycoprotein IIb/IIIa inhibitors should NOT be given, **even after successful crossing and stenting** of the CTO, because a minor wire perforation could reopen and cause delayed pericardial effusion and tamponade.

3.6 TECHNIQUES TO INCREASE GUIDE CATHETER SUPPORT (LARGE BORE GUIDE, ACTIVE GUIDE SUPPORT, GUIDE CATHETER EXTENSIONS, ANCHOR TECHNIQUES)

Strong guide support is essential for achieving high CTO PCI success rates and for maximizing efficiency, irrespective of the crossing technique selected. Techniques to increase guide catheter support include (Fig. 3.27):

---

1. Deep guide catheter intubation (active support).[27]
2. Large and supportive shape guide catheters (passive support).
3. One or multiple buddy wires.[28]
4. Guide catheter extensions (Section 2.3.5) and specialized guidewire support catheters (Sections 2.4.13–2.4.15).
5. Anchor techniques.[29,30]

These techniques can also be used in combination; an example is the Anchor–Tornus technique.[31]

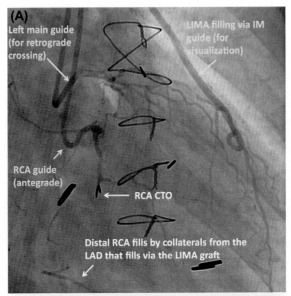

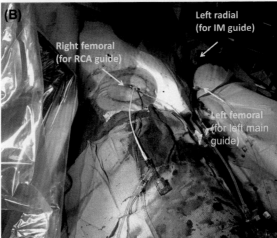

FIGURE 3.26 **Example of triple arterial access.** Bifemoral and left radial access were obtained to enable retrograde crossing of a right coronary artery chronic total occlusion. The right femoral artery was used for engaging the right coronary artery, the left femoral artery for engaging the left main (through which retrograde crossing was performed), and the left radial artery was used for engaging the left internal mammary artery graft (through which visualization of the right posterior descending artery was achieved). *Reproduced from Michael TT, Banerjee S, Brilakis ES. Role of internal mammary artery bypass grafts in retrograde chronic total occlusion interventions.* J Invasive Cardiol *2012;24:359–62, with permission from HMP Communications.*

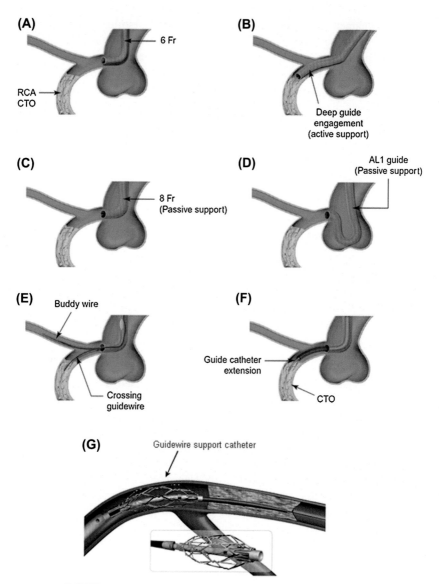

(A)

6 Fr

RCA
CTO

(B)

Deep guide
engagement
(active support)

(C)

8 Fr
(Passive support)

(D)

AL1 guide
(Passive support)

(E)   Buddy wire

Crossing
guidewire

(F)

Guide catheter
extension

CTO

(G)   Guidewire support catheter

FIGURE 3.27   Overview of active and passive guide support techniques.

3.6.1 Active Support

Deep guide catheter insertion can significantly increase backup support but carries a small risk of dissection. Active guide catheter support can be achieved by clockwise rotation for the right coronary artery guide and forward pushing for the left main guide (Fig. 3.27B). It is important to be aware of the various guide tip characteristics from different manufacturers (softer vs. sharper tips, as in sheathless guide catheters). It is also imperative to closely monitor the pressure tracing of the guide and avoid injecting when the pressure waveform is dampened to avoid proximal vessel dissection. Sometimes, aggressive maneuvers for active support may result in seating the guide catheter onto an aortic cusp, impeding opening the valve and/or causing acute aortic regurgitation that can lead to hemodynamic collapse.

3.6.2 Passive Support

Passive guide support can be increased by larger size guides (Fig. 3.27C) or more supportive shapes (such as Amplatz guide catheter, Fig. 3.27D).

3.6.3 Buddy Wires

One or more buddy wires can be advanced in the main vessel or a side branch (Fig. 3.27E), generally using supportive workhorse wires, a technique that usually provides only mild to moderate support. A nonpolymeric stiff wire (such as the BHW, Ironman, or Grand Slam) is preferred for this task, because of better stability as compared with slippery polymer-jacketed wires like the Whisper ES.

3.6.4 Guide Catheter Extensions and Support Catheters

Three guide catheter extensions are available in the United States in 2017: the Guideliner V3 (Vascular Solutions), the Trapliner (Vascular Solutions), and the Guidezilla II (Boston Scientific), which are described in detail in Section 2.3.5. Other guide catheter extensions are available outside the United States, such as the Guidion (IMDS). Use of guide catheter extensions can significantly increase guide catheter support and allow use of smaller caliber guide catheters (Fig. 3.27F).

Four support catheters are available in the United States in 2017: the Multicross, the Centercross (Section 2.4.13), the Prodigy (Section 2.4.14), and the Novacross (Section 2.4.15). These catheters can be considered as guide catheter extensions with extra support from a nitinol cage (for the MultiCross and CenterCross), nitinol wires (Novacross), or a low-pressure balloon (for the Prodigy catheter) (Fig. 3.27G).

3.6.5 Anchor Techniques

Anchor techniques can be useful to increase guide catheter support. The most common is the **side-branch anchoring** technique (Fig. 3.28 A), in which a balloon is inflated in a proximal side branch of the target vessel.[17,31–39] Using a long balloon increases support by increasing the grip of the vessel, especially if the hydrophilic coating is wiped off the surface of the balloon with a wet gauze before insertion. In the **co-axial anchor** technique (Fig. 3.28B), a balloon is inflated in the coronary artery proximal to the occlusion, enhancing the

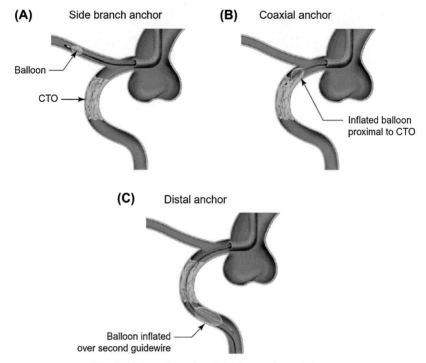

(A) Side branch anchor   **(B)** Coaxial anchor

Balloon

CTO

Inflated balloon
proximal to CTO

(C) Distal anchor

Balloon inflated
over second guidewire

FIGURE 3.28   Illustration of the three anchor techniques.

guidewire penetration capacity.[35] The **distal anchor** technique (Figs. 3.28C and 3.29) is similar to the side-branch anchor, except that the balloon is inflated distal to or at the occlusion within the target artery.[29,40] Two guidewires are required for the distal anchor technique, one to deliver the anchor balloon and a second guidewire (which is pinned by the anchor balloon against the vessel wall) for delivering equipment such as microcatheters, balloons, stents, and guide catheter extensions to the lesion (Fig. 3.29).[34,36,41–43] The **buddy wire stent anchor** technique (Fig. 3.30) can be used if the proximal vessel needs stenting: a buddy wire can be inserted and a stent deployed over this guidewire, effectively trapping the buddy wire, which then provides strong guide catheter support.

3.6.5.1 Limitations of the Anchor Techniques

a. Injury at the site of the anchor balloon inflation, which is usually inconsequential in these small side branches.[29] The risk can be minimized by sizing the balloon 1:1 to the side branch and inflating the anchor balloon at relatively low pressures (4–8 atm).
b. Distal dissection can occur with the distal anchor technique, potentially leading to extensive stenting.

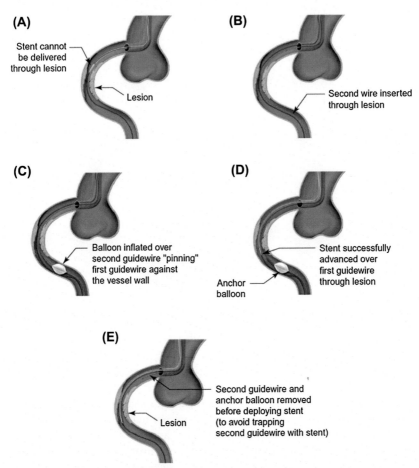

(A) Stent cannot be delivered through lesion

Lesion

(B) Second wire inserted through lesion

(C) Balloon inflated over second guidewire "pinning" first guidewire against the vessel wall

(D) Stent successfully advanced over first guidewire through lesion

Anchor balloon

(E) Second guidewire and anchor balloon removed before deploying stent (to avoid trapping second guidewire with stent)

Lesion

FIGURE 3.29 **Step-by-step illustration of the distal anchor technique.** The distal anchor technique can be useful when difficulty is encountered delivering a stent (or other equipment) through a lesion (A). A second guidewire is inserted next to the initial guidewire (B) and a balloon (which is usually easier to deliver than a stent) is delivered distal to the lesion and inflated, pinning the initial guidewire against the vessel wall (C) and enabling stent delivery over the first guidewire (D). The second guidewire (and balloon) are subsequently withdrawn before stent deployment (E).

c. Larger guide catheters (at least 7 Fr) are needed for delivering a distal anchor balloon and other equipment.[36,42,43]

d. Use of a side branch anchor may modify the proximal vessel anatomy and hinder antegrade wiring.

e. Rarely, ischemia can occur if larger branches are used for anchoring, such as diagonal branches. Allowing intermittent deflation of the anchoring balloon may be necessary to relieve ischemia in such cases.

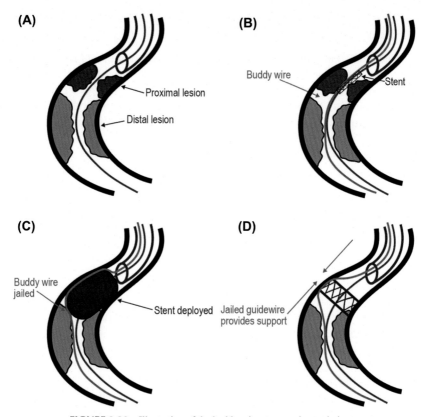

FIGURE 3.30 Illustration of the buddy wire stent anchor technique.

3.7 TRAPPING AND OTHER TECHNIQUES TO REMOVE AN OVER-THE-WIRE SYSTEM OVER A SHORT GUIDEWIRE

There are four techniques for removing or exchanging microcatheters or any over-the-wire system when using a short (180–190 cm) guidewire:

1. Trapping (preferred technique)
2. Hydraulic exchange
3. Use of a guidewire extension
4. Circumcision of the over-the-wire system

3.7.1 Trapping

Trapping is the best technique for removing or exchanging microcatheters or any over-the-wire system when using a short (180–190 cm) or even long (300 cm) guidewire, because it:

a. Minimizes guidewire movement, which can result in guidewire position loss, distal vessel injury, and/or perforation.

b. Minimizes radiation exposure (as fluoroscopy is only needed during the initial phase of catheter withdrawal).

The small inner diameter of a 6 Fr guide catheter prevents utilization of the trapping technique for over-the-wire balloons, the CrossBoss catheter, the Tornus, and the Stingray balloon. However, 6 Fr guides with 0.071 inch inner diameter such as Medtronic Launchers can reliably trap low-profile catheters such as Turnpike, SuperCross, Corsair, Caravel, and Finecross, especially with use of the Trapper balloon.

The TrapLiner catheter (Chapter 2, Section 2.3.5) combines guide extension functionality with a built-in trapping balloon to pin short guidewires. This catheter is available in 6–8 Fr sizes. The advantages of this system include reduced requirement for fluoroscopy to position a trapping balloon and reduced equipment use, as trapping balloons can be difficult to reinsert through the Y-connector after multiple inflations during prolonged procedures.

3.7.1.1 Trapping Technique

Step 1: Withdraw the over-the-wire balloon or microcatheter into the guide catheter just proximal to the position where the trapping balloon will be inflated (Fig. 3.31B)

Step 2: Insert the trapping balloon through the Y-connector, next to (but not over) the guidewire. Keeping the stylet in the end of the balloon catheter can assist in the advancement through the Y-connector.

Type of balloon: Rapid-exchange balloons are preferred because over-the-wire balloons have larger profiles and may not fit into the guide along with other balloons, especially the Stingray balloon or the Venture catheter. The specially designed Trapper balloon (Section 2.12) is currently the preferred trapping balloon.

Size of the balloon: 2.5 mm for 6 Fr guide catheters, 3.0 mm for 7 and 8 Fr guide catheters.

Length: Ideally ≥20 mm (longer balloon length provides more area of contact and pins the wire better, which is especially important for trapping polymer-jacketed wires, which are more slippery).

Caveats: A previously used trapping balloon may get deformed and be difficult to insert through the Y-connector on subsequent attempts. In such cases the trapping balloon should be exchanged for a new one.

Step 3: Advance the trapping balloon to the distal portion of the guide catheter beyond the marker of the microcatheter or over-the-wire balloon, usually at or near the primary curve (Fig. 3.31C).

Caveat: When short guides are used, balloon markers cannot be relied upon to prevent inadvertent advancement of the trapping balloon into the target vessel.

Step 4: Inflate the trapping balloon (Fig. 3.31D).

Pressure: 15–20 atm to provide adequate trapping.

Caveat: Balloon may rupture if inflated next to handmade side holes.

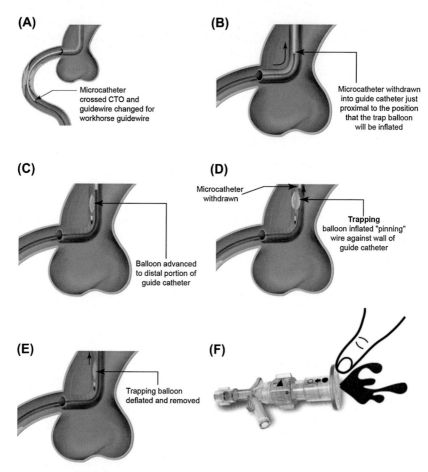

(A)

Microcatheter crossed CTO and guidewire changed for workhorse guidewire

(B)

Microcatheter withdrawn into guide catheter just proximal to the position that the trap balloon will be inflated

(C)

Balloon advanced to distal portion of guide catheter

(D)

Microcatheter withdrawn

Trapping balloon inflated "pinning" wire against wall of guide catheter

(E)

Trapping balloon deflated and removed

(F)

FIGURE 3.31 **Illustration of the trapping technique.** After the chronic total occlusion has been crossed and the crossing guidewire replaced with a workhorse guidewire (A), the microcatheter (or over-the-wire balloon) is withdrawn into the guide catheter (B), the trapping balloon is advanced distal to the microcatheter tip (C) and inflated (D), enabling removal of the microcatheter. The trapping balloon is then deflated and removed (E), followed by back-bleeding of the guide catheter to prevent air embolism (F).

Step 5: Withdraw the over-the-wire balloon and microcatheter from the guide catheter (Fig. 3.31D). Once the trapping balloon is in place and inflated, fluoroscopy is generally not required to maintain distal wire position. Extra care, however, must be utilized when working with polymer-jacketed guidewires, as they have a tendency to slip past the inflated trap balloon if significant negative tension is exerted.

Step 6: Deflate and remove the trapping balloon (Fig. 3.31E).

Step 7: Back-bleed the Y-connector (Fig. 3.31F).

This is a very important step, because often air is entrained into the guide catheter during trapping. If contrast is injected without back-bleeding, coronary air embolization is likely to occur.

Step 8: The microcatheter has been successfully removed while maintaining distal wire position.

Tip

If an anchor balloon is kept in a side branch and its size is appropriate (i.e., ≥2.5 mm in a 6 or 7 Fr guide catheter), this balloon can be deflated and temporarily withdrawn into the guide catheter for trapping. This maneuver saves time and cost.

3.7.2 Hydraulic Exchange (Fig. 3.32)

Hydraulic exchange (also called jet exchange or Nanto technique) is easier to perform, but is less reliable and may not maintain the distal position of the guidewire.[44,45] Hydraulic exchange can be performed through any size guide catheter as follows:

Step 1: Fill an inflating device with normal saline (alternatively a standard mixture of contrast and saline can be used). The inflating device should be filled with the maximum volume possible. Alternatively, a 3 cc syringe can be used instead of an inflating device.

Step 2: Withdraw the microcatheter or over-the-wire balloon as far as possible until the back end of the short guidewire is at the hub of the microcatheter/over-the-wire balloon (Fig. 3.32A).

Step 3: Connect the saline-filled inflating device to the hub of the microcatheter/over-the-wire balloon (wet-to-wet connection) (Fig. 3.32B and C).

Step 4: Inflate the inflating device to 20 atm, while performing fluoroscopy (Fig. 3.32D).

Step 5: Under fluoroscopy, when the inflating device pressure reaches 20 atm, withdraw the microcatheter/over-the-wire balloon (Fig. 3.32E). It is very important that the inflation device pressure is maintained at 20 atm as the microcatheter is withdrawn. Otherwise pressure may be lost and the wire may be withdrawn.

Caveats: Use of the hydraulic exchange should be avoided in CTO PCI in general and especially when using stiff, tapered-tip guidewires or stiff polymer-jacketed guidewires to minimize the risk for inadvertently advancing the guidewire and causing distal guidewire perforation. Hydraulic exchange should also be avoided for removing over-the-wire balloon catheters and the Stingray balloon because of an increased friction between the inner surface of the catheter and the guidewire that may result in loss of wire position.

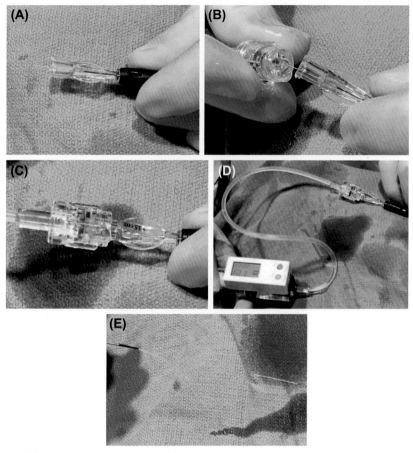

FIGURE 3.32   Illustration of the hydraulic exchange technique. *Courtesy of Dr. William Nicholson.*

3.7.3 Use of a Guidewire Extension (Fig. 3.33)

3.7.3.1 How?

By inserting the back end of the short guidewire into the guidewire extension. Using a guidewire extension can be performed with any size guide catheter.

3.7.3.2 Caveats

1. Each guidewire type has specific guidewire extension that should be available in the cath lab. For example:
   a. Abbott guidewires: DOC guidewire extension (145 cm long)
   b. Asahi guidewires: Asahi guidewire extension (150/165 cm long)
2. The connection should be tightened as much as possible to minimize the risk of the guidewire and guidewire extension coming apart during withdrawal of the microcatheter or over-the-wire balloon.
3. Be careful to avoid kinking or bending of the wire when tightening the connection.

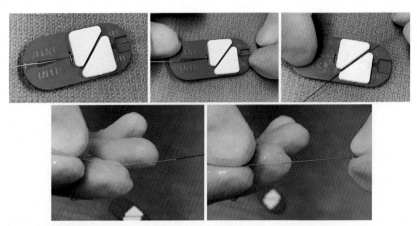

FIGURE 3.33 Using a guidewire extension to remove a microcatheter over a short guidewire. *Courtesy of Dr. William Nicholson.*

3.7.4 Circumcision Technique (Fig. 3.34)

3.7.4.1 How to Perform the Microcatheter Circumcision Technique

(see Online Video "Microcatheter "circumcision" technique").

This is an advanced and cumbersome technique, but can be very useful, especially in cases where the guidewire becomes entrapped within the microcatheter/over-the-wire balloon. The circumcision technique can be performed with any size guide catheter.

> **Step 1**: Withdraw the microcatheter or over-the-wire balloon as far as possible until the back end of the short guidewire is at the hub of the microcatheter/over-the-wire balloon.
>
> **Step 2**: Using a scalpel and a hard surface, a circumferential cut is made as close to the Y-introducer as possible (Fig. 3.34A).
>
> **Step 3**: Once circumferential cutting is complete, the microcatheter/over-the-wire balloon fragment is removed (Fig. 3.34B).
>
> **Step 4**: Steps 1–3 are repeated until the entire microcatheter/over-the-wire balloon is removed.

3.7.4.2 Caveats

1. Cutting should be done very carefully to minimize the risk of injuring the patient or the operator, and to avoid damaging the guidewire, making it difficult, if not impossible, to pass other catheters over it.
2. Consider using a hard surface, such as an upside down saline bowl to perform the circumcision on.
3. The microcatheter/over-the-wire balloon will be destroyed during this maneuver and hence cannot be reused.

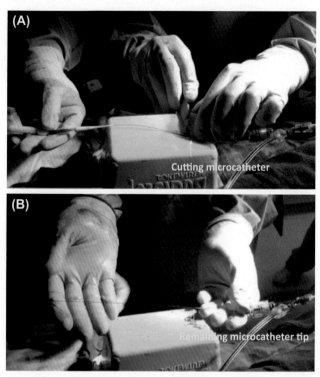

FIGURE 3.34 **Illustration of the circumcision technique for removing a microcatheter over a short guidewire.** (A) A scalpel is used over a hard surface (backstop splash prevention device in this case) to cut the proximal part of the microcatheter. (B) After the microcatheter is cut, the proximal part of the microcatheter is removed and the process is repeated until the entire microcatheter has been removed.

3.8 MONITORING PRESSURE WAVEFORM AND ELECTROCARDIOGRAM

This is a basic rule of cardiac catheterization that also applies to CTO PCI. Continuous monitoring of the pressure waveform and the electrocardiogram can help identify potential problems or complications early, so that preventive or treatment measures can be undertaken. It is recommended to use different colors on antegrade and retrograde guiding catheter hemodynamic tracings to facilitate monitoring during the case.

Electrocardiographic changes of concern include:

1. New ST segment depression
2. New ST segment elevation (Fig. 3.35)
3. Bradycardia
4. Tachycardia
5. Ventricular extrasystoles during wire manipulations
6. Ventricular fibrillation

Pressure waveform changes of concern include:

1. Hypotension (Section 12.1.2.1; a common cause during CTO PCI is prolapse of an Amplatz guide catheter through the aortic valve causing severe aortic regurgitation)
2. Pulsus paradoxus (Fig. 3.35)
3. Hypertension

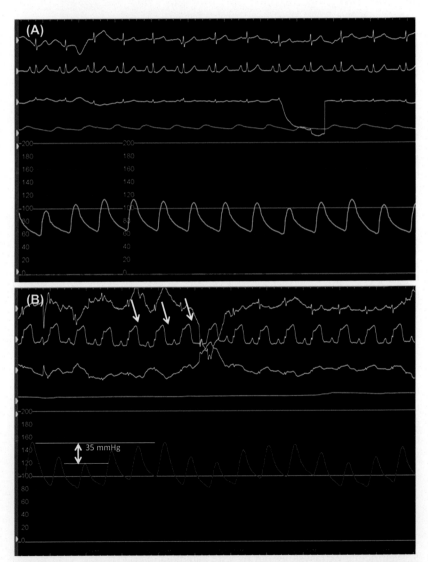

FIGURE 3.35   Electrocardiographic and pressure waveform changes during chronic total occlusion percutaneous coronary intervention. (A) Baseline. (B) ST-segment elevation (*arrows*) and development of 35 mmHg pulsus paradoxus after perforation of side branch in a patient with prior coronary artery bypass graft surgery.

4. Pressure dampening. The waveform of an ostially or proximally occluded vessel may be dampened with guide catheter engagement. This can be managed with the use of side-hole guides (Section 2.3.3) or careful attention to avoidance of injection through dampened guides. Injecting while pressure is dampened can lead to coronary or aortocoronary dissection and/or air embolism.

Tip: Since two guide catheters are used in the majority of CTO PCI cases, color coding the right coronary guide catheter waveform red can facilitate rapid and easy identification of the source of each pressure waveform (Fig. 3.36).

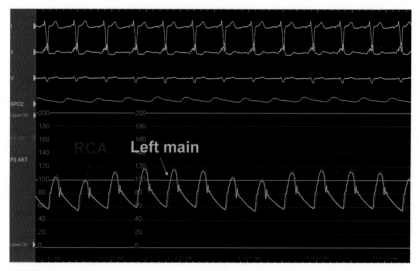

FIGURE 3.36 Suggested color-coding for differentiating the origin of each pressure waveform (*red = right, other color = left main*).

In summary, knowledge and consistent application of the following eight basic techniques can significantly enhance the likelihood of CTO PCI success:
1. Avoiding ad hoc procedures.
2. Performing dual injection in nearly all cases.
3. Carefully studying the target lesion before the procedure.
4. Using long, bifemoral, 8 Fr arterial sheaths.
5. Optimizing anticoagulation.
6. Applying various techniques to increase guide catheter support.
7. Consistently using the trapping technique for equipment exchanges.
8. Constantly monitoring the pressure waveform and the electrocardiogram.

REFERENCES

1. Blankenship JC, Gigliotti OS, Feldman DN, et al. Ad hoc percutaneous coronary intervention: a consensus statement from the Society for Cardiovascular Angiography and Interventions. *Catheter Cardiovasc Interv* 2013;**81**:748–58.
2. Brilakis ES, Banerjee S, Lombardi WL. Retrograde recanalization of native coronary artery chronic occlusions via acutely occluded vein grafts. *Catheter Cardiovasc Interv* 2010;**75**:109–13.
3. Gasparini GL, Oreglia JA, Reimers B. A case of retrograde left main primary percutaneous coronary intervention during cardiogenic shock: the added value of performing coronary chronic total occlusion procedures. *Int J Cardiol* 2016;**215**:396–8.
4. Patel VG, Zankar A, Brilakis E. Use of the retrograde approach for primary percutaneous coronary intervention of an inferior ST-segment elevation myocardial infarction. *J Invasive Cardiol* 2013;**25**:483–4.
5. Deharo P, Strange JW, Mozid A. Primary percutaneous coronary intervention of native chronic total occlusions to treat ST elevation myocardial infarction secondary to acute vein graft occlusion. *Catheter Cardiovasc Interv* 2017;**90**:251–6.
6. Singh M, Bell MR, Berger PB, Holmes Jr DR. Utility of bilateral coronary injections during complex coronary angioplasty. *J Invasive Cardiol* 1999;**11**:70–4.
7. Brilakis ES, Grantham JA, Rinfret S, et al. A percutaneous treatment algorithm for crossing coronary chronic total occlusions. *JACC Cardiovasc Interv* 2012;**5**:367–79.
8. Nicholson WJ, Rab T. Simultaneous diagnostic coronary angiography utilizing a single arterial access technique. *Catheter Cardiovasc Interv* 2006;**68**:718.
9. Morino Y, Abe M, Morimoto T, et al. Predicting successful guidewire crossing through chronic total occlusion of native coronary lesions within 30 minutes: the J-CTO (Multicenter CTO Registry in Japan) score as a difficulty grading and time assessment tool. *JACC Cardiovasc Interv* 2011;**4**:213–21.
10. Nombela-Franco L, Urena M, Jerez-Valero M, et al. Validation of the J-chronic total occlusion score for chronic total occlusion percutaneous coronary intervention in an independent contemporary cohort. *Circ Cardiovasc Interv* 2013;**6**:635–43.
11. Kotsia A, Christopoulos G, Brilakis ES. Use of the retrograde approach for preserving the distal bifurcation after antegrade crossing of a right coronary artery chronic total occlusion. *J Invasive Cardiol* 2014;**26**:E48–9.
12. Brilakis ES, O'Donnell CI, Penny W, et al. Percutaneous coronary intervention in native coronary arteries versus bypass grafts in patients with prior coronary artery bypass graft surgery: insights from the Veterans Affairs Clinical assessment, Reporting, and tracking Program. *JACC Cardiov Interv* 2016;**9**:884–93.
13. Kahn JK, Hartzler GO. Retrograde coronary angioplasty of isolated arterial segments through saphenous vein bypass grafts. *Catheter Cardiovasc Diagn* 1990;**20**:88–93.
14. Mashayekhi K, Behnes M, Akin I, Kaiser T, Neuser H. Novel retrograde approach for percutaneous treatment of chronic total occlusions of the right coronary artery using ipsilateral collateral connections: a European centre experience. *EuroIntervention* 2016;**11**:e1231–6.
15. Dash D. Complications encountered in coronary chronic total occlusion intervention: prevention and bailout. *Indian Heart J* 2016;**68**:737–46.
16. Werner GS, Ferrari M, Heinke S, et al. Angiographic assessment of collateral connections in comparison with invasively determined collateral function in chronic coronary occlusions. *Circulation* 2003;**107**:1972–7.

17. Rathore S, Katoh O, Matsuo H, et al. Retrograde percutaneous recanalization of chronic total occlusion of the coronary arteries: procedural outcomes and predictors of success in contemporary practice. *Circ Cardiovasc Interv* 2009;**2**:124–32.

18. Heil M, Schaper W. Influence of mechanical, cellular, and molecular factors on collateral artery growth (arteriogenesis). *Circ Res* 2004;**95**:449–58.

19. Rentrop KP, Cohen M, Blanke H, Phillips RA. Changes in collateral channel filling immediately after controlled coronary artery occlusion by an angioplasty balloon in human subjects. *J Am Coll Cardiol* 1985;**5**:587–92.

20. Magro M, Schultz C, Simsek C, et al. Computed tomography as a tool for percutaneous coronary intervention of chronic total occlusions. *EuroIntervention* 2010;**6**(Suppl. G):G123–31.

21. Opolski MP, Achenbach S, Schuhback A, et al. Coronary computed tomographic prediction rule for time-efficient guidewire crossing through chronic total occlusion: insights from the CT-RECTOR multicenter registry (Computed Tomography Registry of Chronic Total Occlusion Revascularization). *JACC Cardiovasc Interv* 2015;**8**:257–67.

22. Luo C, Huang M, Li J, et al. Predictors of interventional success of antegrade PCI for CTO. *JACC Cardiovasc Imaging* 2015;**8**:804–13.

23. Ghoshhajra BB, Takx RA, Stone LL, et al. Real-time fusion of coronary CT angiography with x-ray fluoroscopy during chronic total occlusion PCI. *Eur Radiol* 2017;**27**:2464–73.

24. Tanaka Y, Moriyama N, Ochiai T, et al. Transradial coronary interventions for complex chronic total occlusions. *JACC Cardiovasc Interv* 2017;**10**:235–43.

25. Burzotta F, De Vita M, Lefevre T, Tommasino A, Louvard Y, Trani C. Radial approach for percutaneous coronary interventions on chronic total occlusions: Technical issues and data review. *Catheter Cardiovasc Interv* 2014;**83**:47–57.

26. Alaswad K, Menon RV, Christopoulos G, et al. Transradial approach for coronary chronic total occlusion interventions: insights from a contemporary multicenter registry. *Catheter Cardiovasc Interv* 2015;**85**:1123–9.

27. Von Sohsten R, Oz R, Marone G, McCormick DJ. Deep intubation of 6 French guiding catheters for transradial coronary interventions. *J Invasive Cardiol* 1998;**10**:198–202.

28. Burzotta F, Trani C, Mazzari MA, et al. Use of a second buddy wire during percutaneous coronary interventions: a simple solution for some challenging situations. *J Invasive Cardiol* 2005;**17**:171–4.

29. Di Mario C, Ramasami N. Techniques to enhance guide catheter support. *Catheter Cardiovasc Interv* 2008;**72**:505–12.

30. Saeed B, Banerjee S, Brilakis ES. Percutaneous coronary intervention in tortuous coronary arteries: associated complications and strategies to improve success. *J Interv Cardiol* 2008;**21**:504–11.

31. Kirtane AJ, Stone GW. The Anchor-Tornus technique: a novel approach to "uncrossable" chronic total occlusions. *Catheter Cardiovasc Interv* 2007;**70**:554–7.

32. Fujita S, Tamai H, Kyo E, et al. New technique for superior guiding catheter support during advancement of a balloon in coronary angioplasty: the anchor technique. *Catheter Cardiovasc Interv* 2003;**59**:482–8.

33. Hirokami M, Saito S, Muto H. Anchoring technique to improve guiding catheter support in coronary angioplasty of chronic total occlusions. *Catheter Cardiovasc Interv* 2006;**67**:366–71.

34. Matsumi J, Saito S. Progress in the retrograde approach for chronic total coronary artery occlusion: a case with successful angioplasty using CART and reverse-anchoring techniques 3 years after failed PCI via a retrograde approach. *Catheter Cardiovasc Interv* 2008;**71**:810–4.

35. Fang HY, Wu CC, Wu CJ. Successful transradial antegrade coronary intervention of a rare right coronary artery high anterior downward takeoff anomalous chronic total occlusion by double-anchoring technique and retrograde guidance. *Int Heart J* 2009;**50**:531–8.
36. Lee NH, Suh J, Seo HS. Double anchoring balloon technique for recanalization of coronary chronic total occlusion by retrograde approach. *Catheter Cardiovasc Interv* 2009;**73**:791–4.
37. Saito S. Different strategies of retrograde approach in coronary angioplasty for chronic total occlusion. *Catheter Cardiovasc Interv* 2008;**71**:8–19.
38. Surmely JF, Katoh O, Tsuchikane E, Nasu K, Suzuki T. Coronary septal collaterals as an access for the retrograde approach in the percutaneous treatment of coronary chronic total occlusions. *Catheter Cardiovasc Interv* 2007;**69**:826–32.
39. Surmely JF, Tsuchikane E, Katoh O, et al. New concept for CTO recanalization using controlled antegrade and retrograde subintimal tracking: the CART technique. *J Invasive Cardiol* 2006;**18**:334–8.
40. Mahmood A, Banerjee S, Brilakis ES. Applications of the distal anchoring technique in coronary and peripheral interventions. *J Invasive Cardiol* 2011;**23**:291–4.
41. Christ G, Glogar D. Successful recanalization of a chronic occluded left anterior descending coronary artery with a modification of the retrograde proximal true lumen puncture technique: the antegrade microcatheter probing technique. *Catheter Cardiovasc Interv* 2009;**73**:272–5.
42. Mamas MA, Fath-Ordoubadi F, Fraser DG. Distal stent delivery with Guideliner catheter: first in man experience. *Catheter Cardiovasc Interv* 2010;**76**:102–11.
43. Fang HY, Fang CY, Hussein H, et al. Can a penetration catheter (Tornus) substitute traditional rotational atherectomy for recanalizing chronic total occlusions? *Int Heart J* 2010;**51**:147–52.
44. Nanto S, Ohara T, Shimonagata T, Hori M, Kubori S. A technique for changing a PTCA balloon catheter over a regular-length guidewire. *Catheter Cardiovasc Diagn* 1994;**32**:274–7.
45. Feiring AJ, Olson LE. Coronary stent and over-the-wire catheter exchange using standard length guidewires: jet exchange (JEX) practice and theory. *Catheter Cardiovasc Diagn* 1997;**42**:457–66.
46. Joyal D, Thompson CA, Grantham JA, Buller CEH, Rinfret S. The retrograde technique for recanalization of chronic total occlusions: a step-by-step approach. *JACC Cardiovasc Interv* 2012;**5**:1–11.
47. Michael TT, Banerjee S, Brilakis ES. Role of internal mammary artery bypass grafts in retrograde chronic total occlusion interventions. *J Invasive Cardiol* 2012;**24**:359–62.

Chapter 4

Antegrade Wire Escalation: The Foundation of Chronic Total Occlusion Percutaneous Coronary Intervention

Antegrade wire escalation is the simplest and most widely used chronic total occlusion (CTO) crossing technique.[1–3] At least 50% of CTO interventions are currently successfully recanalized using antegrade wire escalation.[4–6] Familiarity and confidence with this technique provides the foundation upon which all other CTO percutaneous coronary intervention (PCI) techniques (antegrade dissection/ reentry and retrograde) are built. Wire escalation may be most helpful in short occlusions (i.e., <20 mm length), longer occlusions of straight segments and/or where a through-and-through microchannel is suspected, and in selected cases of occlusive in-stent restenosis.

Step 1 Selecting a Microcatheter or Over-the-Wire Balloon for Guidewire Support

Goal

Select the equipment most likely to assist with CTO crossing.

How?

A microcatheter or over-the-wire balloon should be used for antegrade crossing in all CTOs (i.e., CTO crossing *should not* be attempted with unsupported guidewires) because such a system:

a. Enhances the wire penetrating capacity (Figure 2.17).

b. Allows wire tip reshaping without losing wire position.

c. Facilitates wire exchanges.

d. Prevents twisting of wires when using the parallel wire technique.

   A microcatheter is preferred by most operators (as described in Chapter 2, Section 2.4), because it:

a. Allows accurate assessment of the microcatheter tip location (because the marker is located at the tip, whereas in 1.20–1.50 mm balloons the marker is located in midshaft and the tip is not angiographically visible).

b. Is more resistant to kinking.

Manual of Chronic Total Occlusion Interventions. http://dx.doi.org/10.1016/B978-0-12-809929-2.00004-1

These advantages are particularly important in cases of **tortuosity** or **poor guide catheter support**, because over-the-wire balloons:

a. Are prone to kinking upon wire removal, thus hindering reliable wire exchanges.[7]
b. Provide less support due to lack of wire braiding.
c. Are more likely to cause proximal vessel injury.[7]

Step 2 Getting to the Chronic Total Occlusion

Goal
Deliver a guidewire and microcatheter/over-the-wire balloon to the proximal CTO cap.

How?
Unless the CTO proximal cap is ostial or very proximal, it should be accessed with a workhorse guidewire advanced through a microcatheter, over-the-wire balloon, or the CrossBoss catheter.

CTO wires with high penetrating power and tapered tips should not be used to traverse the proximal vessel to get to the CTO proximal cap because:

a. They can cause vessel injury, especially in diffusely diseased vessels (Fig. 4.1).
b. The wire bend required to reach the CTO is usually different (much larger) than the wire bend used when entering and crossing the CTO (much smaller) (Fig. 4.2).

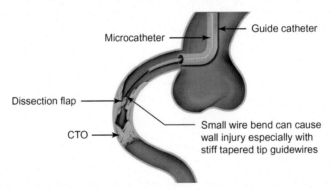

FIGURE 4.1 Illustration of proximal vessel injury during attempts to reach the proximal cap of the chronic total occlusion.

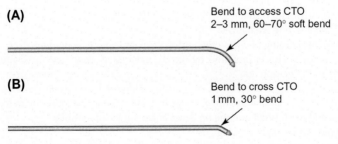

FIGURE 4.2 Example of guidewire bends used to reach the chronic total occlusion (CTO) (A) and traverse a CTO (B).

A soft-tipped, workhorse guidewire should be used to reach the CTO proximal cap, followed by the microcatheter or over-the-wire balloon (Fig. 4.3). The guidewire is then switched for the CTO crossing guidewire through the microcatheter or over-the-wire balloon (Fig. 4.4).

After removing the workhorse guidewire, contrast injection through the microcatheter can sometimes be very useful for clarifying the location and characteristics of the proximal cap.

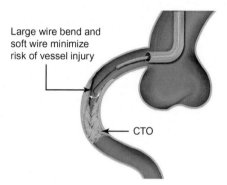

Large wire bend and soft wire minimize risk of vessel injury

CTO

FIGURE 4.3   Reaching the chronic total occlusion (CTO) proximal cap using a soft-tipped guidewire.

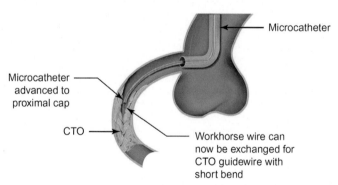

Microcatheter

Microcatheter advanced to proximal cap

CTO

Workhorse wire can now be exchanged for CTO guidewire with short bend

FIGURE 4.4   Exchange of the workhorse guidewire for a chronic total occlusion (CTO) crossing guidewire with a short tip.

Pearls of Wisdom

The best way to prolong (or fail) a case is by taking shortcuts! (William Lombardi, MD)

Success is not the result of big actions; instead it is the result of small steps taken carefully!

Step 3 Selecting a Guidewire for Chronic Total Occlusion Crossing

Goal
Select the most appropriate guidewire for initial antegrade CTO crossing.

How?
Although several coronary guidewires are available for CTO crossing, a simplified selection and escalation scheme is preferred (Fig. 4.5).[7]

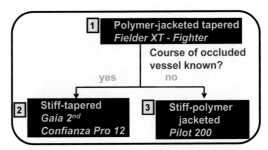

FIGURE 4.5 Illustration of guidewire escalation during antegrade chronic total occlusion crossing attempts.

A detailed description of the guidewires and their properties is presented in Chapter 2, Section 2.5. A tapered, polymer-jacketed wire (such as the Fielder XT, Fielder XT-A or Fighter) is usually used first to track a microchannel (which may sometimes be invisible). This attempt should be brief, unless progress is achieved.

If this wire fails to cross, and the course of the CTO vessel is well understood (especially if the CTO is short), a stiff, tapered guidewire (such as the Gaia 2nd) is preferred. If the course of the CTO is unclear, then a stiff, polymer-jacketed guidewire (such as the Pilot 200) or a composite core, moderate stiffness, nontapered guidewire (such as the Ultimate Bros 3) is preferred, because it is more likely to track the vessel architecture than exit the vessel wall.

Step 4 Guidewire Tip Shaping

Goal
Shape the wire tip to maximize the likelihood of successful CTO crossing.

How?
A small (1 mm long, 30–45 degrees) distal bend (Fig. 4.6) is preferred for CTO crossing because it:
a. Enhances the penetrating capacity of the guidewire.
b. Facilitates entry into microchannels.
c. Reduces the likelihood of deflection outside the vessel architecture or into branches arising within the occlusion.
d. Improves steerability within tight spaces, such as the CTO segment, which would normally straighten larger bends.

Creating such a small bent can only be accomplished by inserting the guidewire through an introducer, rather than using the side of the introducer, as is commonly done for workhorse guidewires (Fig. 4.7).

a. The guidewire is inserted through an introducer with approximately 1 mm protruding through the tip.
b. The guidewire tip is bent by 30–45 degrees (sometimes a syringe is used to bend the tip of very stiff guidewires, such as the Confianza Pro 12 guidewire, as they can puncture the operator's glove).
c. The guidewire tip is inspected to verify optimal shaping.
d. The guidewire is withdrawn into the introducer and advanced into the micro-catheter or over-the-wire balloon (it is best to insert the shaped guidewires into a microcatheter using an introducer to prevent potential tip damage or deformation).

Special care must be taken when shaping composite core wires because if their bend gets too sharp, it is difficult to straighten it and the attempts may damage the tip. In contrast, the polymer-jacketed guidewires are somewhat resistant to shaping and are prone to straightening during manipulations.

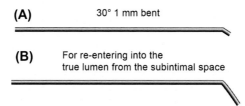

FIGURE 4.6 Illustration of wire tip bends for proximal cap penetration (A) and reentering the true lumen from the subintima (B).

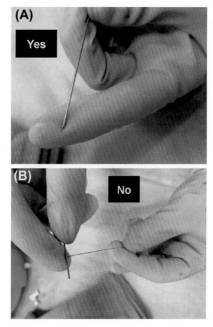

FIGURE 4.7 How to shape a guidewire for chronic total occlusion intervention. The CTO wire should be shaped by inserting it through an introducer (A) and NOT by using the side of the introducer (B).

Step 5 Advancing the Guidewire: Sliding Versus Drilling Versus Penetration

Goal

Cross CTO into the distal true lumen.

How?

Traditionally, three guidewire handling techniques have been described, two (sliding and drilling) used for both non-CTO and CTO lesions and one (penetration) used specifically for CTO lesions. Combinations of these techniques are often used, as part of the simplified wire escalation algorithm shown in Fig. 4.5.

1. **Sliding** is usually the first step in CTO crossing and consists of forward movement of a tapered, polymer-jacketed guidewire (such as the Fielder XT or Fighter wires), aiming to track microchannels within the CTO. The wire is advanced with gentle tip rotation and probing (i.e., a modest controlled drilling movement). These wires provide limited tactile feedback, hence visual assessment of the wire course is important. An apparent deflection of the tip must be avoided because it may lead to subintimal wire entry. If the wire fails to progress within a few minutes, change of guidewire and advancement technique is performed.

2. **Drilling** consists of controlled rotation of the guidewire in both directions. Rotation should be limited to <90 degrees in each direction. Usually guidewires with moderate tip stiffness (3–6 g) are used (such as the Gaia and Ultimate 3 guidewires), followed by escalation to stiffer wires (stiffer wires provide less tactile feedback). A small tip bend is crucial for this technique to avoid the creation of a large subintimal space. Non–polymer-jacketed wires are recommended as first choice (such as the Ultimate 3 g or Miracle 6 g) for this method because of their better tactile feedback.

3. **Penetration** consists of forward guidewire advancement intentionally steering (directing) the wire, not blindly rotating it, usually using a stiff (such as Gaia 2nd or 3rd, Miracle 12, Hornet 14, or Confianza Pro 12) guidewire. The guidewire is used as a needle to penetrate the occlusion. This technique is important for lesions with a calcified, hard to penetrate, proximal cap and for steering though shorter occlusions when the vessel course is well understood.

Crossing the CTO can be divided into three steps.

1. **Crossing the proximal cap.**
   If the proximal cap is heavily calcified, then a highly penetrating wire, such as the Confianza Pro 12, Hornet 14, Stingray, or the Astato 20 guidewire may be necessary to puncture the proximal cap and enable guidewire entrance into the occlusion. This is often done with simultaneous use of techniques to increase guidewire support (Section 9.2 in Chapter 9).

2. **Navigating through the occlusion.**
   After crossing the proximal cap with the microcatheter, the initial highly penetrating guidewire is exchanged for a softer guidewire to track the occluded segment (**escalation–deescalation**). Composite core guidewires, such as the Ultimate 3 and the Gaia series, are often used, as well as polymer-jacketed wires, such as the Fielder XT, Fighter, or Pilot 200.

3. **Distal entry into the lumen.**
   The microcatheter is advanced close to the distal cap and a highly steerable guidewire (such as the Gaia series) is used to direct the wire toward the distal true lumen. Occasionally, stiffer guidewires (such as the Confianza Pro 12) may need to be used if the distal cap is calcified and resistant to penetration. Contralateral injection should be used early to determine whether the guidewire is entering a true or a false lumen.

Wire Advancement Tips and Tricks

1. A microcatheter or over-the-wire balloon should be used to support the guide-wire during crossing attempts to enhance the penetrating capacity of the guide-wire (Section 2.4), and to allow for easy guidewire reshaping and exchanges. The microcatheter should be as close as possible to the tip of the wire, but not too close to avoid biasing the direction of the wire. When significant progress is made with a wire and the operator is confident that the wire is within the vessel (step 6), the microcatheter should be advanced as close to the wire tip as possible to maintain the ground gained and maximize the wire tip penetrating force.

2. Flexibility is important: if no progress is achieved within a few minutes, the wire and wire advancement strategy should be modified.

3. Alternating between approximately orthogonal views frequently prevents inadvertent departure from the intended guidewire course, especially when directed penetration is employed. Vessel wall calcifications can also be very helpful as an aid for determining the vessel course and guidewire location.

4. In case of noncalcified, long lesions and/or tortuosity of the occluded vessel,[8] the course can be determined if the coronary CT images are postprocessed and shown at the same angle as that of the gantry during wire advancement (Section 3.3.6 and Figure 3.25).[9]

Step 6 Assess Wire Position

Goal

Determine the guidewire position after crossing attempts. It is critical to under-stand the guidewire course *before* proceeding with microcatheter or balloon advancement to prevent perforation.

How?

There are three possible outcomes after the guidewire advances through the lesion:

a. Crossing into the distal true lumen.
b. Crossing into the subintimal space.
c. Exiting the vessel architecture (wire perforation).

How Can the Guidewire Location Be Ascertained?

1. **Contralateral injection** is the best method when collaterals arise primarily from the contralateral coronary artery. Obtaining two orthogonal views is piv-otally important, except in very straightforward cases.

2. **Sudden, spontaneous freedom of the wire tip** or proximal end movement as one passes the distal cap gives an important clue that distal true lumen wire position has been achieved. However, when moved forcefully or spun, stiff wires will create sufficient subintimal space to simulate a true lumen position.

3. **Distal wiring with a workhorse guidewire:** if the operator is certain that the guidewire is within the vessel architecture, a microcatheter is advanced over the CTO crossing guidewire, which is exchanged for a workhorse guidewire. Easy advancement of a workhorse wire, especially if branches can be inten-tionally selected, suggests a true lumen position. A coiled spring workhorse wire will not traverse easily in the subintima and will likely coil up or prolapse on itself indicating subintimal position.

4. **Aspirating through the microcatheter:** blood return is suggestive of distal true lumen crossing; however, subintimal hematoma formation can also result in blood return. Aspiration can help reduce the size of subintimal hematoma.

5. Two other methods can be used, but both are suboptimal and both require that the operator is certain (based on dual injection) that the guidewire is *not* outside the vessel architecture:

   • **Intravascular ultrasound,** but advancing the IVUS catheter into the subintimal space can extend the dissection and hinder wire reentry attempts.

   • **Contrast injection** through the microcatheter or over-the-wire balloon. This maneuver will cause staining if the microcatheter is located in the subintimal space, hence its use is *not* recommended).

Assessing Wire Position: Tips and Tricks!

1. During CTO crossing, **equipment should *never* be advanced over a guidewire if the wire is suspected to be outside the vessel architecture** based upon careful angiography or the wire's behavior. Small wire perforations are well tolerated, but catheter perforation can be catastrophic.

2. The position of the guidewire tip should be checked when it reaches 2–3 mm proximal to the distal cap. The guidewire should not be advanced further until after confirmation that it is pointing toward the distal true lumen in two orthogonal views (to reduce the likelihood of subintimal guidewire entry).

3. In case of an exclusively or dominantly ipsilateral collateral supply, introducing an OTW or microcatheter into the donor branch for selective contrast injection is a useful and safe method with very low risk of antegrade hydraulic dissection (Fig. 4.8). A thrombus aspiration catheter can also be used and is more effective for this purpose if the branch is large.

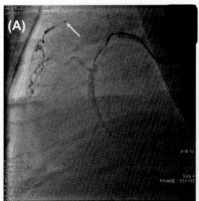 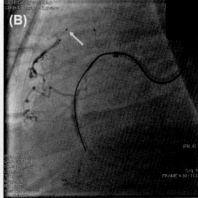

FIGURE 4.8 Contrast injection through an over-the-wire balloon (*arrow*) inserted in an ipsilateral collateral facilitates (A) and confirms (B) chronic total occlusion crossing. *Courtesy of Dr. Imre Ungi.*

Step 7a Wire Crosses Into the Distal True Lumen

(See Online Case 67.)

Goal

Complete CTO PCI with balloon and stents.

How?

1. Advance the microcatheter/over-the-wire balloon into the distal true lumen. Additional measures such as use of a guide catheter extension or balloon anchoring in proximal side branches may be required to assist crossing of the CTO with a microcatheter or over-the-wire balloon, as described in Chapter 8.
2. Remove the CTO crossing guidewire and exchange it for a workhorse guidewire. CTO crossing wires are more likely to cause distal vessel perforation and dissection as compared with workhorse guidewires when used for equipment delivery. Sometimes use of a supportive non–polymer-jacketed wire (such as the BHW, Ironman, or Grand Slam) can facilitate equipment delivery even with limited guide catheter support or in calcified or tortuous vessels.
3. Remove the microcatheter (ideally using the trapping technique to minimize wire motion and use of fluoroscopy, as described in Chapter 3, Section 3.7).
4. Proceed with standard balloon angioplasty and stenting.

Step 7b Wire Enters the Subintimal Space

Goal

Enter into the distal true lumen.

How?

If the guidewire is found to be in the subintimal space when advanced past the distal cap, the following techniques can be used:

a. Bring a microcatheter into the subintimal space and position its tip adjacent to a well-seen segment of true distal lumen. In a projection that provides for good navigation, use **directed penetration** to reenter the lumen. Avoid extending the subintimal track in an uncontrolled fashion.

b. **Subintimal crossing and reentry techniques** (described in Chapter 5).
   A knuckle wire or dissection catheter (i.e., the CrossBoss catheter) can be used to subintimally cross all the way to the distal true lumen, followed by wire-based or device-based (i.e., Stingray-based) reentry attempts. A CrossBoss catheter can extend the subintimal track in a controlled fashion that minimizes the risk for hematoma formation.
       Or,

c. Parallel wire techniques:
   - **Parallel wire**
   - **See-saw**
   - **Dual lumen microcatheter for redirection**
   These three techniques are based on the assumption that the original guidewire will prevent entry of additional guidewires into the subintimal space.

In all these techniques the initial guidewire is left in place and crossing is attempted with a second guidewire that is advanced next to the original guidewire, either through a second microcatheter (see-saw technique), without a second microcatheter (parallel wire technique), or through a dual lumen catheter.

Parallel Wire and See-Saw Technique
(See Online Case 68.)

The parallel wire technique is one of the oldest and most popular techniques for CTO PCI,[10,11] although many operators favor up-front use of subintimal dissection/reentry techniques.

In the parallel wire technique (Fig. 4.9), when the guidewire enters the subintimal space (or occasionally a side branch) it is left in, the microcatheter is removed, and a second guidewire is advanced parallel to the first wire over the microcatheter until it enters into the distal true lumen. In the parallel wire technique a single support catheter is used, whereas in a variation of the parallel-wire technique, called the see-saw technique (Fig. 4.10), two microcatheters (or over-the-wire balloons) are used to support both guidewires. Alternatively, a dual lumen microcatheter such as the TwinPass Torque (see Section 2.4.8) can be used to direct the second guidewire (Fig. 4.11).[12]

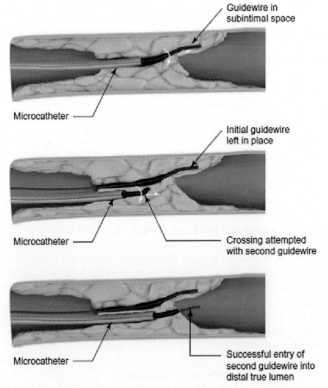

Guidewire in subintimal space

Microcatheter

Initial guidewire left in place

Microcatheter

Crossing attempted with second guidewire

Microcatheter

Successful entry of second guidewire into distal true lumen

FIGURE 4.9 Illustration of the parallel wire technique.

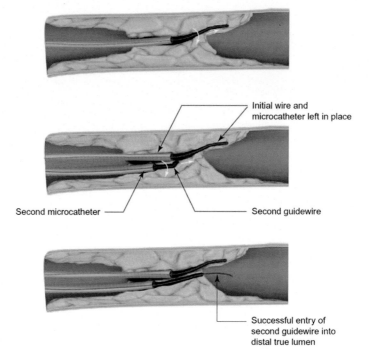

FIGURE 4.10   Illustration of the see-saw technique.

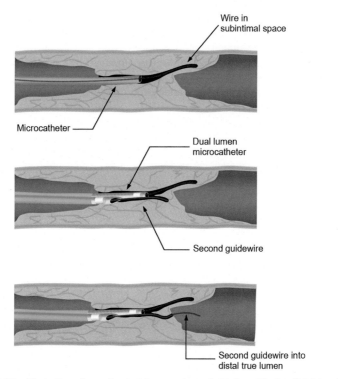

FIGURE 4.11   Illustration of use of a dual lumen microcatheter for achieving distal true lumen entry during antegrade crossing attempts once the first guidewire enters the subintimal space.

Parallel Wire Technique Tips and Tricks

1. Prolonged parallel wire attempts can cause enlargement of the subintimal space and hinder reentry attempts. As a result parallel wire strategies are not favored among many hybrid operators (who prefer to proceed directly to reentry using a dedicated reentry system, as described in Chapter 5).

2. Use of a dual lumen microcatheter is useful because it (a) keeps the position of the first guidewire stable, (b) enhances the second guidewire penetration capacity, (c) straightens the vessel, and (d) allows easier reshaping of the second guidewire.

3. Similarly, use of two microcatheters (see-saw technique) is advantageous, as it allows better support and easier reshaping of the second guidewire.

4. Use of contrast and fluoroscopy may be decreased with parallel wire techniques, as the first guidewire acts as a marker that guides advancement of the second guidewire.

5. The wires most commonly used as second parallel guidewires are stiff and highly torqueable wires, such as the Gaia series, Confianza Pro 12 and Miracle 12, or stiff polymer-jacketed guidewires, such as the Pilot 200. The Gaia family of wires has the advantage of excellent rotation and deflection control and is the preferred second wire for parallel technique.[13]

6. Rotation of the second guidewire should be limited (i.e., the wire should not be spun) to minimize the likelihood of wrapping around the first guidewire. This complication can also be prevented by using a dual lumen microcatheter.

7. Although leaving the first guidewire in place may block entrance of the second guidewire into the same space, occasionally the second guidewire may follow the path of the first guidewire; hence, early redirection of the guidewire within the CTO is important to create a new pathway. Use of intravascular ultrasound in a side branch next to the proximal cap can facilitate determining subintimal or true lumen position of the first guidewire.

8. It may sometimes be difficult to adequately visualize both guidewires, but this can be facilitated by use of orthogonal angiographic views, which also allows understanding of the exact guidewire position during advancement.

9. Occasionally >2 guidewires can be used in parallel wire techniques, but wire visualization can become challenging due to overlap.

10. Sometimes the tip of the second guidewire may require a more acute bend to find the distal true lumen.

11. The parallel wire technique may be particularly useful in cases with intra-occlusion tortuosity, because the first wire stretches the vessel, favorably modifying the curve and making tracking with the second wire easier.

Step 7c Wire Outside the Vessel Architecture

If the guidewire exits the vessel architecture (vessel's adventitia), it should be withdrawn, followed by repeat crossing attempts using the same or a different guidewire. Wire exit without advancing a microcatheter or other equipment is very unlikely to cause frank perforation or tamponade due to the small caliber of the guidewire.

Occasionally the guidewire may enter a side branch and appear as if it has exited the vessel architecture. Therefore, careful assessment of the diagnostic angiogram is critical for understanding the coronary anatomy and guidewire position. If the guidewire is confirmed to be in a side branch, use of intravascular ultrasound can help clarify the course of the main vessel.

REFERENCES

1. Grantham JA, Marso SP, Spertus J, House J, Holmes Jr DR, Rutherford BD. Chronic total occlusion angioplasty in the United States. *JACC Cardiovasc Interv* 2009;**2**:479–86.
2. Morino Y, Kimura T, Hayashi Y, et al. In-hospital outcomes of contemporary percutaneous coronary intervention in patients with chronic total occlusion insights from the J-CTO Registry (Multicenter CTO Registry in Japan). *JACC Cardiovasc Interv* 2010;**3**:143–51.
3. Sianos G, Werner GS, Galassi AR, et al. Recanalisation of chronic total coronary occlusions: 2012 consensus document from the EuroCTO club. *EuroIntervention* 2012;**8**:139–45.
4. Christopoulos G, Karmpaliotis D, Alaswad K, et al. Application and outcomes of a hybrid approach to chronic total occlusion percutaneous coronary intervention in a contemporary multicenter US registry. *Int J Cardiol* 2015;**198**:222–8.
5. Wilson WM, Walsh SJ, Yan AT, et al. Hybrid approach improves success of chronic total occlusion angioplasty. *Heart* 2016;**102**:1486–93.
6. Maeremans J, Walsh S, Knaapen P, et al. The hybrid algorithm for Treating chronic total occlusions in Europe: the RECHARGE registry. *J Am Coll Cardiol* 2016;**68**:1958–70.
7. Brilakis ES, Grantham JA, Rinfret S, et al. A percutaneous treatment algorithm for crossing coronary chronic total occlusions. *JACC Cardiovasc Interv* 2012;**5**:367–79.
8. Luo C, Huang M, Li J, et al. Predictors of interventional success of antegrade PCI for CTO. *JACC Cardiovasc Imaging* 2015;**8**:804–13.
9. Ghoshhajra BB, Takx RA, Stone LL, et al. Real-time fusion of coronary CT angiography with x-ray fluoroscopy during chronic total occlusion PCI. *Eur Radiol* 2017;**27**:2464–2473.
10. Rathore S, Matsuo H, Terashima M, et al. Procedural and in-hospital outcomes after percutaneous coronary intervention for chronic total occlusions of coronary arteries 2002 to 2008: impact of novel guidewire techniques. *JACC Cardiovasc Interv* 2009;**2**:489–97.
11. Mitsudo K, Yamashita T, Asakura Y, et al. Recanalization strategy for chronic total occlusions with tapered and stiff-tip guidewire. The results of CTO new techniQUE for STandard procedure (CONQUEST) trial. *J Invasive Cardiol* 2008;**20**:571–7.
12. Chiu CA. Recanalization of difficult bifurcation lesions using adjunctive double-lumen microcatheter support: two case reports. *J Invasive Cardiol* 2010;**22**:E99–103.
13. Khalili H, Vo MN, Brilakis ES. Initial experience with the Gaia composite core guidewires in coronary chronic total occlusion crossing. *J Invasive Cardiol* 2016;**28**:E22–5.

Chapter 5

Antegrade Dissection/Reentry

Antegrade dissection/reentry is a safe and efficient strategy for crossing long chronic total occlusions (CTOs), as outlined in the hybrid CTO crossing algorithm (Chapter 7). Antegrade dissection takes advantage of the distensibility of the subintimal space for traversing the occlusion rapidly and safely, concentrating subsequent efforts in crossing back into the distal true lumen (reentry). In the past, distal true lumen reentry was problematic because satisfactory tools and techniques were lacking, but dedicated equipment (Stingray balloon and guidewire, Boston Scientific) has significantly facilitated reentry.

5.1 CLARIFYING THE TERMINOLOGY

The terminology utilized in dissection/reentry CTO strategies can be confusing.[1] CTO crossing can occur either in the antegrade or the retrograde direction. In either direction, crossing can be achieved either from true-to-true lumen or by first entering the subintimal space, followed by reentry into the true lumen (dissection/reentry strategies) (Figs. 5.1 and 5.2).[1]

The term subintimal may increase this confusion, as there is typically no intimal layer within the atheroma of a totally occluded artery. Rather, subintimal in CTO percutaneous coronary intervention (PCI) has evolved as a general term

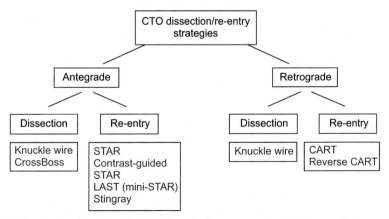

FIGURE 5.1 Classification of the CTO dissection/reentry strategies. *CART*, controlled antegrade and retrograde tracking and dissection; *CTO*, chronic total occlusion; *LAST*, limited antegrade subintimal tracking; *STAR*, subintimal tracking and reentry.

Manual of Chronic Total Occlusion Interventions. https://doi.org/10.1016/B978-0-12-809929-2.00005-3

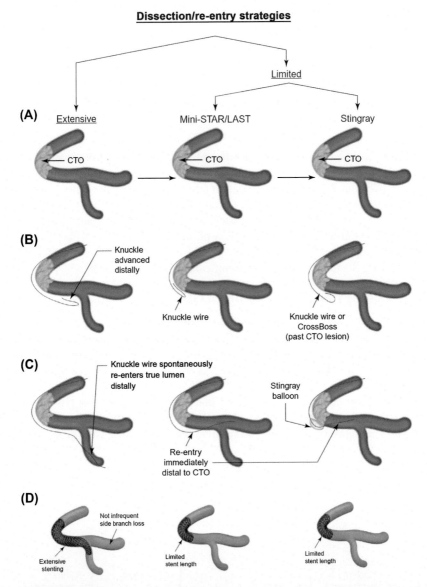

FIGURE 5.2 Illustration of various dissection/reentry strategies for recanalizing chronic total occlusions (CTOs).

that refers to a tissue plane within or beyond the occlusion that may be (1) subintimal, (2) intraplaque, (3) intraadventitial, or (4) combinations thereof, where the location of a tissue plane is related to disease morphology and position along the length of the artery.

Extensive dissection/reentry (subintimal tracking and reentry (STAR) technique), requires stenting of long coronary segments, often sacrifices side branches, and has been associated with poor long-term outcomes with high

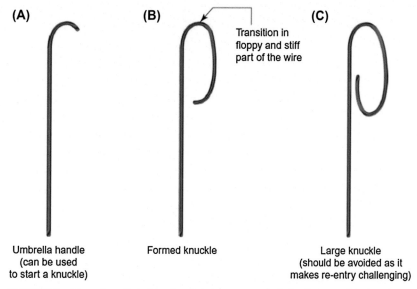

(A)

(B)

Transition in
floppy and stiff
part of the wire

(C)

Umbrella handle
(can be used
to start a knuckle)

Formed knuckle

Large knuckle
(should be avoided as it
makes re-entry challenging)

FIGURE 5.3 Illustration of knuckle wires. It is important to limit the diameter of the knuckle, as large knuckles enlarge the subintimal space and hinder reentry.

rates of in-stent restenosis.[2–4] The goal is always to achieve recanalization with a limited dissection/reentry (using wire-based strategies or dedicated reentry systems, such as the Stingray reentry system Figure 5.2), allowing targeted reentry, side branch preservation, and shorter stent lengths.

In the **antegrade** approach, dissection can be achieved by:

1. **Wire-based strategy (i.e., inadvertent wiring or knuckle wire)**. A knuckle (prolapsed guidewire) is formed by pushing a polymer-jacketed guidewire, (usually Fielder XT or Pilot 50 or 200) until it forms a tight loop at its tip (Fig. 5.3). The knuckle is then advanced subintimally through the occlusion. Compared to trying to advance the tip of a wire, advancing a knuckle is much faster, safer (the tight loop minimizes the risk of vessel perforation), and less likely to enter side branches.
2. **Catheter-based strategy, using the CrossBoss catheter**.[5]

In the **antegrade** approach, reentry can be achieved by:

1. **Wire-based strategies (not recommended)**.
   a. Continuing to advance the knuckled guidewire until it spontaneously reenters the true lumen (usually at a distal bifurcation). This is the STAR technique that was invented by Antonio Colombo.[2] A modification of the STAR technique called the contrast-guided STAR or Carlino technique, named after its inventor, uses subintimal contrast injection through a microcatheter inserted into the proximal cap to create/visualize a dissection plane and facilitate guidewire advancement (Fig. 5.4).[6] However, the STAR technique (1) often results in side branch loss, (2) is less predictably successful, and (3) has high reocclusion rates (likely due to long stent length and limited vessel

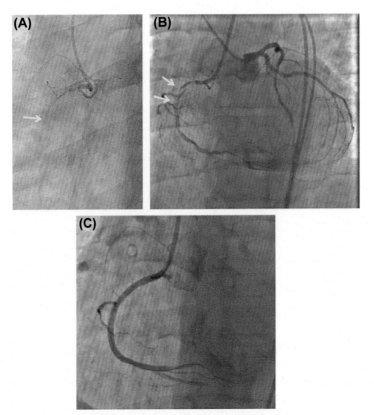

FIGURE 5.4   Contrast-guided subintimal tracking and reentry. Chronic total occlusion of the proximal right coronary artery (*arrow*, A), treated with injection of contrast via a microcatheter resulting in subintimal contrast entry (*arrows*, B) into the distal true lumen, with successful recanalization after stenting (C). *(Reproduced with permission from Michael TT, Papayannis AC, Banerjee S, Brilakis ES. Subintimal dissection/reentry strategies in coronary chronic total occlusion interventions.* Circ Cardiovasc Interv *2012;5:729–38.)*

outflow). It is rarely used as a definitive technique, but may be employed as a last-ditch effort, especially in the right coronary artery.[3]

   b. Reentering the true lumen as early as possible after the occlusion with a guidewire distal to the occlusion, which can be achieved by the **mini-STAR**[7] or the limited antegrade subintimal tracking (**LAST**)[8] technique (described in detail in Section 5.4, Step 4.2). These techniques, however, tend to have lower success rates because of difficulty in reliably reentering the true lumen, often due to extensive uncontrolled dissection with subintimal hematoma formation and true lumen compression.

2. **Dedicated reentry systems (recommended).**
   a. Using the Stingray (Boston Scientific) balloon and guidewire.[9,10]

In the **retrograde approach**, dissection is usually performed using a knuckle wire and reentry is achieved using the techniques described in Chapter 6, Step 7.2.

5.2 WHEN SHOULD ANTEGRADE DISSECTION/REENTRY BE USED?

Antegrade dissection/reentry can be used (Fig. 5.5):

1. After failure of antegrade wire escalation (inadvertent subintimal wire crossing) or failure of the retrograde approach.
2. As the initial crossing strategy (primary dissection/reentry).

   Good candidate lesions for **primary dissection/reentry** strategy are those with:

1. Well-defined proximal cap.
2. Large-caliber distal vessel.
3. No large branches within the CTO or more importantly at the distal cap.
4. Lack of good interventional collaterals.
5. ≥20 mm length.

5.2.1 The Antegrade Dissection/Reentry Debate

The optimal role and timing of antegrade dissection/reentry in CTO PCI has been a subject of debate. Hybrid operators (Chapter 7) favor early application of antegrade dissection/reentry to increase success rates and improve efficiency of the procedure, while keeping the risk low. Other operators argue that dissection/reentry should only be used as a last resort after other crossing strategies fail.

Several studies have shown that more complex lesions are more likely to require use of advanced crossing techniques (i.e., antegrade dissection/reentry and the retrograde approach; Fig. 6.44).[11,12] However, antegrade dissection/reentry carries lower risk than the retrograde approach,[13] hence for many operators it is the preferred initial advanced crossing technique in challenging cases.

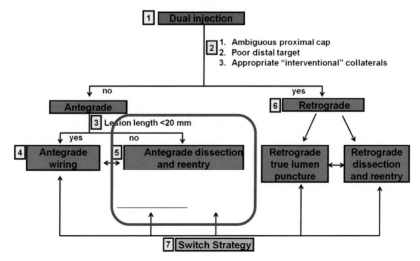

FIGURE 5.5   When to use antegrade dissection and reentry.

Moreover, several studies have shown similar restenosis rates with antegrade dissection/reentry and antegrade wiring. Antegrade dissection reentry can also provide unique solutions to anatomic challenges, such as proximal cap ambiguity (move-the-cap techniques, such as balloon-assisted subintimal entry and scratch-and-go", Section 9.1), wire uncrossable lesions (Carlino technique, Section 9.2), balloon uncrossable lesions (subintimal lesion modification or subintimal distal anchor, Section 8.1), and crossing of in-stent restenosis (using the CrossBoss catheter, Section 9.7, Online Case 19). However, the cost of antegrade dissection/reentry equipment can be high, limiting adoption in some areas.

If technically and economically feasible and among operators with experience in the techniques, **limited** antegrade dissection/reentry (i.e., use of the Stingray system for reentry close to the distal cap) should be the preferred advanced crossing technique for complex lesions. In contrast, **extensive** dissection/reentry techniques (such as STAR; see Section 5.6.1) have been associated with high restenosis and reocclusion rates[3,4] and should be rarely used, except potentially as a final bail-out maneuver.

5.3 HOW TO USE THE CROSSBOSS CATHETER

The CrossBoss catheter is used according to the following four steps (Fig. 5.6).

Step 1 CrossBoss Delivery to the Proximal Cap (Fig. 5.7)

Unless the CTO proximal cap is ostial or very proximal, it should be accessed with a workhorse guidewire advanced through a microcatheter, over-the-wire balloon, or the CrossBoss catheter itself, as described in Step 4.2.

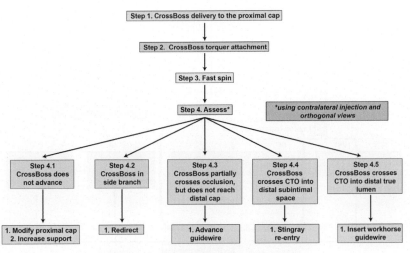

FIGURE 5.6 How to use the CrossBoss catheter.

The CrossBoss catheter is then advanced into the proximal cap and the guidewire is retracted within the CrossBoss catheter (but not removed, as it may help prevent blood entry and thrombus formation within the CrossBoss catheter lumen) (Fig. 5.7).

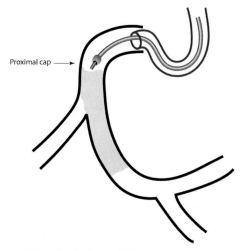

Proximal cap

FIGURE 5.7   CrossBoss delivery to the proximal cap.

Step 2 CrossBoss Torquer Attachment (Fig. 5.8)

The CrossBoss catheter torquer is positioned 2–3 cm (two to three finger widths) proximal to the Y-connector and tightened.

This is done to limit potentially excessive forward movement of the CrossBoss catheter (so-called CrossBoss jump) during catheter spinning. As the CrossBoss catheter engages and penetrates tissue, at times the device stores torsional energy and has the propensity to jump during advancement and navigation.

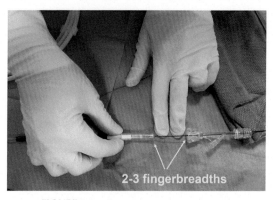

2-3 fingerbreadths

FIGURE 5.8   CrossBoss torquer attachment.

Step 3 Fast Spin (Fig. 5.9)

The CrossBoss catheter is rotated using the **fast-spin** technique and gentle forward pressure.

The Y-connector is held between the small finger and the palm of the left hand and the torque device is rotated using the index finger and thumb of both hands.

The catheter can be spun by hand in either direction, rotating **as fast as possible (until crossing or significant operator discomfort!)**. Faster spinning decreases friction and increases the likelihood of advancement and crossing. During catheter advancement it is important to keep the CrossBoss torquer device close (two finger-breadths) to the hemostatic valve, to prevent excessive forward movement of the CrossBoss.

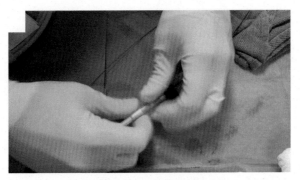

FIGURE 5.9 Fast-spin technique.

Step 4 Assess (Fig. 5.10)

The result of the fast-spin technique is assessed using contralateral injection (antegrade injections should not be performed after antegrade dissection/reentry is started to minimize the risk of extending a proximal dissection) and orthogonal views (to detect possible entry of the CrossBoss catheter into a side branch). There are five possible outcomes of the fast-spin technique.

4.1. CrossBoss Fails to Advance (Fig. 5.11)

Failure of the CrossBoss to advance can be due to poor guide catheter support or a hard, calcified proximal cap. Potential solutions include:

1. Increase guide catheter support (for example by using a more supportive guide catheter, a side-branch anchor technique, or a guide catheter extension as described in detail in Chapter 3, Section 3.6).

2. In patients with hard, calcified proximal cap a stiff guidewire, such as the Confianza Pro 12, Hornet 14, or Astato 20 can be used to puncture the proximal cap (should not be advanced >5–10mm to prevent vessel perforation). This wire should be immediately withdrawn and the CrossBoss advanced by itself, using the fast-spin technique. Alternatively, a polymer-jacketed wire

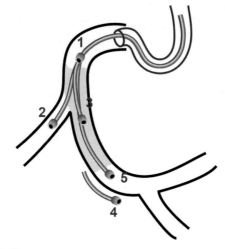

FIGURE 5.10   Assessment of the CrossBoss catheter position.

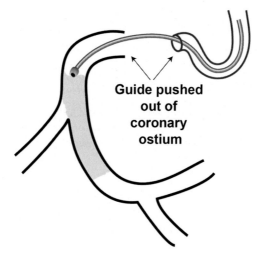

FIGURE 5.11   CrossBoss fails to advance.

can be used to create a knuckle and subsequently advanced, followed by CrossBoss crossing of the final CTO segment.

3. Change to a guidewire crossing strategy.

4.2. CrossBoss Enters a Side Branch (Fig. 5.12)

Although the CrossBoss catheter is highly unlikely to exit the vessel adventitia, due to its blunt trip, it can enter side branches. If undetected, entry of the CrossBoss into a side branch can be a catastrophic complication, as continued advancement can make it exit the side branch, causing a large and often challenging-to-treat perforation.

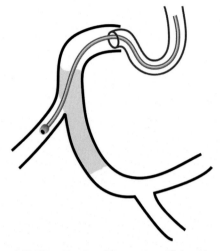

FIGURE 5.12   CrossBoss enters a side branch.

Detection: Side-branch course of the CrossBoss catheter is detected by using imaging in various projections and contralateral injection. Side-branch course is suspected when the CrossBoss catheter is not dancing in sync with the CTO target vessel.

Management: The CrossBoss catheter is retracted and redirected, usually using a knuckled polymer-jacketed guidewire or less commonly a stiff guidewire (such as the Gaia 2nd or Confianza Pro 12), which are less likely to enter the side branches.

4.3. CrossBoss Partially Crosses the Occlusion (Fig. 5.13)

Partial CrossBoss crossing can be due to intraocclusion calcification and/or tortuosity or due to interaction with a previously deployed stent.

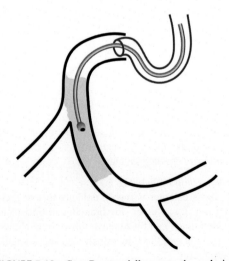

FIGURE 5.13   CrossBoss partially crosses the occlusion.

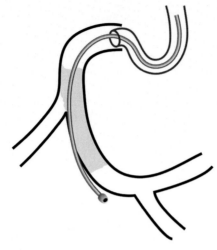

FIGURE 5.14 CrossBoss crosses into subintimal space distal to the distal cap.

Management: The CrossBoss catheter is retracted and redirected, usually using a knuckled polymer-jacketed guidewire or less commonly a stiff guidewire (such as the Gaia 2nd or Confianza Pro 12).

4.4. CrossBoss Crosses Into Subintimal Space Distal to the Distal Cap (Fig. 5.14)

- Crossing of the CrossBoss catheter in the subintimal space distal to the distal cap creates favorable conditions for reentry, as it has a low profile and hence does not enlarge the subintimal space.
- Reentry is optimally performed using the Stingray balloon (see Section 5.5) without reentry attempts with guidewires, as the latter may enlarge the area of dissection and cause large subintimal hematomas, which in turn can compress the distal true lumen, and hinder reentry attempts.
- The CrossBoss should be removed over a stiff, straight, nonlubricious guide wire, such as a Miracle 12, using the trapping technique to prevent wire movement and maintain distal position without enlarging the dissection.
- After subintimal crossing with the CrossBoss it is useful to disconnect the contrast-containing syringe from the antegrade guide catheter manifold (or cover the manifold) to minimize the risk of hydraulic dissection (Fig. 5.15). Inadvertent contrast injection could enlarge the subintimal space and hinder reentry attempts.

4.5. CrossBoss Crosses Into Distal True Lumen (Fig. 5.16)

The CrossBoss catheter may cross into the distal true lumen in approximately one-third of cases (Fig. 5.17). A workhorse guidewire is then inserted into the distal true lumen and the CrossBoss catheter is removed (ideally using the trapping technique to minimize the risk of losing guidewire position), followed by standard balloon angioplasty and stenting.

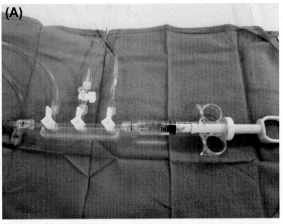

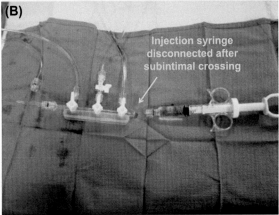

Injection syringe
disconnected after
subintimal crossing

FIGURE 5.15 Example of disconnecting the injection syringe from the manifold (panel B; panel A shows the connected manifold) after antegrade subintimal crossing to prevent inadvertent contrast injection that could enlarge the subintimal space.

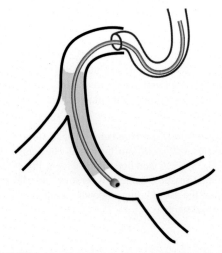

FIGURE 5.16 CrossBoss crosses into distal true lumen.

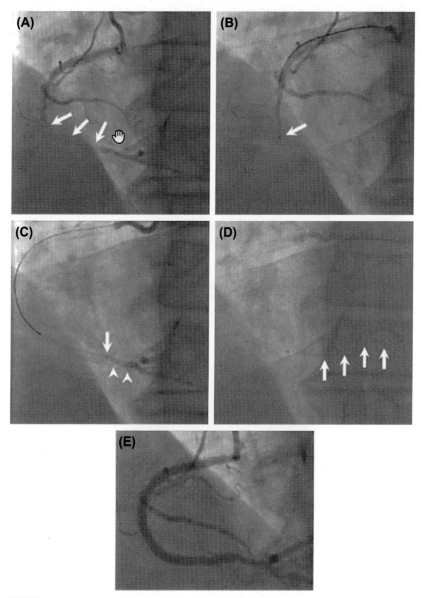

FIGURE 5.17 Example of true-to-true lumen crossing using the CrossBoss catheter. Previously failed chronic total occlusion of a right coronary artery (*arrows* in A) with CrossBoss catheter tip (*arrows* in B and C) crossing to the distal true lumen (*arrowheads* in C), facilitating distal wire placement (*arrows* in D), predilation, and final result (E). (*Reproduced with permission from Whitlow PL, Burke MN, Lombardi WL, et al. Use of a novel crossing and re-entry system in coronary chronic total occlusions that have failed standard crossing techniques: results of the FAST-CTOs (Facilitated Antegrade Steering Technique in Chronic Total Occlusions) trial.* JACC Cardiovasc Interv *2012;5:393–401.*)

5.4 HOW TO USE A KNUCKLE WIRE FOR CROSSING

Step-by-Step Knuckle Wire Crossing (Fig. 5.18)

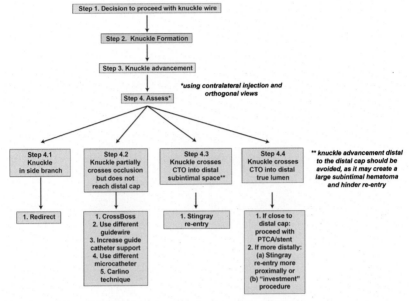

FIGURE 5.18   Step-by-step knuckle wire crossing.

Step 1 Decision to Proceed With a Knuckle Wire (Fig. 5.19)

The most common scenario for using a knuckled guidewire is when the antegrade (or retrograde) guidewire enters into the subintimal space. Alternative strategies to

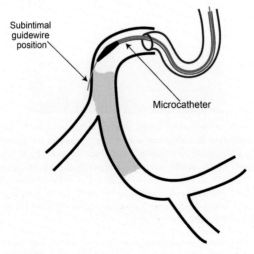

FIGURE 5.19   Decision to proceed with a knuckle wire.

knuckled guidewire are (1) redirecting the initial guidewire into the true lumen, (2) using a parallel wire technique (Chapter 4, Step 7b), or (3) retrograde guidewire crossing (Chapter 6).

Subintimal Crossing: Knuckle Wire Versus CrossBoss

Two techniques can be used for subintimal CTO crossing: (1) wire-based (knuckle wire technique) and (2) catheter-based (CrossBoss device).

The advantages of the CrossBoss catheter over a knuckle wire are the following:

1. It creates a smaller and more controlled subintimal dissection space (Fig. 5.20), enabling a more predictable and controlled reentry into the distal true lumen.

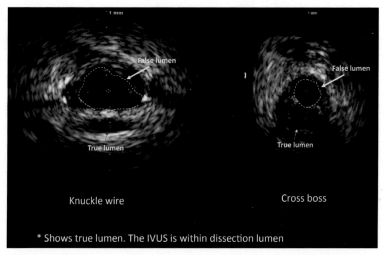

FIGURE 5.20 Comparison dissection created using a knuckled guidewire (larger size) and created with a CrossBoss catheter (smaller size). *(Courtesy of Dr. Craig Thompson.)*

2. The relatively stiff CrossBoss catheter tends to advance along a longitudinal path parallel to the artery axis, whereas guidewires sometimes wrap around the artery circumference (in a barber-pole fashion; Fig. 5.21), potentially hindering subsequent advancement of other devices and reentry into the distal true lumen.

The advantages of the knuckled guidewire are that:

1. It is cheaper.
2. It may be less likely to enter into side branches, especially when larger knuckles are formed.
3. It may be easier to advance around areas of severe calcification and/or tortuosity.

In some cases both the CrossBoss catheter and a knuckle wire may be used (knuckle-Boss technique) to navigate beyond side branches or advance beyond calcific or tortuous anatomy.

FIGURE 5.21 Antegrade intimal plaque tracking versus subintimal tracking. Antegrade intimal plaque tracking (*solid line*) and subintimal tracking (*dotted line*). Once the wire migrated into the subintimal space, the wire easily advances into the subintimal space and wraps around the intimal (in a barber-pole fashion). It is difficult for the subintimal wire to cross into the distal true lumen because of lower resistance to advance into subintimal space coupled with increased resistance to traversing the plaque. *(Reproduced with permission from Sumitsuji S, Inoue K, Ochiai M, Tsuchikane E, Ikeno F. Fundamental wire technique and current standard strategy of percutaneous intervention for chronic total occlusion with histopathological insights.* JACC Cardiovasc Interv *2011;4:941–51, Elsevier.)*

Step 2 Knuckle Formation (Fig. 5.22)

How to Knuckle a Wire

1. Advance microcatheter into subintimal space. More supportive microcatheters such as the Corsair and TurnPike may provide stronger support for forming a knuckle as compared with lower profile microcatheters, such as the FineCross or Caravel.
2. Use a polymer-jacketed wire such as Fielder XT, Fighter or Pilot 200—softer wires usually form smaller, tighter loops.
3. Create an umbrella bend using the introducer (this is not necessary, as a loop is likely to form anyway with aggressive pushing).
4. Push *without* spinning to minimize risk for wire fracture or entanglement.
5. Do not be afraid to **push hard**!
6. A loop usually forms at the junction of the radiopaque and the stiffer radiolucent parts of the wire (Fig. 5.3).
7. Try to keep the knuckle small; you may need to withdraw into the microcatheter and readvance if the knuckle size becomes too large. Alternatively, the microcatheter can be advanced trying to keep the tip of the catheter as close as possible to the leading edge of the knuckle.
8. If a loop does not form, withdraw and readvance.

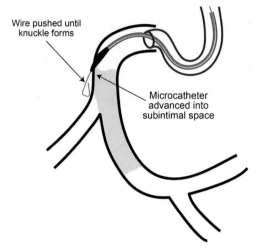

Wire pushed until
knuckle forms

Microcatheter
advanced into
subintimal space

FIGURE 5.22   Knuckle formation.

Precaution: Ensure the knuckle is started within the subintimal space—do not start to knuckle if there are concerns that the wire is located outside the vessel architecture, as this may cause a catastrophic perforation.

Troubleshooting: Unable to Form a Knuckle
Solutions
1. Confirm microcatheter position in the subintimal space within the vessel architecture, using orthogonal projections.
2. Try a different guidewire. For example, change between soft, tapered, polymer-jacketed guidewires (such as Fielder XT or Fighter) and stiff, polymer-jacketed guidewires (such as Pilot 200 or Gladius).
3. Try a different microcatheter.
4. Reposition the microcatheter more proximal or more distal within the occlusion.
5. Use the Carlino technique to start a dissection plane that can subsequently be entered by the guidewire.

Step 3 Advance the Knuckle (Fig. 5.23)
1. Advancement is made in intermittent, forceful forward pushes.
2. After confirmation of subintimal position alongside the vessel wall, the micro-catheter is advanced closer to the tip of the knuckled guidewire.

Troubleshooting: Unable to Advance Knuckle
Causes
1. Calcification and tortuosity
2. Poor guide catheter support

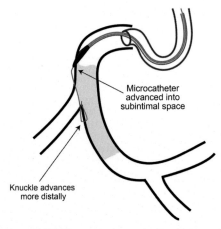

FIGURE 5.23   Advance the knuckle.

Solutions

1. **Increase support**:
   a. Advance microcatheter to the tip of the knuckle
   b. Use a supportive guide catheter, such as Amplatz Left 1 for right coronary artery CTOs
   c. Guide catheter extensions
   d. Side branch anchor
   e. CenterCross and NovaCross catheter
   f. Coaxial anchor (the knuckle is advanced through an inflated over-the-wire balloon)
   g. Power knuckle: a second guidewire is advanced adjacent to the microcatheter and a 1:1 sized balloon is inflated over the second guidewire, anchoring the microcatheter against the vessel wall (see Online Cases 57, 82, and 83)
   h. Use a different guidewire (e.g., Pilot 200 is stiffer than the Fielder XT and forms bigger knuckles, whereas the Fielder XT is smaller and forms tighter knuckles)

Step 4 Assess the Knuckle Position (Fig. 5.24)

The result of knuckle advancement is assessed using contralateral injection and orthogonal views. There are four possible outcomes with knuckle advancement.

4.1. Knuckle Enters a Side Branch (Fig. 5.25)

The knuckle wire is extremely unlikely to exit the vessel adventitia, due to its blunt bend, but can enter side branches, such as acute marginal branches during right coronary artery CTO PCI.

Detection: Side-branch course of the knuckled guidewire is detected by using imaging in various projections and contralateral injection. Side branch course is suspected when the knuckled guidewire is not dancing in sync with the CTO target vessel.

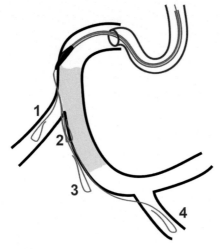

FIGURE 5.24   Assess the knuckle position.

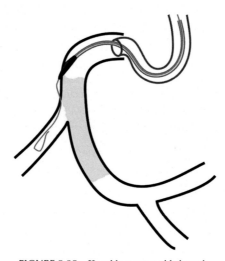

FIGURE 5.25   Knuckle enters a side branch.

Management: The knuckled guidewire is retracted and redirected. Using a stiffer polymer-jacketed guidewire (such as the Pilot 200 or Gladius) can create larger loops, which are less likely to enter side branches.

4.2. Knuckle Partially Crosses the Occlusion but Does Not Reach the Distal Cap (Fig. 5.26)

Sometimes the knuckled guidewire cannot cross to the distal cap, due to calcification, poor guide catheter support, or other reasons.

Detection: Contralateral injection shows intraocclusion location of the guidewire, with the loop dancing in sync with the vessel.

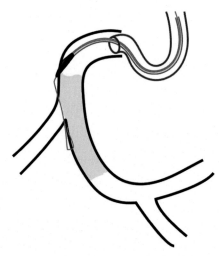

FIGURE 5.26   Knuckle partially crosses the occlusion but does not reach the distal cap.

Management: Before proceeding with aggressive knuckling advancement maneuvers it is critical to ascertain (using contralateral injection) that the guidewire is within the vessel architecture and not within a side branch or in the pericardium. After confirmation that the knuckled guidewire is within the occlusion and not within a branch, the following actions can be taken:

1. Exchange the knuckled guidewire for a CrossBoss catheter. Knuckles (especially large ones) can cause extensive dissection and subintimal hematoma that can impair reentry attempts. This can be prevented by crossing the distal CTO segment with the CrossBoss catheter ("finish with the Boss"), or by starting a knuckle in a more proximal location in the CTO target vessel.
2. Use a different polymer-jacketed guidewire to form a knuckle, such as the Pilot 200 or Gladius.
3. Increase guide catheter support, for example by using a side branch anchor or a guide catheter extension.
4. Use a more supportive microcatheter, such as the Corsair, Turnpike, or Turnpike Spiral.
5. Use the Carlino technique (Section 5.6.3): advance the microcatheter as far as possible within the occlusion and inject 0.5–1 cc of contrast under cine. The contrast injection can modify the plaque, facilitating guidewire crossing (see Online Case 13, 27, 37, 45, 48, 49, 57, 60, 62, 84).

4.3. Knuckle Crosses the Chronic Total Occlusion Into the Distal Subintimal Space (Fig. 5.27)

This is the intended outcome of knuckle guidewire advancement. After confirmation that the guidewire is indeed in the distal subintimal space (by using contralateral injection and verifying that the wire is "dancing in sync" with the distal vessel), the next step is to achieve reentry into the distal true lumen.

Caution: Too distal advancement of the knuckled guidewire should be avoided, as it may result in side branch loss and longer stent length with higher restenosis rates. The goal is to reenter the distal true lumen as close as possible to the distal cap, so as to minimize the length of dissection.

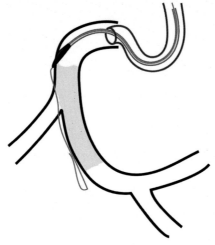

FIGURE 5.27   Knuckle crosses the chronic total occlusion into the distal subintimal space.

Management: There are three broad categories or techniques to allow crossing into the distal true lumen: (1) using the Stingray balloon and guidewire (Section 5.5), (2) using wire-based reentry techniques (Section 5.6), and (3) using the parallel wire technique (Chapter 4, Step 7b).

Most hybrid operators favor immediate use of the Stingray system for achieving reentry because (1) use of guidewire for reentry can cause a large subintimal hematoma that can hinder reentry and (2) it is faster and more efficient. Wire-based reentry techniques should in general be avoided, due to unpredictability of reentry and risk of causing hematoma. Use of the parallel wire technique (leaving the knuckle in place and advancing a second guidewire proximal in the occlusion) is an acceptable option provided that guidewire manipulation in the distal subintimal space is minimized.

4.4. Knuckle Enters the Distal True Lumen (Fig. 5.28)
Infrequently, the knuckled guidewire may enter into the distal true lumen, as demonstrated by contralateral injection. Most commonly this occurs in a distal branch, which is suboptimal, as stenting may result in loss of side branches proximal to the reentry location decreasing the benefit of revascularization; moreover, long stent length is associated with very high restenosis rates. This is the original STAR technique (Section 5.6.1) that should be used only as a last resort when every other attempt has failed.

4.5. Management of Distal Knuckle Crossing (Fig. 5.29)
If knuckle reentry is achieved close to the distal cap, balloon angioplasty and stenting are performed. If, however, reentry is more distal beyond major side branches (which occurs much more frequently) stenting should be avoided, as it may occlude more proximal side branches. Instead, more proximal reentry should be attempted using the Stingray system or an investment procedure can be performed (balloon angioplasty of the dissected segment without stent implantation) (see Online Case 8 and 9).[12] Repeat angiography is performed in 2–3 months, often showing restoration of antegrade flow, allowing stenting without compromising all side branches. Maintenance of patency is more likely if there is good antegrade flow after subintimal ballooning.[14]

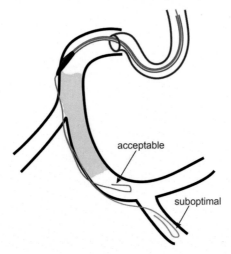

FIGURE 5.28 Knuckle enters the distal true lumen.

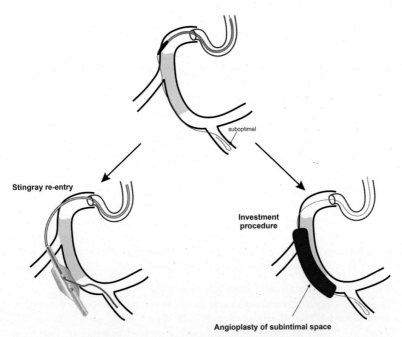

FIGURE 5.29 Management of distal knuckle crossing.

5.5 HOW TO REENTER INTO DISTAL TRUE LUMEN USING THE STINGRAY SYSTEM

After subintimal CTO crossing has been achieved, reentry into the distal true lumen can be achieved using (1) dedicated reentry systems, such as the Stingray system or (2) guidewire-based techniques. Stingray-based techniques are more reproducible and reliable and are the current standard of care for reentry.

Step 1 Preparation of the Stingray Balloon (Fig. 5.30)

Stingray preparation steps (Fig. 5.30) include:
1. Attach a new, completely dry, three-way stopcock to the end of the Stingray balloon port (Fig. 5.30A).

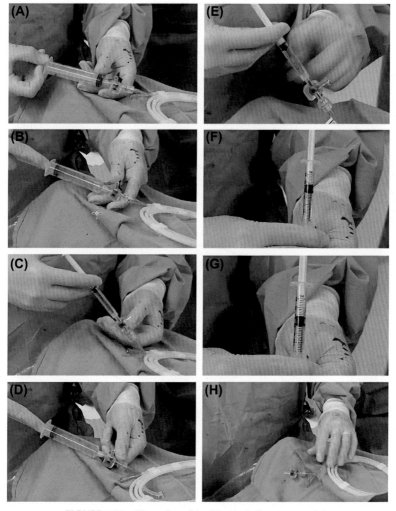

FIGURE 5.30   Illustration of the Stingray balloon preparation.

2. Using a new, completely dry 20 cc luer-lock syringe, aspirate negative 2–3 times (Fig. 5.30B) and turn the stopcock each time, keeping the balloon port closed (Fig. 5.30C), to retain vacuum in the Stingray balloon.
3. Remove the 20 cc syringe, replacing it with a 3 cc luer-lock syringe that contains 100% contrast (Fig. 5.30D).
4. Flush contrast through a three-way stopcock, "priming the pump," ensuring that there are no air bubbles in the stopcock (Fig. 5.30E).
5. Open the stopcock to the syringe—the plunger will advance by 2–3 mm (Fig. 5.30F is before opening stopcock, Fig. 5.30G is after opening stopcock). The contrast syringe can remain attached to the Stingray balloon until ready to connect with the indeflator.
6. The Stingray balloon is now ready for use (Fig. 5.30H).

Step 2 Delivery of the Stingray Balloon to the Reentry Zone (Fig. 5.31).

Stingray balloon delivery is usually easy if the CrossBoss catheter or a knuckle wire was used to create the dissection. Occasionally during attempts to deliver the Stingray balloon, the guidewire may enter the distal true lumen.

- Stingray balloon delivery can be facilitated by using a supportive guidewire for delivery, such as a 300 cm long Miracle 12 guidewire.
- It is recommended that the wire trapping technique be used to avoid inadvertent movement of the exchange wire. The original Stingray balloon required an 8 Fr guide catheter for trapping, whereas the Stingray LP balloon can be trapped in a 7 Fr guide catheter.

If Stingray balloon delivery is challenging the following can be done:

1. Use the Stingray LP (low profile) balloon instead of the original Stingray balloon, as the latter has a lower profile and better crossability.

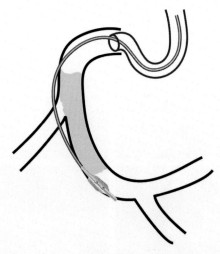

FIGURE 5.31 Delivery of the Stingray balloon to the reentry zone.

2. Predilate the subintimal space at the CTO segment with the smallest possible balloon, such as a 1.25 or 1.5 mm balloon).

3. Increase guide catheter support, for example by using a side branch anchor or a guide catheter extension. However, trapping will not be feasible with guide catheter extensions unless a Trapliner is used.

Step 3 Reentry Into the Distal True Lumen

After the Stingray balloon is delivered to the reentry zone, it is inflated at 2–4 atm. Upon inflation the balloon assumes a flat shape and self-orients, with one surface of the balloon facing the true lumen and the other facing the adventitia.

Reentry can be achieved using either (1) angiographic guidance using only a stiff guidewire (stick-and-drive technique); (2) angiographic guidance using initially a stiff guidewire to puncture the subintimal space followed by exchange to a polymer-jacketed guidewire (stick-and-swap technique); or (3) without angiographic guidance (double-blind stick-and-swap technique). The double-blind stick-and-swap technique is currently preferred by many CTO operators, as it minimizes use of contrast and radiation, while still maintaining high success rates.[15]

3.1. Stingray Reentry: Stick-and-Drive Technique

3.1.1. Obtain Optimal Angiographic View (Fig. 5.32)

Contralateral injection is performed using orthogonal angiographic projections to select the optimal view for reentry (Fig. 5.33). The ideal view is the one in which the Stingray balloon is seen as one line located at the side of the vessel lumen (Fig. 5.33B). The side view is necessary to determine in which direction the wire leaves the Stingray balloon, so as to direct it toward the vessel lumen.

3.1.2. Advance Stingray Wire (Fig. 5.34)

To reenter the distal true lumen the Stingray guidewire is advanced through the Stingray balloon side ports under fluoroscopic guidance. The Stingray balloon has three exit ports: two of them are 180 degrees apart on the flat surface of the Stingray balloon for vessel reentry, and the third is the end hole. The proximal exit port is proximal to the two markers and the distal exit port is between the two markers. If the wire enters the exit port that faces away from the true lumen, it is withdrawn and redirected into the other exit port facing the true lumen.

Tip: Apart from the Stingray guidewire, other wires can be used for distal true lumen entry, such as the Gaia 2nd guidewire (which comes with a premade bend that is similar to the bend of the Stingray guidewire).

3.1.3. Confirm Distal True Lumen Wire Position (Fig. 5.35)

Once the wire is exiting from the correct port, it is advanced without rotation so as to puncture back into the true lumen. This often creates a "pop" sensation. Contralateral injection is now helpful to determine whether true lumen reentry has been achieved. If this is the case and the distal vessel is not severely diseased and the lumen is large, the Stingray guidewire can be rotated 180 degrees and advanced further down into the vessel. This is also called the stick-and-drive method.

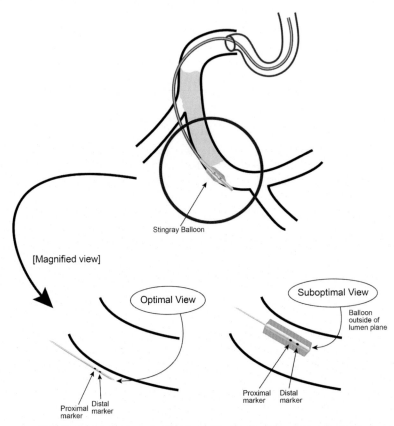

FIGURE 5.32 Obtain optimal angiographic view for Stingray-based reentry.

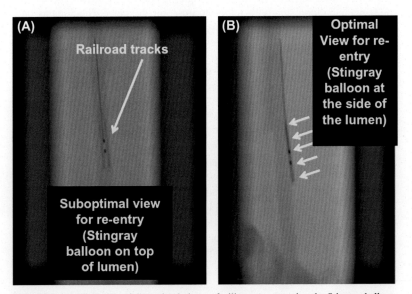

FIGURE 5.33 Selection of the optimal view to facilitate reentry using the Stingray balloon.

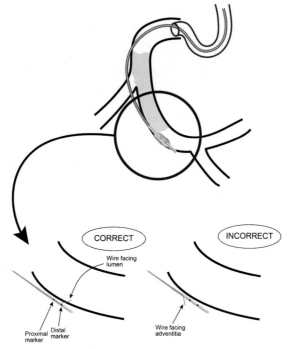

FIGURE 5.34 Advance the Stingray wire.

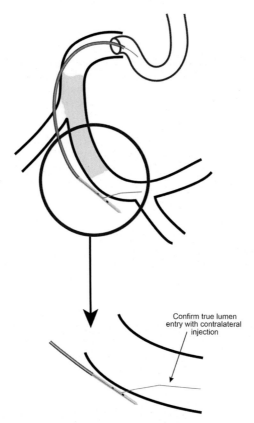

FIGURE 5.35 Confirm distal true lumen wire position.

Tip: If uncertain, the wire position can be confirmed using intravascular ultrasound.

see Online Case 35.

3.1.4. Remove Stingray Balloon and Exchange for a Workhorse Guidewire

The Stingray balloon is best removed using the trapping technique, to minimize the risk of losing the distal wire position or causing distal vessel injury. The original Stingray balloon requires an 8 Fr guide catheter for trapping, whereas the Stingray LP balloon can be trapped within a 7 Fr guide catheter (or a Trapliner). After removal of the Stingray balloon a microcatheter is advanced over the guidewire that has reentered the vessel to allow exchange for a workhorse guidewire (using a workhorse guidewire during balloon and stent delivery minimizes the risk for distal vessel perforation).

An example of successful Stingray-based reentry is demonstrated in Fig. 5.36.

3.2. Stick-and-Swap Technique (Fig. 5.37)

See online cases 2, 16, 17, 25, 34, 35, 72, 83, 97.

3.2.1. Obtain Optimal Angiographic View

Do this as described for the stick-and-drive technique.

3.2.2. Advance Stingray Wire

This is performed as described in Section 3.1.2.

3.2.3. Swap

After the Stingray (or occasionally Gaia or Hornet guidewire) is advanced through the side port of the balloon it is rotated 180 degrees and withdrawn to enlarge the wire exit connection and is then removed from the Stingray balloon. Another guidewire (usually stiff, polymer-jacketed, such as the Pilot 200) is advanced through the same Stingray balloon exit port and rotated until it tracks the course of the vessel. The Pilot 200 guidewire should be shaped similar to the Stingray guidewire with a 1 mm 30 degrees distal bend. If more bend is created on the wire the chances of entering the side port significantly decrease.

3.2.4. Confirm Distal True Lumen Position

This is done as described in Section 3.1.3.

3.2.5. Remove Stingray Balloon and Exchange for a Workhorse Guidewire

This is done as described in Section 3.1.4.

3.3. Double-Blind Stick-and-Swap Technique

(Online cases 13, 30, 32, 43, 45, 48, 79, 80, 82, 85, 92, 93, 94, 100)

3.3.1. Advance Stingray Wire

The Stingray (or occasionally another stiff guidewire, such as the Gaia 2nd) is advanced through both side exit ports of the Stingray balloon, for approximately 3–5 mm. Occasionally, hard pushing of the guidewire is needed to achieve advancement.

Tip: There is occasionally concern among operators that advancing the guide-wire toward the adventitia can cause perforation. However, guidewire advancement alone through the adventitia without advancing a balloon or microcatheter will almost never cause coronary perforation.

3.3.2. Swap

The Stingray guidewire is removed and a polymer-jacketed guidewire (such as Pilot 200) is advanced through each side port in random order. The wire is advanced rotating, observing for smooth guidewire movement down the vessel. Difficulty with wire

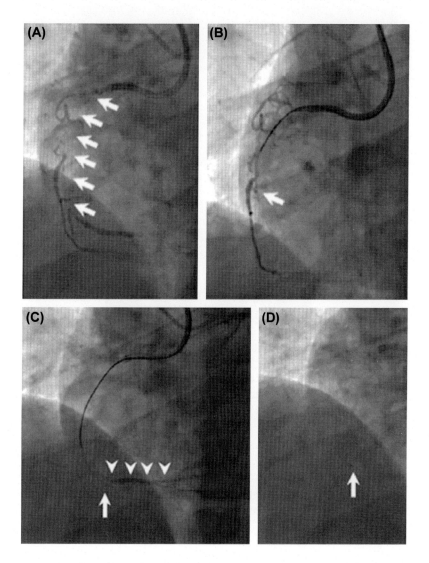

FIGURE 5.36 Successful reentry into the distal true lumen using the Stingray balloon and wire. Chronic total occlusion of a right coronary artery (*arrows*, A) with CrossBoss catheter tip (*arrows*, B, C) crossing into the false lumen (true lumen; *arrowheads*, C) followed by Stingray catheter placement (*arrow*, D), Stingray guidewire reentry (*arrow*, E) through the Stingray catheter into the distal true lumen (*arrows*, F), predilation, and final result (G). *(Reproduced with permission from Whitlow PL, Burke MN, Lombardi WL, et al. Use of a novel crossing and re-entry system in coronary chronic total occlusions that have failed standard crossing techniques: results of the FAST-CTOs (Facilitated Antegrade Steering Technique in Chronic Total Occlusions) trial. JACC Cardiovasc Interv 2012;5:393–401.*

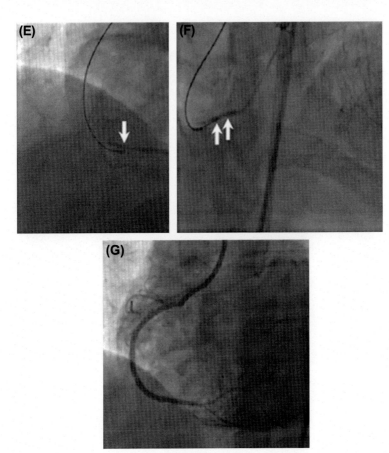

FIGURE 5.36 cont'd.

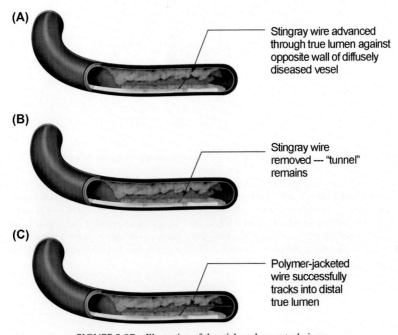

(A) Stingray wire advanced through true lumen against opposite wall of diffusely diseased vessel

(B) Stingray wire removed --- "tunnel" remains

(C) Polymer-jacketed wire successfully tracks into distal true lumen

FIGURE 5.37 Illustration of the stick-and-swap technique.

advancement and/or buckling of the guidewire usually suggests subintimal or sub-adventitial guidewire position. The wire should be withdrawn and either advanced again through the same port or advanced through the contralateral side exit port.

3.3.3. Confirm Distal True Lumen Position
This is done as described in Section 3.1.3.

3.3.4. Remove Stingray Balloon and Exchange for a Workhorse Guidewire
See Online Case 48. This is done as described in Section 3.1.4.

Stingray Troubleshooting
1. **Unable to advance the Stingray balloon to the reentry zone**.
   If the Stingray cannot be delivered to the reentry zone, the following steps could be performed:
   a. Modify the subintimal channel by:
      i. Balloon angioplasty with a small (1.2–1.5 mm) balloon or Threader.
      ii. Advancing a microcatheter, such as Corsair and Turnpike to the reentry zone.
   b. Use techniques to enhance guide catheter support, such as anchoring techniques or guide catheter extensions (Chapter 3. Section 3.6).
2. **Poor visualization of the Stingray balloon and the distal true lumen**.
   Potential Solutions
   a. Double-blind stick-and-swap technique.
   b. Meticulous preparation of the Stingray balloon.
   c. Orthogonal views to identify optimal reentry projection: the goal is to achieve a sideways projection (Fig. 5.33B) and avoid a railroad projection (Fig. 5.33A).
   d. Magnified views (discouraged during the other parts of the CTO PCI procedure as they can significantly increase radiation dose).
3. **Unable to reenter diffusely diseased distal vessel**.
   Reentry from the subintimal space into the distal true lumen can be difficult in patients with small, diffusely-diseased distal vessels, because the reentry lumen is small and the Stingray wire may go through the true lumen into the opposite vessel wall (Fig. 5.37A).
   Solutions
   a. Change the reentry site by moving the Stingray balloon to a healthier, straighter, and larger vessel segment (the horizontal part of the distal right coronary artery is usually preferable for reentry). This technique is called **bobsled** (see Online Case 16, 48, 79, 93).
   b. Use the **stick-and-swap** or double-blind stick-and-swap technique, as described in Fig. 5.37.
4. **Compression of distal true lumen by hematoma**.
   Prevention of subintimal hematoma is key to successful reentry. Use of the CrossBoss catheter (rather than a knuckle wire) for dissection (especially in the distal segment of the occlusion) can reduce the risk compared with a knuckle wire. This strategy is called "finish with the Boss." If a hematoma develops, then aspiration of the hematoma can be attempted, either through the Stingray balloon itself, or ideally through another microcatheter or over-the-wire balloon advanced next to the Stingray balloon. This is called the subintimal trans-catheter withdrawal (**STRAW**) technique (Figs. 5.38 and 5.39) (Online Cases 48 and 79).[16] This technique can only be performed through an 8 Fr guide

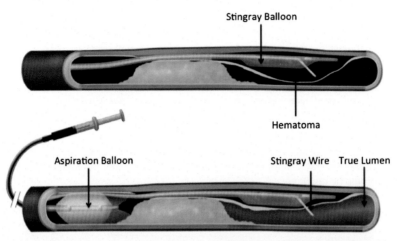

FIGURE 5.38 Illustration of the subintimal transcatheter withdrawal (STRAW) technique. (A) Subintimal hematoma that compressed the distal true lumen, hindering reentry attempts with the Stingray balloon and guidewire. (B) Insertion of a second over-the-wire balloon advanced into the subintimal space, through which aspiration is performed, decompressing the hematoma and allowing reexpansion of the distal true lumen, hence facilitating reentry.[16]

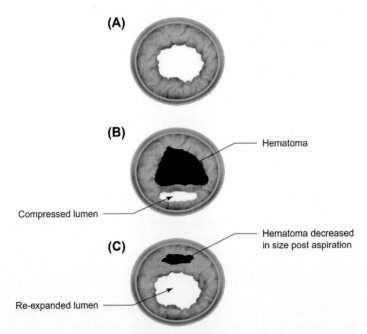

FIGURE 5.39 Cross-sections of the vessel illustrating the subintimal transcatheter withdrawal technique. (A) The vessel distal to the chronic total occlusion. (B) The effect of hematoma formation in the subintimal space and compression of the distal true lumen. (C) Reexpansion of the distal true lumen after aspiration of the hematoma, which facilitates guidewire reentry.

catheter. A modified STRAW technique can be performed by advancing a guide extension, such as a Guideliner catheter (Vascular Solutions), into the vessel and aspirating at the manifold. The STRAW technique is most effective if the proximal vessel is occluded (e.g., with a balloon) to prevent continuing expansion from proximal blood flow.[16]

Sometimes a subintimal hematoma can cause distal true lumen compression even after successful reentry and stent implantation. In such cases use of a cutting balloon may be useful to decompress the hematoma and restore antegrade flow.

5. **Distal vessel calcification**.
Reentry can be challenging in densely calcified vessels where the Stingray wire slides parallel to the true lumen in the subintimal space rather than puncturing through. Potential solutions include:
   a. Use a different stiff, tapered, highly penetrating guidewire, such as the Gaia 2nd or 3rd (which is preshaped, in a shape similar to the shape of the Stingray wire), the Confianza Pro 12 guidewire (Asahi Intecc), or the Hornet 14 (Boston Scientific) to make a puncture through the calcified vessel wall.
   b. Attempt reentry into a more distal location (by advancing the Stingray balloon— "bobsled technique ").
   c. Place a secondary bend proximally on the Stingray or other wire to contact a different location for reentry.
   d. Retrograde crossing, if reentry fails.

6. **Occlusion of side branch at distal cap**.
See Online Case 25 and 100.
   Problem
   Reentry into one of the branches at the distal CTO cap can result in occlusion of the other branch.
   Solutions
   a. Reenter proximal to the bifurcation, which is usually accomplished using a stiff, penetrating, guidewire.
   b. Enter one vessel and refenestrate the other, preferably using the Stingray system (see Online Cases 25 and 80). Alternatives include reentry using a stiff guidewire (Online Case 100), or changing to a mini-STAR technique. Sometimes, retrograde crossing may be needed to restore patency of the other vessel (Fig. 5.40).

7. **Inability to advance a Stingray wire through a Stingray balloon**.
   Causes
   a. Severe vessel tortuosity.
   b Kinking of Stingray balloon shaft.
   Solutions
   a. Try to straighten the Stingray balloon if possible.
   b. Try to advance a polymer-jacketed guidewire first, such as the Pilot 200.
   c. Use a different stiff guidewire, such as the Gaia 2nd or Confianza Pro 12.
   d. Do *not* push aggressively, as the guidewire can perforate the Stingray balloon lumen.
   e. If all else fails, it may be necessary to remove the Stingray balloon and wire the lesion again.

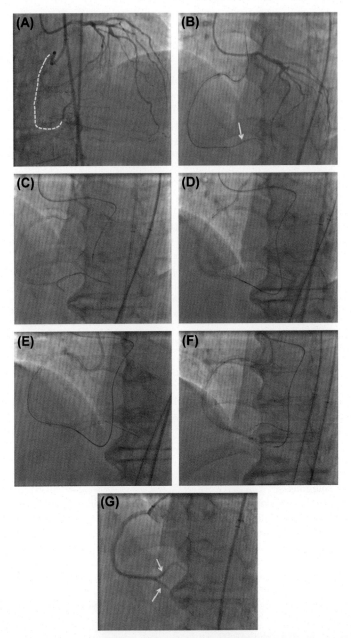

FIGURE 5.40 **Example of using the retrograde approach for preserving both branches of a bifurcation at the distal chronic total occlusion (CTO) cap.** Percutaneous coronary intervention of a long (*dotted line*, A) right coronary artery CTO was attempted. Antegrade crossing into the right posterolateral vessel (PLV; *arrow*, B) was achieved using a CrossBoss catheter, a Pilot 200, and Confianza Pro 12 guidewire. Antegrade flow was restored into the PLV after balloon predilation, but several attempts to antegradely wire the right posterior descending artery (PDA) failed. Retrograde crossing into the right PDA via a septal collateral was then performed (C). After stenting into the PLV antegradely, retrograde crossing into the distal right coronary artery was performed with a Confianza Pro 12 guidewire (D, E), followed by wire externalization and PDA stenting (F) using a mini-crush technique. An excellent final angiographic result was achieved with preservation of flow into both PDA and PLV (*arrows*, G).

5.6 NON-STINGRAY REENTRY TECHNIQUES

Although use of the Stingray balloon and guidewire is the preferred, more-controlled reentry technique, if it is not available or if it fails to achieve reentry, other reentry techniques can be used.

5.6.1 Subintimal Tracking and Reentry

Step 1: The reentry guidewire (usually polymer-jacketed, such as Pilot 50 or 200 or Fielder XT or FC) is inserted within the lesion and a loop is formed at the distal tip.

Step 2: The knuckled guidewire is advanced until it spontaneously reenters into the true lumen (usually at a distal bifurcation).[7] This may be easier to accomplish than more limited wire reentry or Stingray reentry techniques.

What Can Go Wrong?

   a. Extensive shearing and occlusion of side branches, resulting in limited outflow. The STAR technique should NOT be used in the left anterior descending artery, as it may lead to significant loss of the septal and diagonal branches (and may also preclude future coronary bypass graft surgery due to extensive stent implantation).
   b. Inability to reenter the distal true lumen. If this occurs then use of the Stingray system may facilitate reentry, although reentry will likely continue to be challenging because of the extensive dissection caused by the STAR technique.
   c. High restenosis and reocclusion rates (likely due to the long stent length and limited outflow, although studies are limited by presenting data with both drug-eluting and bare-metal stents and considering TIMI 2 flow post-PCI as successful procedure):
      i.   52% need for target vessel revascularization in the original Colombo series[2]
      ii. 54% restenosis with contrast-guided STAR[4]
      iii. 57% reocclusion rate in an Italian registry[3]

In summary, the STAR technique should be used only as a last resort option in CTO PCI due to loss of side branches and high restenosis rates.

5.6.2 Limited Antegrade Subintimal Tracking (LAST) or Mini-STAR (Fig. 5.41)

Step 1: The antegrade microcatheter is advanced into the CTO lesion.
Step 2: The reentry guidewire is selected. In mini-STAR,[7] the Fielder FC or XT wire is used. In LAST, a Pilot 200 or a Confianza Pro 12 guidewire is used.
Step 3: The reentry guidewire is shaped to facilitate reentry, as follows (Fig. 5.41):

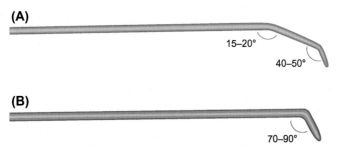

FIGURE 5.41   Illustration of the wire bends used for the mini-STAR (A) and the limited antegrade subintimal tracking (B) techniques.

> **Mini STAR**: A 40–50 degrees curve is created 1–2 mm proximal to the tip and a second 15–20 degrees curve is created 3–5 mm proximal to the tip.
>
> **LAST**: A 90 degrees curve is created 2–3 mm proximal to the tip.

Step 4: The reentry wire is manipulated until reentry is achieved into the distal true lumen. This can be occasionally facilitated by use of the Venture wire control catheter[17] or a dual lumen microcatheter.

What Can Go Wrong?

a. Failure to reenter. Other techniques (such as the Stingray system and retrograde crossing) can be used in these cases.
b. Distal true lumen compression by subintimal hematoma. The STRAW technique (Figs. 5.38 and 5.39) can be used to decompress the hematoma.
c. Wire perforation alone will only rarely cause tamponade, but it is important to detect it early to prevent advancement of equipment, such as balloons and microcatheters, which will enlarge the exit point and may cause a catastrophic perforation.
d. Distal reentry, which can lead to similar consequences as the STAR technique (loss of side branches and high restenosis and reocclusion rates).

Because of all the preceding limitations and because upfront use of the LAST technique can create subintimal hematoma, hindering reentry, use of the LAST technique is currently discouraged by most operators.

5.6.3 Carlino Technique

See Online Cases 13, 27, 37, 45, 48, 49, 53, 60, 62 and 84.

Historical Perspective

Mauro Carlino is a highly creative interventional cardiologist from Milan, Italy who pioneered the technique of using contrast to cause subintimal dissection. In the original version of the technique a significant amount of contrast was used, essentially performing a variation of the STAR technique (i.e., causing extensive dissection with the goal of distal true lumen reentry). As with the original STAR technique, Carlino's original technique had increased risk for complications, such as perforation (storm-cloud dissection) and high restenosis rates. The technique was subsequently modified to its current form, in which only a small volume of contrast is injected, limiting the extent of dissection and associated risk. The context of use also changed: instead of use as a standard upfront technique, the modified Carlino technique is currently used for specialized indications, such as:

1. Wire uncrossable lesions (see Chapter 9, Section 9.2)
2. Proximal cap ambiguity (see Chapter 9, Section 9.1)

Modified Carlino Technique: Step-by-Step Description

Step 1: An antegrade microcatheter (or over-the-wire balloon) is advanced as far as possible into the proximal cap of the CTO lesion. A microcatheter is preferred because it has the radiopaque marker at its tip (in contrast to small balloons that have radiopaque markers in the middle of their shaft).

Step 2: The coronary guidewire is removed and a small syringe is attached to the microcatheter.

Step 3: A small amount of contrast (0.5–1.0 mL) is injected *gently* into the occlusion, under cine-angiography.

Step 4: The course of the contrast is assessed. There are four possible outcomes:

a. **Tubular dissection:** implies position of microcatheter within the vessel architecture.

b. **Storm-cloud dissection:** implies position of microcatheter within a small branch. The microcatheter should be withdrawn and redirected.

c. **Patchy appearance:** implies position of microcatheter within vessel architecture, indicating patches of loose tissue adjacent to a highly calcific occlusion.

d. **Dissection into distal true lumen,** creating a pathway that can be subsequently tracked by a guidewire (often a polymer-jacketed guidewire, such as Pilot 200 or a composite core guidewire, such as Gaia 2nd).

REFERENCES

1. Michael TT, Papayannis AC, Banerjee S, Brilakis ES. Subintimal dissection/reentry strategies in coronary chronic total occlusion interventions. *Circ Cardiovasc Interv* 2012;**5**:729–38.
2. Colombo A, Mikhail GW, Michev I, et al. Treating chronic total occlusions using subintimal tracking and reentry: the STAR technique. *Catheter Cardiovasc Interv* 2005;**64**:407–11 . discussion 12.
3. Valenti R, Vergara R, Migliorini A, et al. Predictors of reocclusion after successful drug-eluting stent-supported percutaneous coronary intervention of chronic total occlusion. *J Am Coll Cardiol* 2013;**61**:545–50.

4. Godino C, Latib A, Economou FI, et al. Coronary chronic total occlusions: mid-term comparison of clinical outcome following the use of the guided-STAR technique and conventional anterograde approaches. *Catheter Cardiovasc Interv* 2012;**79**:20–7.

5. Whitlow PL, Burke MN, Lombardi WL, et al. Use of a novel crossing and re-entry system in coronary chronic total occlusions that have failed standard crossing techniques: results of the FAST-CTOs (Facilitated Antegrade Steering Technique in Chronic Total Occlusions) trial. *JACC Cardiovasc Interv* 2012;**5**:393–401.

6. Carlino M, Godino C, Latib A, Moses JW, Colombo A. Subintimal tracking and re-entry technique with contrast guidance: a safer approach. *Catheter Cardiovasc Interv* 2008;**72**:790–6.

7. Galassi AR, Tomasello SD, Costanzo L, et al. Mini-STAR as bail-out strategy for percutaneous coronary intervention of chronic total occlusion. *Catheter Cardiovasc Interv* 2012;**79**:30–40.

8. Lombardi WL. Retrograde PCI: what will they think of next? *J Invasive Cardiol* 2009;**21**:543.

9. Werner GS. The BridgePoint devices to facilitate recanalization of chronic total coronary occlusions through controlled subintimal reentry. *Expert Rev Med Dev* 2011;**8**:23–9.

10. Brilakis ES, Badhey N, Banerjee S. "Bilateral knuckle" technique and Stingray re-entry system for retrograde chronic total occlusion intervention. *J Invasive Cardiol* 2011;**23**:E37–9.

11. Christopoulos G, Wyman RM, Alaswad K, et al. Clinical utility of the Japan-Chronic total occlusion score in coronary chronic total occlusion interventions: results from a multicenter registry. *Circ Cardiovasc Interv* 2015;**8**:e002171.

12. Wilson WM, Walsh SJ, Yan AT, et al. Hybrid approach improves success of chronic total occlusion angioplasty. *Heart* 2016;**102**:1486–93.

13. Stetler J, Karatasakis A, Christakopoulos GE, et al. Impact of crossing technique on the incidence of periprocedural myocardial infarction during chronic total occlusion percutaneous coronary intervention. *Catheter Cardiovasc Interv* 2016;**88**:1–6.

14. Visconti G, Focaccio A, Donahue M, Briguori C. Elective versus deferred stenting following subintimal recanalization of coronary chronic total occlusions. *Catheter Cardiovasc Interv* 2015;**85**:382–90.

15. Christopoulos G, Kotsia AP, Brilakis ES. The double-blind stick-and-swap technique for true lumen reentry after subintimal crossing of coronary chronic total occlusions. *J Invasive Cardiol* 2015;**27**:E199–202.

16. Smith EJ, Di Mario C, Spratt JC, et al. Subintimal TRAnscatheter withdrawal (STRAW) of hematomas compressing the distal true lumen: a novel technique to facilitate distal reentry during recanalization of chronic total occlusion (CTO). *J Invasive Cardiol* 2015;**27**:E1–4.

17. Badhey N, Lombardi WL, Thompson CA, Brilakis ES, Banerjee S. Use of the venture wire control catheter for subintimal coronary dissection and reentry in chronic total occlusions. *J Invasive Cardiol* 2010;**22**:445–8.

18. Sumitsuji S, Inoue K, Ochiai M, Tsuchikane E, Ikeno F. Fundamental wire technique and current standard strategy of percutaneous intervention for chronic total occlusion with histopathological insights. *JACC Cardiovasc Interv* 2011;**4**:941–51.

Chapter 6

The Retrograde Approach

6.1 HISTORICAL PERSPECTIVE

The retrograde technique differs from the standard antegrade approach in that the occlusion is approached from the distal vessel, advancing a wire against the original direction of blood flow (i.e., retrograde).[1,2] The guidewire is advanced into the artery distal to the occlusion through either a bypass graft or through collateral channels. This approach differs from the antegrade approach, in which all equipment is inserted only proximal to the occlusion and travels in the same direction as the original arterial flow (i.e., antegrade).

The retrograde chronic total occlusion (CTO) percutaneous coronary intervention (PCI) technique was first described by Kahn and Hartzler in 1990, who performed balloon angioplasty of a left anterior descending artery (LAD) CTO via a saphenous vein graft (SVG).[3] In 1996 Silvestri et al. reported retrograde stenting of the left main artery via an SVG.[4] In 2006 Surmely, Tsuchikane, Katoh et al. first reported retrograde crossing via septal collaterals,[5] starting the modern era of the retrograde techniques through septal[5–10] and epicardial[11] collaterals as well as through arterial bypass grafts.[10] The introduction of specialized equipment and further refinements of the technique started in Japan[12–14] with rapid adoption both in Europe[15–18] and in the United States.[19–20]

6.2 ADVANTAGES OF THE RETROGRADE APPROACH

Crossing in the retrograde direction can sometimes be easier than antegrade crossing because the distal cap:

1. Is usually easier to enter than the proximal cap, as it is more frequently tapered.[21]
2. Is often softer than the proximal cap, likely because of exposure to lower filling pressure.
3. Is less frequently anatomically ambiguous.

Moreover, the antegrade approach may not be feasible or desirable in some CTOs, for example ostial and stumpless CTOs; CTOs with ambiguous proximal cap; CTOs with bifurcation at the distal cap; long and tortuous CTOs; CTOs previously attempted and failed antegradely; failed CTOs with extensive dissection; or CTOs with diffuse disease distal to the occlusion. In cases where antegrade wiring is challenging because of ambiguous course, the retrograde wire can help

Manual of Chronic Total Occlusion Interventions. https://doi.org/10.1016/B978-0-12-809929-2.00006-5

direct the antegrade wire and ensure that the latter is in the structure of the vessel proper. Retrograde CTO PCI might also be advantageous in patients with severe renal insufficiency and clear retrograde channels, because most steps required to complete the retrograde approach require limited contrast injection.

6.3 SPECIAL EQUIPMENT

In addition to the standard equipment needed for the antegrade approach, the retrograde approach requires specialized equipment: short guides, specialized microcatheters (150–155 cm long), and long guidewires for externalization, such as the R350 and RG3 wires, as described in Chapter 2.

1. **Short guide catheters and guide catheter extensions**
   The standard guide catheter length is 100 cm (shaft length, although the length from the hub to the guide tip is approximately 106 cm).[1] If standard guide catheters are used for the retrograde approach, equipment may not be long enough to reach the lesion retrogradely (Online Case 99): the retrograde microcatheter might not reach the antegrade guide catheter, and wires advanced retrogradely may be too short to be externalized, especially with retrograde crossing through epicardial collaterals or bypass grafts. Utilizing a shorter guide catheter extends the reach of balloons, wires, and microcatheters advanced retrogradely, as the length of the catheter outside the body has been decreased by the shorter guide length. Shorter guide catheters (usually 90 cm guides are used) are commercially available, but if they are not locally available, any guide can be shortened using an interposition segment of sheath, as described in Section 2.3.2.[22] Another option is to use a guide catheter extension deeply intubated into the CTO target vessel, which is especially useful for retrograde crossing through epicardial collaterals or (left or right) internal mammary artery grafts.

2. **Microcatheters**
   Several microcatheters are available for the retrograde approach, such as the Corsair, Caravel, Turnpike, Turnpike LP, Finecross, Nhancer ProX, and Micro 14 (Section 2.4). Only long (150 cm, except for the Nhancer and Micro 14, which are 155 cm long) microcatheters should be used for retrograde crossing. Larger microcatheters, such as the Corsair and Turnpike, facilitate collateral crossing, providing collateral dilation at the same time, and have good penetration power into the distal cap, yet may be challenging to deliver through small and tortuous collaterals. Lower profile microcatheters, such as the Caravel, Turnpike LP, Finecross, and Micro 14, may be easier to deliver, especially through small caliber epicardial collaterals, but provide less support for crossing complex (such as heavily calcified) lesions.

3. **Externalization guidewires**
   Dedicated externalization guidewires are currently available (RG3 and R350) (Section 2.5.6) and should be used for externalization whenever possible. These wires are long (330 and 350 cm, respectively), thinner than

standard guidewires, and have hydrophilic coating over more than half of their shafts. The tip of the wire should not be bent, to facilitate antegrade equipment loading after externalization.

If a dedicated externalization guidewire is not available or cannot be used (e.g., when the guidewire crosses into the antegrade guide catheter but the microcatheter cannot be advanced into the antegrade guide catheter), a Rotawire (BostonScientific) or Viperwire Advance (CSI), which are 330 and 325 cm, respectively in length, can be used although they are delicate and prone to kinking. A standard length (300 cm) guidewire can often be used for externalization, especially if short guide catheters or guide catheter extensions are being used. However, externalization of standard guidewires is more challenging and requires more time and more force, potentially leading to compression of the heart with hypotension, bradycardia, and occasionally asystole. Lubricating the microcatheter with Rotaglide may facilitate externalization of such wires.

4. **Collateral crossing guidewires**

Preferred wires for septal crossing are composite core soft guidewires (such as the Sion, Suoh 03, and Samurai RC) or soft, polymer-jacketed wires such as the Fielder FC, Fielder XT-R, and Sion Black (Section 2.5.4). The best tip bend is very short (1 mm) and quite shallow (20–30°, although some operators use 90 degrees bends) to allow for tracking very small, tortuous collaterals.

6.4 STEP-BY-STEP DESCRIPTION OF THE PROCEDURE

Step 1 Decide That Retrograde Is the Next Step

Goal

Decide when the retrograde approach should be used. If retrograde crossing is selected the activated clotting time (ACT) should be kept >350 s.

How?

A. Appropriate collaterals exist.

   And

B. There is local experience and expertise in the retrograde technique.

   And

C1. The antegrade approach fails.

   Or

C2. As the initial crossing strategy (primary retrograde) in the following cases:

   1. Ambiguous proximal cap or stumpless occlusions.
   2. Ostial occlusions.[23–24]
   3. Long occlusions.
   4. Severe proximal tortuosity or calcification.
   5. Small or poorly visualized distal vessel.
   6. CTO vessels that are difficult to engage, such as anomalous coronary arteries.[23,25]
   7. Occlusion involving a major distal bifurcation.

Step 2 Selecting the Collateral

Goal
Select the collateral(s) that will be used for the retrograde approach.

How?
The usual preference order for selecting a retrograde collateral channel is bypass graft, septal, then epicardial. Given high risk for dissection causing ischemia and hemodynamic compromise (see Online Case 46), left internal mammary artery grafts should only be used as the last resort for retrograde access (see Online Cases 29 and 46) with strong consideration of using prophylactic hemodynamic support. The advantages and disadvantages of each collateral channel are shown in Fig. 6.1.[1] The classification and optimal angiographic views for evaluating collateral vessels are discussed in detail in Section 3.3.5.

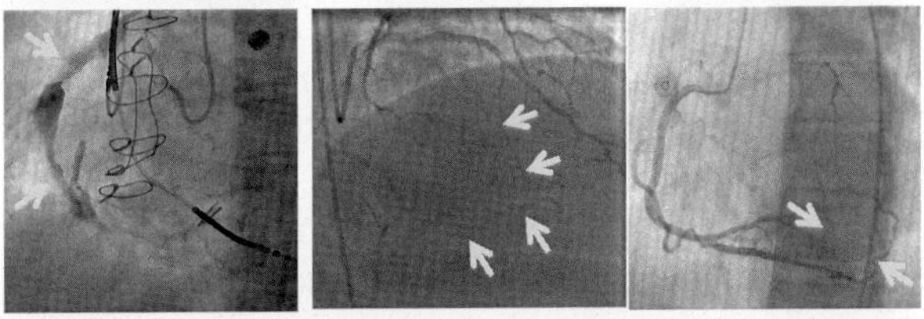

	Bypass	Septal	Epicardial
Tortuosity	+	++	+++
Tamponade risk	+	+	+++
Wiring difficulty	+	++	+++
Able to dilate	Yes	Yes	No

FIGURE 6.1 Comparison of advantages and disadvantages of various collateral vessels that can be used for retrograde CTO interventions. *Reproduced with permission from Brilakis ES, Karmpaliotis D, Patel V, Banerjee S. Complications of chronic total occlusion angioplasty. Interv Cardiol Clin 2012;1:373–89.*

Bypass grafts Online cases 10, 16, 29, 50, 57, 61, 81, 85, 87, 96, 102, 103 are large and easy to wire (Fig. 6.2), but are infrequently available: they were used in 19% of retrograde CTO PCI in prior coronary artery bypass graft (CABG) surgery patients in one series.[26] Although CABG surgery causes scarring of the pericardium it does not eliminate the likelihood of free pericardial effusion (or even worse, loculated pericardial effusion[27-30]) and tamponade in case of perforation during CTO PCI.[31] Even acutely occluded SVGs (Online Cases 87 and 103) can serve as conduits to the distal arterial segment of chronically occluded native coronary arteries.[32] Bypass grafts may tent the vessel to which they are anastomosed, potentially changing the expected course of the native coronary vessel.

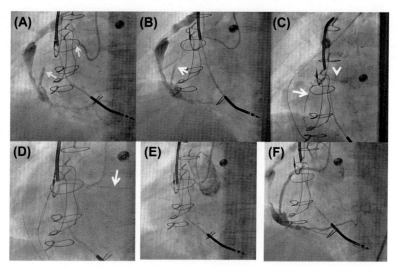

FIGURE 6.2 Illustration of a retrograde intervention of the native right coronary artery (*arrows*, panel A) through a degenerated and aneurysmal saphenous vein graft. After a failed antegrade attempt for CTO crossing (*arrow*, panel B), a guidewire was advanced retrogradely into the distal right coronary artery via a saphenous vein graft with support of a Venture catheter (*arrow*, panel C). A knuckle (*arrowhead*, panel C) was formed on the retrograde guidewire and advanced toward the proximal right coronary artery. After inflation of a 3.0 mm antegrade balloon in the proximal right coronary artery a Confianza Pro 12 guidewire (*arrow*, panel D) was advanced retrogradely into the aorta (reverse controlled antegrade and retrograde tracking and dissection (CART) technique), followed by the retrograde microcatheter. An RG3 guidewire was snared and externalized through a JR4 guide catheter (panel E), followed by antegrade delivery of drug-eluting stents over the externalized guidewire and restoration of the right coronary artery patency (panel F). *Reproduced with permission from Brilakis ES, Grantham JA, Thompson CA, et al. The retrograde approach to coronary artery chronic total occlusions: a practical approach.* Catheter Cardiovasc Interv *2012;79:3–19.*

There is currently controversy as to whether patent but degenerated SVGs used as retrograde conduit for CTO PCI should be coiled after successful completion of CTO PCI.[26] Coiling could stop competitive flow through the native stented segment and possibly decrease the risk of subsequent stent occlusion or thrombosis (see Online Case 10).

Retrograde crossing via internal mammary artery grafts, such as the left internal mammary artery (LIMA), should be done with extreme caution and only by experienced retrograde operators, as it can cause hemodynamic collapse, even without injury of the graft, often due to straightening of the graft tortuosity by the wires and microcatheters (see Online Cases 29, 37, 46, 50, 57).

Wiring of septal collaterals (Figs. 6.3 and 6.4)[6] is preferred over wiring of epicardial collaterals, mainly because the risk of tamponade following channel perforation is substantially lower in septal collaterals as compared with epicardial collaterals.[33] Injury or perforation of a septal collateral is less likely to cause acute myocardial infarction, myocardial hematoma,[9] or tamponade[33] compared with perforation of an epicardial collateral. Treatment of collateral perforation is presented in Chapter 12. In addition, septal collaterals are usually less tortuous

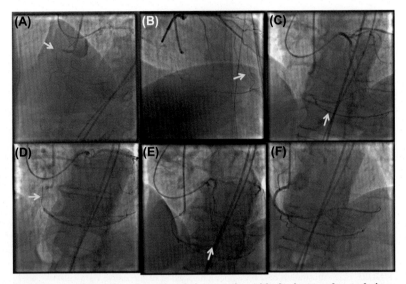

FIGURE 6.3 Illustration of retrograde intervention with the just-marker technique (described in detail in step 7.3 of this chapter). A proximal right coronary artery chronic total occlusion (CTO) (*arrow*, panel A) could not be crossed antegradely. Left main injection demonstrated a septal collateral branch (*arrow*, panel B) that was successfully crossed with a Fielder FC guidewire (*arrow*, panel C), which was then advanced to the occlusion site (*arrow*, panel D). Using the retrograde wire as a marker, a Confianza Pro 12 wire was advanced antegradely through the CTO (*arrow*, panel E), followed by successful stenting of the right coronary artery (panel F). *Reproduced with permission from Brilakis ES, Grantham JA, Thompson CA, et al. The retrograde approach to coronary artery chronic total occlusions: a practical approach.* Catheter Cardiovasc Interv *2012;79:3–19.*

than epicardial channels and are less likely to cause ischemia during crossing, as multiple septal collaterals usually exist.

Selecting the shorter collateral is preferred because (1) it provides better support and (2) it minimizes the risk of not being able to reach the target lesion. However, if a septal collateral enters the vessel close to the distal cap, there may not be enough space to allow for delivery of a wire and a microcatheter into the distal true lumen; using a collateral that enters the vessel more distally is preferred in such cases. Collaterals with corkscrew morphology and >90 degrees angle with the recipient vessel may be challenging or impossible to wire,[12] whereas nontortuous, large collaterals (CC1 or CC2 by the Werner classification[34] as described in Section 3.3.2.1) are the easiest to wire. Often invisible collaterals (CC 0 by the Werner classification) can be successfully crossed using the surfing technique.[35]

It is generally easier to advance a wire through a septal collateral from the LAD to the right coronary artery (RCA) than from the RCA to the LAD, because the RCA ends of septal collaterals usually have more acute turns at their origins and more tortuosity in their lower courses (Fig. 6.4B and C).[22]

Finally, **epicardial collaterals** (Fig. 6.5) (see Online Cases 13, 37, 38, 54, 56, 62, 63, 64, 88, 93, and 97) are the least preferred for retrograde CTO PCI,[7,36] because they are usually more tortuous than septal collaterals[21,22] and

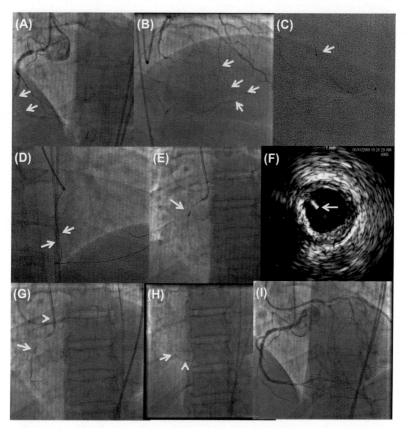

FIGURE 6.4 Illustration of retrograde intervention with wire externalization.
Antegrade crossing attempts of a mid-right coronary artery (RCA) chronic total occlusion (CTO) (*arrows*, panel A) failed due to the presence of a large side branch at the occlusion site. Injection of the left main demonstrated a large, tortuous septal collateral branch (*arrows*, panel B) filling the RCA. Selective injection through a Finecross catheter (*arrow*, panel C) highlighted the collateral vessel course. Kissing wire attempts after retrograde and antegrade wire (*arrows*, panel D) subintimal advancement in the mid RCA failed. After retrograde puncture with the wire, intravascular ultrasonography (*arrow*, panel E) demonstrated that the retrograde wire was located in the proximal true lumen (*arrow*, panel F). The retrograde guidewire was trapped into the antegrade guide (*arrowhead*, panel G) followed by retrograde balloon dilatation (*arrows*, panel G) of the CTO. After externalization of the retrograde guidewire a balloon was advanced antegradely (*arrow*, panel H), while a retrograde balloon (*arrowhead*, panel H) covered the intraseptal portion of the wire. After implantation of multiple drug-eluting stents the RCA patency was restored (panel I). *Reproduced with permission from Brilakis ES, Grantham JA, Thompson CA, et al. The retrograde approach to coronary artery chronic total occlusions: a practical approach.* Catheter Cardiovasc Interv *2012;79:3–19.*

their perforation can lead to rapid tamponade, especially in patients with an intact pericardium. In patients with prior CABG surgery, epicardial collateral perforation can lead to hematoma and localized tamponade, not accessible with pericardiocentesis (Section 12.1.1.2.5). Moreover, if epicardial collaterals are the only source of collateral blood flow and they become occluded

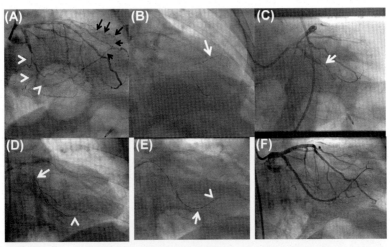

FIGURE 6.5 Example of retrograde chronic total occlusion (CTO) percutaneous coronary intervention via an epicardial collateral. Coronary angiography demonstrating a CTO of the second obtuse marginal branch (*arrowheads*, panel A), which was filling via an epicardial collateral from the second diagonal branch (*arrows*, panel A). The epicardial collateral was successfully wired with a Fielder FC wire (*arrow*, panel B) through a Finecross catheter. The retrograde guidewire formed a knuckle (*arrow*, panel C) and was advanced retrogradely in the subintimal space proximal to the occlusion. After failure of the controlled antegrade and retrograde tracking and dissection (CART) and reverse CART techniques, the antegrade guidewire formed a knuckle and was advanced parallel to the retrograde guidewire into the subintimal space distal to the occlusion (*arrowhead*, panel D). The retrograde guidewire knuckle is shown by an *arrow* in panel D. Reentry into the true lumen distal to the occlusion was achieved with a Stingray wire (*arrowhead*, panel E) through a Stingray balloon (*arrow*, panel E). Stenting restored the patency of the second obtuse marginal branch (F). *Reproduced with permission from Brilakis ES, Badhey N, Banerjee S. "Bilateral knuckle" technique and Stingray re-entry system for retrograde chronic total occlusion intervention. J Invasive Cardiol 2011;23:E37–9.*

during CTO PCI, acute ischemia and myocardial infarction may occur. Sometimes, due to severe tortuosity the control of the guidewire can be limited and advancement challenging or impossible (see Online Case 62). Despite these limitations, with increasing experience and improvements in the retrograde equipment (wires and microcatheters), the use of epicardial collaterals (including ipsilateral epicardial collaterals[37,38]) has been increasing.

In patients with **ipsilateral** collaterals (such as septal collaterals from the proximal into the distal LAD–Figs. 6.6 and 6.7 and Online Case 51; or diagonal or obtuse marginal left-to-left collaterals–Online Cases 56, 66 and 88; or right-to-right atrial collaterals-Online case 101) dual injection may not be required.[39,40]

- A special challenge with ipsilateral collaterals is that the retrograde wire often takes a fairly sharp turn to return into the proximal vessel, which can lead to kinking and difficulty advancing equipment,[39] or more importantly, to collateral rupture (which may occur more frequently with ipsilateral than contralateral collaterals).

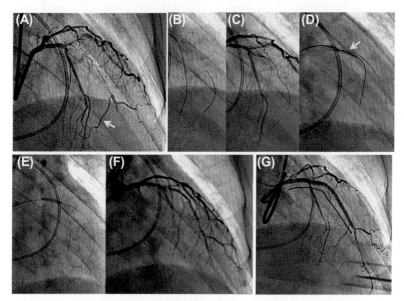

FIGURE 6.6   **Example of retrograde chronic total occlusion (CTO) percutaneous coronary intervention of a left anterior descending artery using the retrograde approach via a septal–septal channel.** Coronary angiography demonstrating a mid-left anterior descending artery (LAD) CTO with the distal vessel filling via an ipsilateral septal collateral (*arrow*, panel A). Crossing of the ipsilateral collateral was achieved with a Pilot guidewire (panel B), followed by retrograde crossing into the proximal LAD using the reverse controlled antegrade and retrograde tracking and dissection technique (panels C, D). The CTO was predilated (panel E), followed by antegrade wiring (panel F) and stenting with an excellent final result (panel G). *Modified with permission from Utsunomiya M, Mukohara N, Hirami R, Nakamura S. Percutaneous coronary intervention for chronic total occlusive lesion of a left anterior descending artery using the retrograde approach via a septal-septal channel.* Cardiovasc Revasc Med *2010;11:34–40.*

- If the CTO is successfully wired through the collateral, then a second ipsilateral guide catheter may be beneficial for trapping or externalizing the wire, because if the retrograde wire is inserted into the antegrade guide catheter, equipment delivery is more difficult through the same guide catheter. Equipment delivery is easier using a ping-pong technique, in which engagement of the target vessel is alternated between the two guide catheters (Fig. 6.8) (Online Case 51 and 101).[41]

Invisible Collaterals

Some patients may appear to have only epicardial collaterals, but if those collaterals become occluded, then septal collaterals may also appear (recruitable collaterals). Selective injection, the so-called tip injection, of the septal perforator branches (through an over-the-wire balloon, or through a microcatheter) may also allow visualization of previously invisible collaterals. Another option for crossing invisible septal collaterals is with the surfing technique, in which septal collaterals are probed and crossed without contrast injection.[35] This technique increases

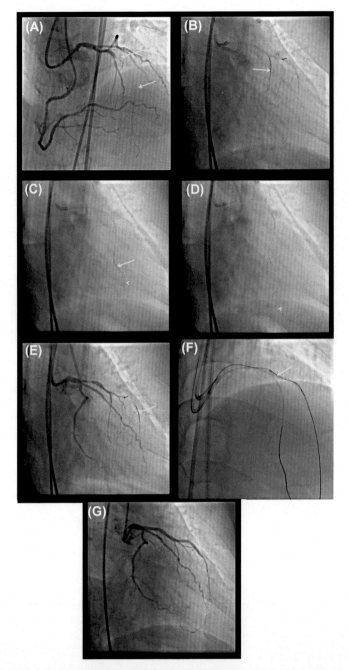

FIGURE 6.7 **Illustration of retrograde percutaneous coronary intervention of a left anterior descending artery chronic total occlusion (CTO) using the retrograde approach with a Suoh 03 wire via a septal–septal channel.** Mid-left anterior descending artery (LAD) CTO (*arrow*, panel A) with ambiguous proximal cap. Retrograde crossing was attempted via ipsilateral septal collaterals. Retrograde injection via a septal collateral (*arrow*, panel B) allowed advancement of a microcatheter (*arrow*, panel C) and Suoh 03 guidewire (*arrowhead*, panels C, D) that successfully crossed into the distal true lumen (*arrow*, panel E). Using the reverse controlled antegrade and retrograde tracking and dissection technique (panel F), the LAD CTO was successfully recanalized (panel G). *Courtesy of Dr. Masahisa Yamane.*

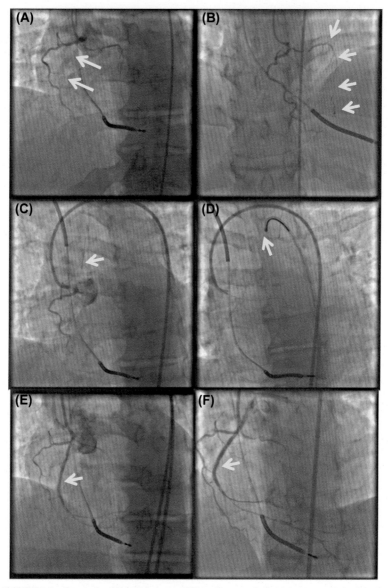

FIGURE 6.8   Example of the ping-pong guide catheter technique for retrograde chronic total occlusion (CTO) percutaneous coronary intervention via an ipsilateral collateral. Online Case 101 Coronary angiography demonstrating a proximal right coronary artery CTO due to in-stent restenosis (*arrows*, panel A), with an ipsilateral atrial collateral (*arrows*, panel B). Retrograde CTO crossing with wire exiting into the aorta (panel C) followed by successful snaring through a second guide catheter (panel D) and stenting (panels E, F) using the two guide catheters in a ping-pong fashion. *Reproduced with permission from Brilakis ES, Grantham JA, Banerjee S. "Ping-pong" guide catheter technique for retrograde intervention of a chronic total occlusion through an ipsilateral collateral.* Catheter Cardiovasc Interv *2011;**78**:395–9.*

the success rates of collateral crossing, but has the limitation that sometimes the collaterals are too small to be tracked by microcatheters.

Rarely collateral vessels may not be apparent during diagnostic angiography. For example, an isolated conus branch can occasionally supply collaterals to an occluded LAD territory and has been used for retrograde PCI[42] or for facilitating antegrade crossing (see Online Case 2) in such cases.

Step 3 Getting to the Collateral

Goal
Advance a wire and microcatheter into the target collateral vessel.

How?
A. Use a workhorse guidewire to minimize the risk for proximal vessel injury.
B. Larger, double bends on the workhorse guidewire are often needed to get to the collateral (Fig. 6.9). After microcatheter advancement the wire is exchanged for a collateral crossing guidewire with a small distal bend.

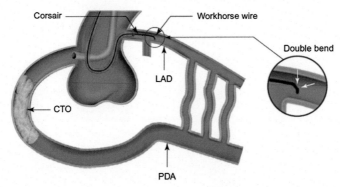

FIGURE 6.9 Illustration of double bend wire shaping for entering a septal collateral.

C. For collaterals with an acute takeoff from the parent vessel consider using the Venture catheter (Fig. 6.10), an angulated microcatheter (such as the SuperCross), or a dual lumen catheter (such as the Twin Pass Torque) to enter the collateral.
   - **Caution**: Wire trapping for removal of the over-the-wire Venture catheter without losing wire position cannot be performed in <8 Fr guides (due to large profile of the Venture catheter).

What Can Go Wrong?
A. **Injury (such as dissection) of the donor vessel**, while trying to enter the collateral (see Section 12.1.1.1.1) (Online Case 97). This can be a catastrophic complication, leading to rapid hemodynamic collapse, and requires immediate treatment (usually with stenting). Stenting of proximal vessel lesions should be considered prior to retrograde crossing to minimize the risk of proximal vessel

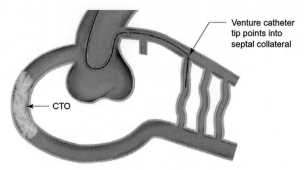

FIGURE 6.10   Illustration of septal collateral branch wiring using a Venture catheter.

dissection (jailing of a septal collateral usually allows wiring of the collateral branch and subsequent equipment delivery through the stent struts). Also **use of a safety wire in the donor vessel** can facilitate donor guide catheter engagement, straighten the artery potentially facilitating wiring of collaterals, and provide access to the vessel in case of a complication, such as dissection or thrombosis (Online Case 22).

Step 4 Crossing the Collateral With a Guidewire

Goal
Cross the collateral with a guidewire.

How?
The technique varies depending on the type of collateral used (septal, epicardial, or SVG).

Step 4a Septal Collateral
See Online Cases 5, 7, 9, 12, 17, 18, 20, 22,23, 28, 31, 32, 33, 36, 40, 41, 42, 43, 44, 53, 57, 58, 59, 60, 64, 65, 69, 70, 71, 74, 77, 78, 79, 84, 90.

Once the microcatheter is inserted into the septal collateral (Fig. 6.11), the workhorse wire is removed and exchanged for a wire with a low tip load that

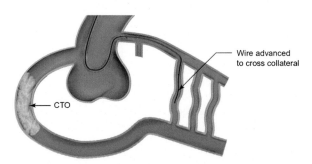

FIGURE 6.11   Illustration of septal collateral crossing.

is highly torquable, most commonly a Sion wire. For septal crossing the wire is shaped with a very small (1–2) mm 30 degrees bend at the tip using the wire introducer. This shape allows tortuosity to be navigated and the wire to be directed away from small side branches.

There are two techniques for subsequent crossing: surfing[35] and contrast-guided.

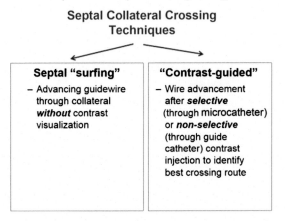

Septal Collateral Crossing Techniques

Septal "surfing"	"Contrast-guided"
– Advancing guidewire through collateral *without* contrast visualization	– Wire advancement after *selective* (through microcatheter) or *non-selective* (through guide catheter) contrast injection to identify best crossing route

Septal Surfing Facts

1. Introduced by Dr. George Sianos.
2. The guidewire (Sion is more commonly used currently) is **advanced rapidly with simultaneous rotation** until it either buckles or advances into the distal target vessel. If the wire buckles it is withdrawn and redirected.
3. Septal surfing can be a very **efficient** crossing method.

Septal Surfing: Tips and Tricks

1. Surfing should *never* be done in epicardial collaterals, because of high risk for perforation.
2. If the wire repeatedly takes the same unsuccessful course, *retract further back* before readvancing to select alternate route.
3. Do *not* push hard and stop immediately when you feel resistance! Force will increase the risk of collateral injury without increasing crossing success.
4. The odds of successful wiring are usually higher in proximal, straighter septals.
5. Septal collaterals are usually straight in their upper half (LAD side), then bow toward the apex and turn again into the posterior descending artery (Fig. 6.12).
   Therefore right anterior oblique (RAO) cranial is the best projection for initial wiring, and RAO caudal for entering into the posterior descending artery.

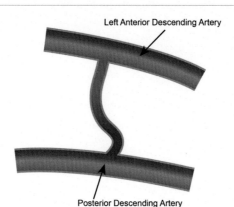

FIGURE 6.12   Right anterior oblique caudal view of septal collaterals.

6. Septal collaterals from the most proximal LAD tend to connect to the right posterolateral branch, whereas more distal septals connect to the posterior descending artery. Very distal septals may connect to a right ventricular branch.

Contrast-Guided Septal Crossing

1. Use a 3 mL luer-lock syringe with 100% contrast. Medallion syringes (Merit Medical) are more resistant to breaking during forceful injection.
2. First aspirate until blood enters the syringe (to avoid air embolization and to ensure that the microcatheter is not against the septal collateral vessel wall). If no blood can be aspirated pull back the microcatheter for a short distance until blood can be aspirated. This prevents air embolism and reduces the risk of hydraulic dissection.
3. Perform cine-angiography while gently injecting contrast with the 3 mL syringe.
4. Flush the microcatheter before reinserting the guidewire (to minimize subsequent stickiness).
5. If a continuous connection to the distal vessel is observed, reattempt crossing through that connection.
6. Do *not* pan to avoid change in collateral road mapping.
7. Consider RAO caudal projection to evaluate the length and tortuosity of the distal part of the septal collateral. Also left anterior oblique projections can be useful if there is limited progress with the RAO views.
8. A description of various types of septal collaterals is shown in Fig. 6.13.

Continued

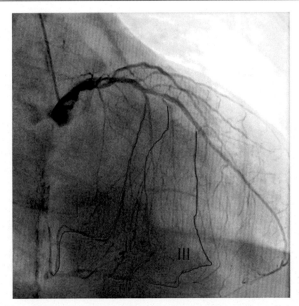

FIGURE 6.13 **Overview of septal collateral anatomy. Three types are described.** (I) Proximal Septal: They often connect to the posterolateral system and have a partial epicardial course. (II) Mid Septal: They generally connect to the PDA and are often tortuous before entering into the posterior descending artery. (III) Distal Septal: Attention is needed because crossing can create high shear stress during externalization. *Courtesy of Dr. Kambis Mashayekhi.*

9. Flush the catheter with saline to avoid wire sticking. The remaining contrast will exit the microcatheter before the saline completely flushes the lumen. This is an extra collateral shot that can be used for viewing the collateral in another projection during cine-angiography.

What Can Go Wrong?
1. Collateral dissection (in most cases further attempts to cross may be performed via a different collateral).
2. Collateral perforation (Section 12.1.1.2.5), which is nearly always benign and only causes localized staining; however, there are reported cases of septal hematoma formation and/or perforation into the pericardium causing hemodynamic compromise.[43,44]
3. Guidewire entrapment: to prevent this complication do not allow big (>1.5 mm) and acute (>75 degrees) bends to form at the tip of the guidewire during attempts for retrograde crossing of the septal collateral.[45] The wire should not be overtorqued by continuous spinning in one direction.
4. Microcatheter tip fracture and entrapment. If the tip of the microcatheter is stuck avoid turning it multiple times; instead withdrawal, followed by rotation is preferable.

Step 4b Epicardial Collateral
See Online Cases 13, 36, 37, 38, 54, 56, 62, 63, 64, 66, 88, 93, 97, 101.

Epicardial Collateral Crossing Tips and Tricks

1. Perform injection through a microcatheter to visualize the collateral vessel course. Be certain that the microcatheter is not wedged (blood aspiration through microcatheter) and inject gently to avoid collateral damage.
2. The Sion Black, Suoh 03, Sion, and Fielder XT-R wires perform best in crossing epicardial collaterals.
3. Advance the wire first, then follow with the microcatheter—never let the microcatheter advance ahead of the guidewire.
4. The microcatheter will straighten tortuosity and allow subsequent advancement.
5. Rotate the wire (do not push) in tortuous segments. Crossing may be easier during diastole, when the angle between collateral turns is wider (Fig. 6.14).

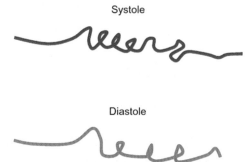

Systole

Diastole

FIGURE 6.14 **Illustration of changes in epicardial collateral channel angulation during the cardiac cycle.** In tortuous epicardial channels, wire crossing through the spiraling segments is the key to success. When the tip gets caught in the curve, quick torque of the wire tends to slide a little on a wider angle in diastole. Therefore, timely torqueing is necessary to go through the spiral segment of the channel.

6. Once the wire reaches the distal true lumen it is advanced to the distal cap before following with the microcatheter.
7. Sometimes crossing epicardial collaterals can prove impossible due to severe tortuosity.

Epicardial Collateral Crossing Facts

1. Epicardial collateral crossing should always be performed using contrast guidance (i.e., no surfing).
2. Orthogonal injections are important to determine the collateral vessel course.
3. Marked tortuosity and small collateral size reduces the likelihood of successful collateral vessel crossing.

What Can Go Wrong?

1. Ischemia of the myocardium supplied by the collateral, especially if there are no other collaterals supplying the same territory. Ischemia can cause arrhythmias and/or hypotension.
2. Collateral perforation that can cause tamponade. In contrast to prior beliefs, perforation in patients with prior CABG surgery can be *more* dangerous than perforation in patients with intact pericardium, as it can lead to loculated effusions compressing various cardiac structures[28] (such as the left atrium[27,29,46] or the right ventricle[30]) that cannot be drained with pericardiocentesis (Section 12.1.1.2.5).
3. Collateral dissection (in most cases further attempts to cross can be performed via a different collateral).
4. Guidewire entrapment: To prevent this, the operator should not allow a loop or knuckle to form at the tip of the guidewire during attempts for retrograde crossing of the collateral (see Online Case 13), although occasionally tiny loops at the tip of a polymer-jacketed guidewire can help cross tortuous collaterals.

Step 4c Bypass Graft (Online cases 10, 16, 29, 50, 57, 61, 81, 85, 87, 96, 102, 103).

Bypass Grafts for Retrograde CTO PCI: Tips and Tricks

1. Both arterial grafts and SVGs (either patent or occluded) can be used for retrograde CTO PCI.
2. There is a risk for perforation and distal embolization (with either patent or occluded SVGs).
3. Internal mammary artery (IMA) bypass grafts are the least preferred bypass grafts for retrograde wiring, because insertion of equipment in the graft could result in pseudolesion formation and even antegrade flow cessation,[47] and because injury of the IMA graft might have catastrophic consequences. Moreover crossing collaterals through IMA grafts can cause IMA dissection, especially in tortuous IMA grafts and when rotating the microcatheter through the collateral. The Caravel microcatheter that is designed to cross the collaterals without rotation is preferred in this setting. If the left IMA (LIMA) to LAD is the only available collateral donor, a mechanical circulatory support device should be strongly considered before crossing the collateral (see Online Case 46).
4. One of the major challenges of retrograde wiring through bypass grafts is navigating severe angulation at the distal anastomosis (see Online Cases 4 and 47). This can be overcome by several techniques, such as using:
   a. polymer-jacketed guidewires and preshaped microcatheters, such as the SuperCross microcatheters[48]
   b. the hairpin (also called reversed) guidewire technique (Section 9.6.2),[49,50] or
   c. the Venture deflectable tip catheter.[51,52]
5. After a native coronary CTO is recanalized, coiling of the SVG may be considered (to minimize risk for subsequent distal embolization and to decrease competitive flow with risk of stent thrombosis), although this approach is controversial.

Step 5 Confirm Guidewire Position Within the Distal True Lumen

Goal

Confirm that the retrograde guidewire has crossed through the collateral into the distal true lumen.

How?

By injecting contrast through the retrograde guide catheter in two orthogonal projections (Figs. 6.15 and 6.16).

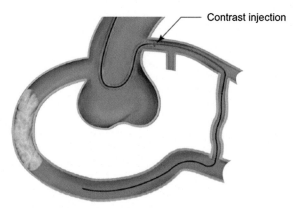

FIGURE 6.15 Illustration of confirmation distal true lumen positioning of the retrograde guidewire, which is an imperative step before advancing the microcatheter through the collateral.

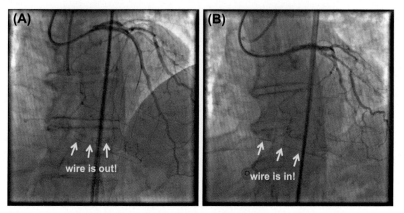

FIGURE 6.16 Example of extraluminal retrograde guidewire position (panel A). The wire was repositioned achieving intraluminal position (panel B). (A) Wire located outside the distal true lumen—the microcatheter should not be advanced with the guidewire in this position. (B) Wire located inside the distal true lumen—the microcatheter can now be safely advanced over the wire.

Distal Wire Position Confirmation

1. Angiographic confirmation of the distal guidewire position should always be done in orthogonal projections *before* advancing the microcatheter through the collateral (to prevent collateral rupture if the wire has exited the vessel) (Fig. 6.16).
2. Possible wire positions.
   - Septum (no crossing achieved)
   - Distal true lumen
   - Cavity (suspected if the wire starts making large back and forth movements); this occurs commonly and is almost always benign
   - Pericardium
   - Nonseptal collateral (occasionally a collateral may appear to be a septal in one projection but may in reality be epicardial; such collaterals have higher risk of rupture; a classic example is an acute marginal collateral supplying the RCA from the distal LAD). Obtaining an orthogonal view can help clarify the collateral type and course.
3. The microcatheter should be advanced only if the guidewire is located inside the distal true lumen. In all other cases the wire should be retracted and redirected.
4. Septal staining is almost always benign and does not cause tamponade (but can cause cardiac biomarker elevation) (Fig. 6.17).

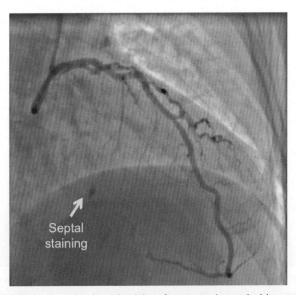

Septal
staining

FIGURE 6.17 Example of septal staining after retrograde septal wiring attempts.

Step 6 Crossing the Collateral With the Microcatheter

Goal

To advance the microcatheter into the distal true lumen.

How?

After distal true lumen guidewire position is confirmed, the guidewire is advanced as far as possible towards the distal CTO cap (or deeply in another distal branch, such as the posterolateral branch) to provide sufficient backup for retrograde microcatheter advancement (Fig. 6.18).

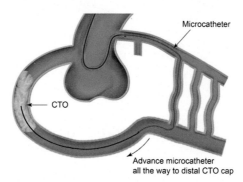

Microcatheter

CTO

Advance microcatheter
all the way to distal CTO cap

FIGURE 6.18   Advancing the microcatheter into the distal true lumen.

What to Do If the Microcatheter Will Not Advance Through the Collateral

(See Online Cases 23, 59.)

1. Rapid clockwise and counterclockwise rotation of the microcatheter (for Corsair and Turnpike), using both hands. Rotate no more than 10 turns in one direction before releasing, to prevent damage to the microcatheter.
2. Increase retrograde guide catheter support, either with active support (forward push of left-sided guides or clockwise rotation of right-sided guides), or by using additional extra support wires, the side branch anchor technique, or a guide catheter extension, such as the GuideLiner and Guidezilla.
3. Try a different microcatheter (lower profile microcatheters, such as the Caravel, Turnpike LP, Finecross, and Micro 14 may be more likely to cross than the Corsair and Turnpike; see Section 2.4).
4. Try a short (135 cm long) microcatheter, which allows for more transmission of torque.
5. Dilate a septal collateral with a small (1.0–1.5 mm) balloon at low pressure (2–4 atm). Epicardial collaterals should never be dilated.
6. Try a new microcatheter (sometimes the microcatheter can become sticky with prolonged use).
7. If the retrograde guidewire is located adjacent to an antegrade subintimal guidewire, the retrograde guidewire can be anchored by inflating a balloon over the antegrade guidewire (Fig. 6.19). Anchoring can provide enough support to advance the microcatheter through the collateral.

Antegrade anchor

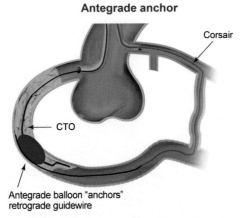

FIGURE 6.19   Anchoring the retrograde guidewire using an antegrade balloon.

8. If the retrograde microcatheter is close to the distal vessel, the retrograde guidewire can be exchanged for a stiffer guidewire, such as the Pilot 200, that can provide more support for advancing the microcatheter.

Alternatively, if the microcatheter cannot cross the collateral:

1. Attempt to cross the CTO with the retrograde guidewire (more likely to be successful in short, noncalcified occlusions).
2. Use the retrograde guidewire as just-marker for antegrade crossing.
3. Try to use another collateral. Leaving the first guidewire in place will increase support for the new guidewire.

What Can Go Wrong?

1. Ischemia can occur, if most or all of the CTO target vessel perfusion comes from the wired collateral. This is most likely to occur with epicardial collaterals, as there are usually multiple septal collaterals. Mild chest discomfort, however, is common during retrograde crossing.
2. Retrograde guide position loss, usually with excessive advancement of the retrograde microcatheter in an effort to overcome resistance. To prevent this careful attention should be paid to the retrograde guide catheter position.
3. Injury of the donor vessel (especially if there is excessive back-and-forth movement of the retrograde guide catheter). In cases of retrograde intervention of RCA CTOs, inserting a safety guidewire in both the LAD and the circumflex will prevent the guide catheter from diving deeply into the LAD and potentially causing dissection.
4. Donor vessel thrombosis (particularly in long retrograde cases and if the donor vessel is diseased). To avoid this, the ACT should be maintained at >350s.

Step 7 Crossing the Chronic Total Occlusion

Goal

Cross the CTO with a guidewire.

How?

Once the collateral branch has been successfully wired and the retrograde microcatheter advanced to the distal cap, there are three ways to cross the lesion:[1]

1. Retrograde crossing of the CTO into the proximal true lumen (retrograde true lumen puncture or "true-to-true").
2. Retrograde dissection/reentry techniques (the reverse CART is the most commonly used technique).
3. Antegrade crossing of the CTO (using the kissing wire or the "just-marker" technique).

All techniques require excellent guide support, which can be achieved using the techniques described in Section 3.6.[53]

Step 7, Part 1 Retrograde True Lumen Puncture

Retrograde true lumen puncture can be achieved in approximately 20%–40% of retrograde CTO PCIs (Figs. 6.4 and 6.20).[12]

The wire that crossed the collateral is advanced to the CTO distal cap, followed by advancement of the microcatheter for additional support. The CTO is then crossed from the distal into the proximal true lumen, either with the same guidewire (as the distal CTO cap may be softer and more tapered than the proximal cap) or with a stiffer guidewire.[1]

Several maneuvers can be used to enhance the chance of crossing, such as inflating a retrograde balloon for more support (coaxial anchor) and using a stiffer, tapered tip, and/or polymer-jacketed wire. However, some argue that use of highly penetrating guidewires (such as the Confianza Pro 12) should be avoided because if retrograde perforation occurs it may be difficult to control. Antegrade intravascular ultrasonography (IVUS) can also facilitate directing the retrograde guidewire into the proximal true lumen (Fig. 6.4F).[54,55]

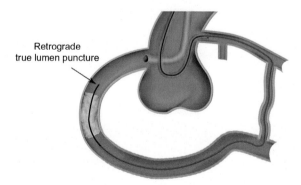

Retrograde true lumen puncture

FIGURE 6.20   Illustration of retrograde true lumen puncture.

What Can Go Wrong?

1. While direct retrograde wire crossing can be an efficient method in short CTO segments, attempts to cross long, tortuous, heavily calcified, or ambiguous CTO segments with stiff guidewires should be avoided due to the significant risk of perforation.
2. Subintimal crossing into the left main coronary artery in an ostial or near-ostial LAD CTO can result in compromise of the circumflex artery or the left main coronary artery and should be avoided. To prevent this complication a guide catheter extension should be advanced into the LAD or circumflex and used as the target for the retrograde guidewire.

Step 7, Part 2 Retrograde Dissection/Reentry

If the CTO subintimal space is entered during manipulation of the antegrade, retrograde, or both guidewires, reentry into the true lumen and CTO crossing can be achieved by two techniques (Fig. 6.21):[1,2]

1. Inflating a balloon over the retrograde guidewire, followed by advancement of the antegrade guidewire into the distal true lumen (CART).[5]
2. Inflating a balloon over the antegrade guidewire, followed by advancement of the retrograde guidewire into the proximal true lumen (reverse CART).

Several variations of the CART techniques have been reported, such as the IVUS-guided CART,[22] the GuideLiner-assisted reverse CART, the stent reverse CART, the confluent balloon[56] technique (Fig. 6.21), and the contemporary reverse CART.

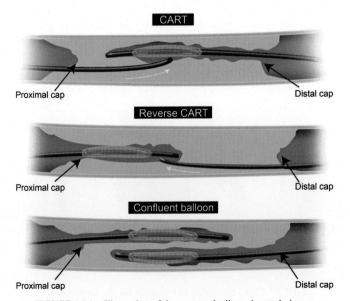

FIGURE 6.21   Illustration of the retrograde dissection techniques.

Step 7, Part 2.1 Controlled Antegrade and Retrograde Tracking and Dissection Technique

See Online Cases 102, 103 First described by Katoh in 2006,[5] the CART technique is based on the principle of creating a subintimal space (ideally confined to the CTO segment) that is known to communicate with the true lumen (Fig. 6.22). The subintimal space is enlarged by inflating a balloon inserted over the retrograde wire.[5] While the balloon is being deflated, an antegrade wire is directed toward the balloon, crossing the lesion and entering

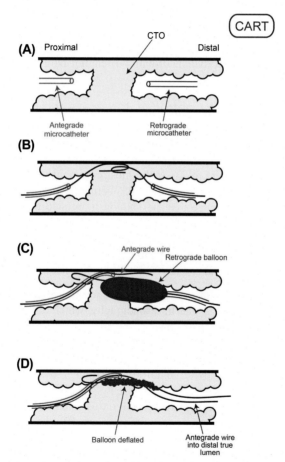

FIGURE 6.22   Illustration of the controlled antegrade and retrograde tracking and dissection (CART) technique. (A) Delivery of an antegrade microcatheter to the proximal cap and a retrograde microcatheter to the distal cap. (B) Subintimal crossing with both the antegrade and retrograde guidewires. (C) Inflation of a balloon in the subintimal space over the retrograde guidewire advancing a guidewire over the antegrade microcatheter. (D) Deflation of the balloon followed by advancement of the antegrade guidewire into the distal true lumen.

the path taken by the balloon and wire. Usually 2.5–3.0 mm diameter, long, over-the-wire balloons are used for the CART technique. After CTO crossing with an antegrade guidewire, balloon angioplasty and stenting is performed in a standard manner.

This technique is often limited by the ability of the balloon to cross the collateral vessel and a higher risk of collateral channel damage and associated complications. In the past balloon dilatation of the septal collaterals was mandatory. In modern days use of the novel microcatheters often allows for balloon crossing without further channel dilation. However, the need for retrograde balloon advancement has been largely replaced by the reverse CART technique and is used only in cases in which the retrograde equipment is not long enough to reach the antegrade guiding catheter (mainly in patients with long epicardial connections and very enlarged hearts), or in cases where the antegrade equipment (microcatheter/balloon) cannot be advanced to the site of wire overlap.

Step 7, Part 2.2 Reverse Controlled Antegrade and Retrograde Tracking and Dissection Technique

See Online Cases 1, 5, 36, 58, 59, 60, 61, 62, 64, 65, 66, 70, 78, 79, 84, 85, 86, 87, 88, 90.

The reverse CART (Fig. 6.23) is the most commonly used technique for retrograde CTO PCI.[20,57] The reverse CART technique is similar to the CART technique, with the difference that the balloon is inflated over the antegrade guidewire, creating a space into which the retrograde guidewire is advanced. It is usually easier to perform reverse CART and make the connection in a straight section of the vessel rather than at a bend.

Failure of the CART and reverse CART technique can be due to:

1. Use of **undersized balloons** (most common reason). Balloon sizing can be facilitated using intravascular ultrasonography, which allows measurement of the media-to-media dimensions.
2. Placement of the antegrade and retrograde guidewire **in different spaces** (one wire in the intima/true lumen and one in the subintimal space) (Figs. 6.24 and 6.25) with inability to connect the different spaces. This challenge can be overcome by creating a dissection in the intimal guidewire, for example by using balloon-assisted subintimal entry or the scratch-and-go technique (see Online Case 59). Another solution for connecting different spaces is use of a cutting balloon (if the antegrade guidewire in located within the true lumen and the retrograde wire in the false lumen—cutting balloon inflation in a false lumen should be avoided; Fig. 6.26).[58]

After retrograde guidewire crossing, **wire externalization is performed in most cases** (step 8, described next). Other treatment options include antegrade wiring and retrograde stent delivery.

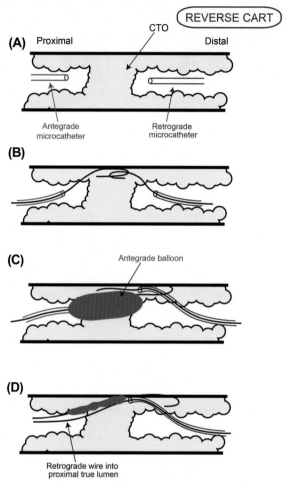

FIGURE 6.23   Illustration of the reverse CART technique. (A) Delivery of an antegrade microcatheter to the proximal cap and a retrograde microcatheter to the distal cap. (B) Subintimal crossing with both the antegrade and retrograde guidewires. (C) Inflation of a balloon in the subintimal space over the antegrade guidewire. (D) Deflation of the balloon followed by advancement of the retrograde guidewire into the proximal true lumen.

Step 7, Part 2.3 Variations of the Reverse Controlled Antegrade and Retrograde Tracking and Dissection Technique

Several modifications of the reverse CART technique have been developed.[1,2]

Step 7, Part 2.3.1 GuideLiner-Reverse Controlled Antegrade and Retrograde Tracking and Dissection

See Online Cases 12, 36, 60, 70, 86, 90.

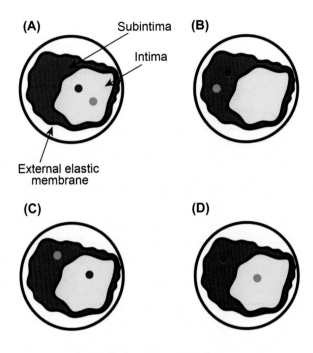

FIGURE 6.24 Antegrade and retrograde guidewire positions as assessed by IVUS in reverse CART. (A) Antegrade and retrograde guidewires are both within intimal plaque. This is the ideal scenario to make a connection, after antegrade balloon dilation in the chronic total occlusion body. If needed, retrograde puncture of intimal plaque with a stiffer wire could be performed. (B) Antegrade and retrograde guidewires are both within the subintimal space. This is another ideal condition in which it is easy to create a connection in the same space after balloon dilation. (C) Antegrade guidewire in intimal plaque but retrograde guidewire in subintimal space. This is a very complex situation in which it is crucial to create a medial disruption with proper balloon sizing to create a connection between the two guidewires. In case of failure, it may be possible to advance the antegrade wire distally to enter the subintimal space and create the previous condition (subintimal–subintimal). (D) Antegrade wire in subintimal space but retrograde wire in intimal plaque, often very calcified. This is the most complex situation because antegrade balloon dilation usually enlarges the subintimal space (increasing intramural hematoma) with low probability of creating a connection between the two guidewires. In this situation, the connection is usually achieved by pushing the retrograde wire in the subintimal space (usually with retrograde knuckle technique). In such a complex case, a possible less-used alternative is retrograde balloon dilation (original CART) to create medial dissection and facilitate antegrade guidewire connection with the retrograde guidewire. *CART*, controlled antegrade retrograde tracking; *IVUS*, intravascular ultrasound. *Modified with permission from Galassi AR, Sumitsuji S, Boukhris M, et al. Utility of intravascular ultrasound in percutaneous revascularization of chronic total occlusions: an overview. JACC Cardiovasc Interv 2016;9:1979–91, Elsevier.*

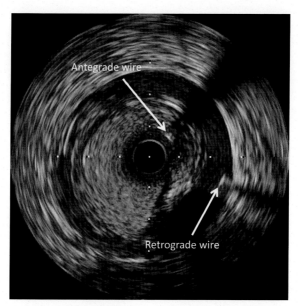

FIGURE 6.25 **Example of the cause of reverse CART failure.** Intravascular ultrasound showing that the antegrade guidewire is within the true lumen, whereas the retrograde guidewire is in the false lumen (scenario C in Fig. 6.24A). This challenge was overcome by using the scratch-and-go technique to advance the antegrade guidewire into the subintimal space (see Online Case 59). *Courtesy of Dr. Scott Harding.*

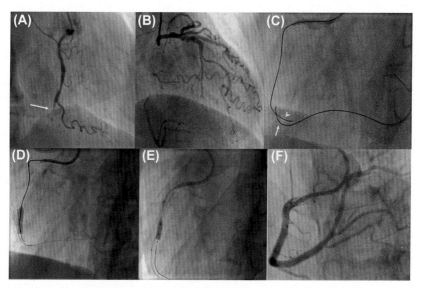

FIGURE 6.26 **Use of a cutting balloon to facilitate reverse controlled antegrade and retrograde tracking and dissection (CART).** (A) Chronic total occlusion of the distal right coronary artery (*arrow*). (B) The posterior descending artery was filling via septal collaterals. (C) Retrograde crossing into the distal true lumen with a Fielder XT wire (*arrow*). A stiff antegrade wire is advanced within the plaque proximally (*arrowhead*). (D) Reverse CART attempts using a 4 mm balloon failed. (E) A 3.5 mm cutting balloon was inflated in the mid-right coronary artery enabling successful reverse CART. (F) Successful recanalization of the right coronary artery after stenting. *Reproduced with permission from Alshamsi A, Bouhzam N, Boudou N. Cutting balloon in reverse CART technique for recanalization of chronic coronary total occlusion.* J Invasive Cardiol *2014;26:E115–6.*

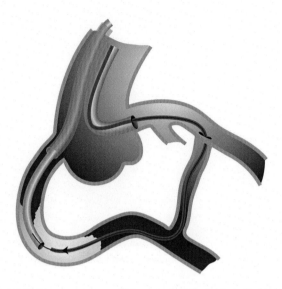

FIGURE 6.27   GuideLiner reverse controlled antegrade and retrograde tracking and dissection.

In the **GuideLiner-reverse CART** technique (Fig. 6.27), a **guide catheter extension** (GuideLiner or Guidezilla) is advanced over the antegrade guidewire to form a proximal target for the retrograde guidewire and facilitate entry into the guide catheter.[59] The larger the size of the guide catheter extension the easier the entry of the retrograde guidewire will be (see Online Case 70). This technique can be used when there is dissection or diffuse disease in the vessel proximal to the connection site or when the retrograde microcatheter is too short to reach the antegrade guide catheter. This technique is also of particular importance for ostial LAD and ostial circumflex CTOs to minimize the risk for left main dissection during retrograde crossing attempts.

Step 7, Part 2.3.2 Intravascular Ultrasound–Guided Reverse Controlled Antegrade and Retrograde Tracking and Dissection

The **IVUS-guided reverse CART** technique is used when there has been difficulty making the connection between the antegrade and retrograde spaces in reverse CART. An IVUS catheter is advanced over the antegrade wire. Although any IVUS catheter can be used, an IVUS catheter with the imaging transducer at its tip, such as the Eagle Eye (Volcano) is preferred. IVUS examination allows determination of whether or not a connection has already been made, definition of the location of the antegrade and retrograde systems in the vessel, and precise sizing of the vessel and the balloon (to maximize the space created for reentry without risking vessel rupture).[13,60]

If IVUS demonstrates that a connection between the retrograde and antegrade spaces has already been made then the problem is usually dissection, recoil following ballooning, tortuosity, or a combination of these factors in the vessel proximal to the connection, preventing passage of the retrograde wire. The solution in such cases is usually performing GuideLiner reverse CART, where a guide catheter extension is delivered just proximal to where the connection has been made, thereby eliminating the problems in the proximal vessel.

If there is no connection visible on IVUS and the antegrade and retrograde wires are in the same space (intimal or subintimal) or if the antegrade wire is intimal and the retrograde wire is subintimal, then the solution is to undertake further dilatation with an appropriately sized balloon advanced over the antegrade wire. If this fails, then the site at which the reverse CART is being attempted should be moved proximally or distally. However, if the retrograde wire is located in the intima and the antegrade wire is subintimal, then reentry can be very challenging (Fig. 6.24). Further balloon dilatation will not facilitate connection. In this situation a stiff wire such as a Confianza Pro 12 should be advanced retrogradely to puncture from through the intimal into the antegrade subintimal space. IVUS can be used to guide this process. Alternatively a knuckled Pilot 200 wire can be advanced from the retrograde microcatheter and usually will enter the subintimal space, allowing a connection to be made.

Step 7, Part 2.3.3 Deflate, Retract, and Advance into the Fenestration Technique

In the **Deflate, Retract, and Advance into the Fenestration (DRAFT)** technique (Fig. 6.28), the antegrade balloon is withdrawn by one operator while the other operator advances the retrograde guidewire into the guide catheter.[61]

Step 7, Part 2.3.4 Stent Reverse Controlled Antegrade and Retrograde Tracking and Dissection

In the **stent reverse CART** technique a stent is placed in the proximal true lumen into the subintimal space to facilitate retrograde wiring into the stent (Fig. 6.29).

The stent reverse CART is used to overcome disease or dissection in the proximal vessel and to provide a large target for the retrograde wire, but has many limitations. First, if the retrograde wire enters through the side of the stent rather than the distal end of the stent the retrograde microcatheter may not advance, and if it does, it will deform the stent, hindering subsequent antegrade equipment delivery. Second, if the stent is placed before a connection is established between the antegrade and retrograde spaces, the retrograde wire may not be able to puncture into the stent. Third, the site of the reverse CART cannot be moved proximally because of the stent

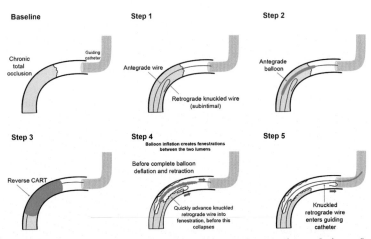

DRAFT
(Deflate, Retract and Advance into the Fenestration Technique)

FIGURE 6.28 **Deflate, retract, and advance into the fenestration technique.** Step 1: Antegrade and retrograde guidewires in the subintimal space, during reverse controlled antegrade and retrograde tracking and dissection. Step 2: A balloon is advanced over the antegrade guidewire. Step 3: Inflation of the antegrade balloon (which is sized 1 to 1 with the vessel). Steps 4 and 5: Balloon inflation over the antegrade guidewire creates multiple fenestrations and subsequent connections between the subintimal space and the true lumen, so that when quick advancement of the retrograde guidewire is performed before complete balloon deflation and retraction, the retrograde knuckled wire can easily penetrate the true lumen. Step 6: The balloon and retrograde guidewire enter the antegrade guide catheter. *Courtesy of Dr. Mauro Carlino.*

and if no connection is made thrombosis of the proximal stent is likely to occur, which may compromise proximal branches covered by the stent. Therefore it is recommended that IVUS be performed prior to stent placement to ensure there is a connection between the antegrade and retrograde space and to allow appropriate stent sizing. Once the stent has been placed and the retrograde wire has been advanced through the stent, further IVUS should be performed over the antegrade wire to ensure central passage of the retrograde wire through the stent. Because of these inherent problems, stent reverse CART has largely been replaced by the GuideLiner reverse CART technique.

Step 7, part 2.3.5 Confluent Balloon Technique

In another variation called the **confluent balloon technique**, both antegrade and retrograde balloons are inflated simultaneously in a kissing fashion to cause the subintimal space to become confluent, allowing wire passage through the CTO (Fig. 6.30 and Online Case 96).[56,62]

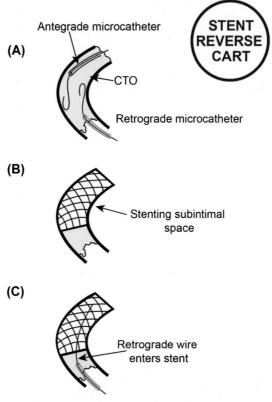

(A)

Antegrade microcatheter

STENT REVERSE CART

CTO

Retrograde microcatheter

(B)

Stenting subintimal space

(C)

Retrograde wire enters stent

FIGURE 6.29   **Stent reverse controlled antegrade and retrograde tracking and dissection technique.** (A) Subintimal location of both antegrade and retrograde guidewires. (B) Stenting over the antegrade guidewire into the subintimal space. (C) Retrograde guidewire crossing into the stented segment.

Step 7, Part 2.3.6 Contemporary Reverse Controlled Antegrade and Retrograde Tracking and Dissection

In the **contemporary reverse CART** technique, a small (usually 2.0–2.5 mm in diameter) antegrade balloon is used to minimize the size of dissection and vessel injury. The retrograde guidewire (usually a composite core guidewire with excellent handling characteristics, such as the Gaia family of wires) is advanced, aiming toward the antegrade balloon.

Step 7, Part 2.3.7 Antegrade Balloon Puncture

Wu et al. created a modification of the reverse CART technique, in which the antegrade balloon remains inflated during retrograde crossing attempts and is punctured by the retrograde guidewire, which is then advanced while the punctured antegrade balloon is retracted under fluoroscopy.[55]

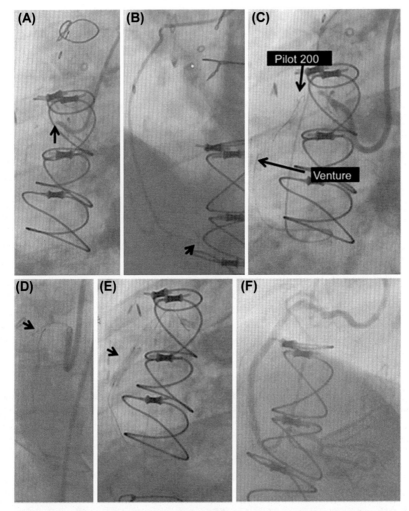

FIGURE 6.30   Illustration of the confluent balloon technique (Online Case 96). Chronic total occlusion (CTO) of the proximal right coronary artery (*arrow*, panel A) with filling of the right posterior descending artery via a diffusely diseased saphenous vein graft with a distal anastomotic lesion (*arrow*, panel B). Using a Venture catheter and a Pilot 200 wire formed into a knuckle (*arrow*, panel C), the CTO was crossed subintimally. A CrossBoss catheter was used for antegrade crossing (*arrow*, panel D), followed by inflation of two 2.5 mm balloons, one advanced over the antegrade and one advanced over the retrograde guidewire (*arrow*, panel E) (confluent balloon technique). The retrograde guidewire successfully crossed into the antegrade guide catheter and was externalized, followed by stent implantation and an excellent result (panel F). *Reproduced with permission from Michael TT, Papayannis AC, Banerjee S, Brilakis ES. Subintimal dissection/ reentry strategies in coronary chronic total occlusion interventions.* Circ Cardiovasc Interv 2012;5: 729–38.

Contrast injections through the guide should not be performed after any attempts for reverse CART, since this will enlarge and extend the localized dissection downstream. To prevent this, the injection syringe can be disconnected from the manifold (as illustrated in Fig. 5.15 of Chapter 5).

What Can Go Wrong?

1. As with all dissection strategies (antegrade and retrograde), side branches at the area of dissection may become occluded, with consequences dependent on the size of the supplied territory (see Section 12.1.1.1.3).
2. Antegrade injections must be avoided as they can cause extensive hydraulic dissection down to the distal vessel.
3. Vessel perforation, if the subintimal balloon is oversized (although most commonly it is undersized). IVUS can help optimize the subintimal balloon size.

Step 7, Part 3   Antegrade Crossing

Just-marker (Fig. 6.31 and Online Case 104) is the simplest (but least reliable) form of the retrograde technique: the retrograde wire is advanced to the distal cap and acts as a marker of the distal true lumen, serving as a target for the antegrade wire.[53] This allows continuous visualization of where the distal true lumen is located, without contrast injections.

Kissing wire entails manipulation of both the antegrade and retrograde wires in the CTO until the ends of the wires meet; the antegrade wire then follows the channel made by the retrograde wire into the distal true lumen (Fig. 6.32).[4,53]

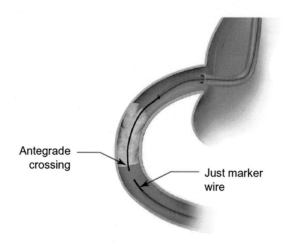

Antegrade crossing

Just marker wire

FIGURE 6.31   Illustration of the just-marker technique.

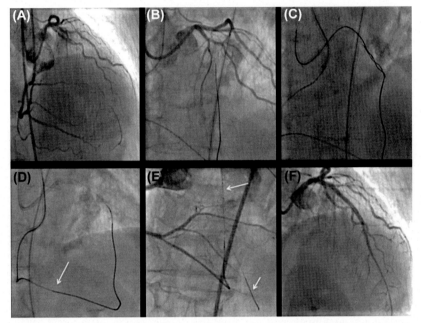

FIGURE 6.32 Long chronic total occlusion (CTO) of the left anterior descending artery (LAD) (panel A) with contralateral collaterals from the right coronary artery. A Sion black guidewire was advanced retrogradely within a Corsair microcatheter through septal collaterals from the posterior descending artery toward the LAD. A Sion blue guidewire was advanced antegradely into the second septal branch of the LAD at the proximal cap of the LAD CTO (panel B). The kissing wire technique was performed with an antegrade Gaia 2nd and a retrograde Ultimate 3g guidewire (panel C). The antegrade Gaia 2nd guidewire supported by an antegrade Finecross microcatheter intubated the tip of the retrograde Corsair (*arrow*, panel D). Thereafter, a dual-lumen Crusade microcatheter (Kaneka, Tokyo, Japan) was advanced antegradely to the mid LAD together with a second antegrade Sion black guidewire. The Sion black guidewire was then advanced into the distal LAD (*white arrows*, panel E). An excellent final result was achieved after drug-eluting stent implantation (panel F). *Courtesy of Dr. Kambis Mashayekhi.*

Step 8 Wire Externalization

Goal
Externalize the retrograde guidewire, in order to use it as rail to advance balloons and stents in an antegrade direction, followed by safe removal of the externalized equipment.

This step is only applicable to cases in which the CTO is crossed in the retrograde direction (retrograde true lumen puncture and reverse CART). If the CTO is crossed in the antegrade direction this step is not needed.

How?
Two options are available for retrograde wire externalization depending on whether the retrograde guidewire enters the antegrade guide or not: (1) wiring the antegrade guide and (2) snaring.

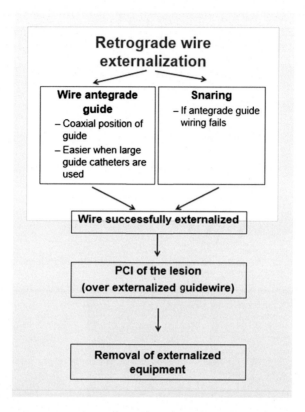

Wiring the antegrade guide is simpler and preferable and may be facilitated by advancing a guide catheter extension (such as GuideLiner or Guidezilla) through the antegrade guide catheter. Wiring the antegrade guide catheter may not always be possible, especially in the following circumstances:

1. Aorto-ostial CTOs.
2. Large vessel caliber.
3. Non coaxial guide positioning.

Option A: Retrograde Guidewire Enters the Antegrade Guide Catheter

Wiring the antegrade guide catheter with the retrograde guidewire (Figs. 6.33 and 6.34) is the simplest technique to externalize a guidewire and should be the first choice whenever possible. After the retrograde guidewire enters the antegrade guide catheter, wire externalization is performed in a step-by-step approach, as follows (Figs. 6.35 and 6.36):

1. After the retrograde wire (that crossed the CTO) enters the antegrade guide, a trapping balloon is inflated within the antegrade guide next to the wire to facilitate advancement of the retrograde microcatheter into the antegrade guide catheter (Fig. 6.35A).

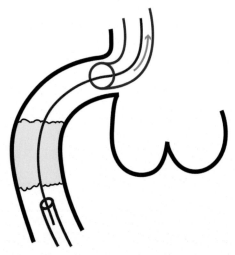

FIGURE 6.33   Advancing the retrograde guidewire into the antegrade guide catheter.

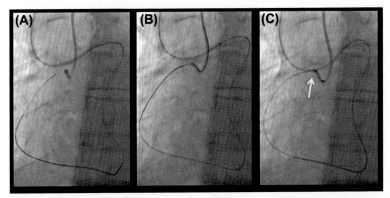

FIGURE 6.34   Conventional wire externalization. A retrograde microcatheter was advanced toward the proximal cap of the right coronary artery chronic total occlusion (panel A). The retrograde guidewire entered the antegrade guide catheter (panel B), followed by advancement of the retrograde microcatheter into the antegrade guide (*white arrow*, panel C). After retracting the retrograde guidewire, an externalization wire was advanced retrogradely via the externalization route within the retrograde microcatheter toward the antegrade guide catheter (not shown). *Courtesy of Dr. Kambis Mashayekhi.*

2. The retrograde guidewire is removed after the trapping balloon is deflated, while the retrograde microcatheter remains within the antegrade guide catheter (Fig. 6.35B).

3. The wire to be externalized (RG3 or R350) is inserted and pushed through the microcatheter (Fig. 6.35C).

4. The antegrade Y-connector is disconnected from the guide catheter and a finger is placed over the antegrade guide catheter hub, until the retrograde guidewire is felt tapping on the finger (Fig. 6.35D).

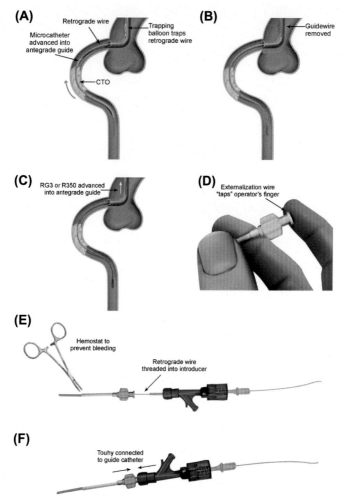

FIGURE 6.35   Step-by-step illustration of the retrograde wire externalization technique.

5. After placing a hemostat over the antegrade guide shaft to minimize bleeding, a wire introducer is inserted through the antegrade Y-connector and the retrograde guidewire tip is threaded through the introducer (Fig. 6.35E).
6. The antegrade Y-connector is reconnected to the guide catheter hub, **without flushing** (to avoid an antegrade hydraulic dissection) (Fig. 6.35F).
7. The retrograde guidewire is pushed until 20–30 cm has exited through the Y-connector.
8. If the tip of the externalized guidewire is damaged, it can be cut off to facilitate loading of balloons/stents or other equipment. This may, however, result in guidewire coil unraveling.

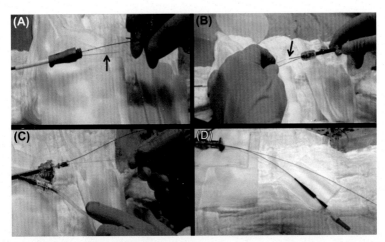

FIGURE 6.36 Illustration of the wire externalization process outside of the body. Externalization was performed through a single guide catheter in a patient with ipsilateral collaterals. The RG3 externalization wire is externalized through the antegrade guiding catheter (*black arrow*, panel A) (6 French). The externalization wire is advanced through the distal end of the introducer needle (*black arrow*, panel B), which in turn has been placed within the antegrade Y-connector. After reconnecting the Y-connector with the antegrade guide catheter the guidewire externalization can be safely performed (panel C). To avoid losing the retrograde end of the externalization wire, a torque device was attached to its tip, in front of the hub of the retrograde microcatheter (panel D). *Courtesy of Dr. Kambis Mashayekhi.*

9. To prevent losing the retrograde ending of the externalization wire during PCI, a torque device should be attached to the retrograde guide end of the externalized wire, so that it is not inadvertently pulled through the retrograde microcatheter (Figure 6.36D).

Option B: Snaring the Retrograde Guidewire (Online Cases 18, 23, 37, 58, 77, 101).
If wiring the antegrade guide catheter with the retrograde wire fails, then wire snaring can be performed (as described in Section 2.7.1). If the retrograde microcatheter successfully crossed the lesion into the aorta, then an externalization wire (R350, RG3) can be advanced through it and snared. If not, then the wire used for retrograde lesion crossing can be externalized, or if it is not long enough it can then be exchanged for a long externalization wire. Short guidewires (180–190 cm) should never be docked with an extension wire followed by an externalization attempt, because of the danger of the connection becoming detached, resulting in collateral channel and vessel injury, or wire loss.

1. Snare preparation (illustrated in Section 2.7.1).
   • Of the commercially available snares, the 27–45 or 18–30 mm EN Snare has three loops facilitating capture of the retrograde guidewire and is preferred to single loop snares, such as the Microvena Amplatz Goose Neck snares and microsnares and the Micro Elite snare.
   • The snare is removed from the package and pulled back into the snare introducer.
   • The snare delivery catheter is discarded (the guide catheter is used for snare delivery).

- The snare is inserted into the antegrade guide by inserting the introducer through the Y-connector (a Co-pilot or a Guardian is preferred to minimize bleeding).
- If snaring is performed through a guide engaging the RCA, it is preferable to use a JR4 guide instead of an Amplatz guide (which is used for most antegrade CTO attempts), as the JR4 guide catheter poses less risk for ostial RCA dissection.

2. The snare is advanced out of antegrade guide and opened (Fig. 6.37A).
3. The retrograde guidewire is advanced through the snare (Fig. 6.37B).

   It is preferable to snare the guidewire you plan to externalize (RG3 or R350), if possible. This is dependent on getting the microcatheter through the CTO segment. The radiopaque portion of these wires is snared, followed by a careful sweep into the antegrade guide catheter.

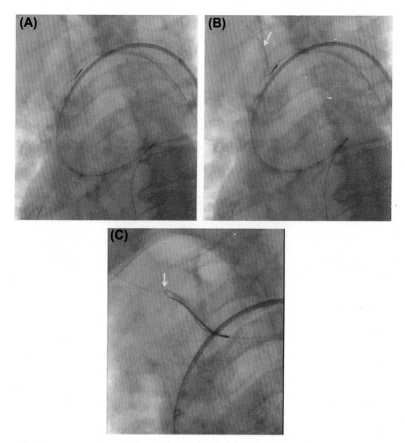

FIGURE 6.37   Illustration of snaring a retrograde guidewire. (Panel A) Illustration of a three-loop snare advanced through the guide catheter and deployed in the ascending aorta. (Panel B) The retrograde guidewire (*arrow*) is advanced through one or more of the snare loops. (Panel C) The snare is withdrawn into the guide, capturing the retrograde guidewire.

If it is not possible to advance the microcatheter through the CTO and into the aorta, then a standard 300 cm long wire may need to be snared. Care must be taken when snaring to avoid fracture or unraveling of the distal part of the guidewire. The ideal snaring location is immediately proximal to the radiopaque portion of the wire.

4. The snare is pulled back, capturing the retrograde guidewire (*arrow*, Fig. 6.37C).

What Can Go Wrong?

   a. Snaring the distal flexible portion of the retrograde wire can result in wire fracture,[63,64] although this is very unlikely with currently used externalization wires (RG3 and R350).
   b. The snared wire may unravel (which is why snaring should be performed under continuous fluoroscopic observation).

5. The retrograde guidewire is pushed through the retrograde microcatheter (while applying gentle traction on the snare) until it exits from the antegrade guide Y-connector (Fig. 6.38).

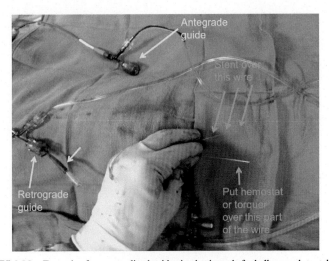

FIGURE 6.38 Example of an externalized guidewire that is ready for balloon and stent delivery.

If the wire tip is deformed it is then cut to facilitate loading of equipment over the externalized portion of the guidewire and a torque device is attached on the retrograde side of the wire to prevent the distal tip from inadvertently slipping into the microcatheter.

What to Do If You Don't Have a Commercial Snare

If a commercial snare is not available, a homemade snare (KAM-snare) can be created (Fig. 6.39).[65] It consists of a wire loop being trapped by an inflated balloon at the distal portion of a guide catheter extension.

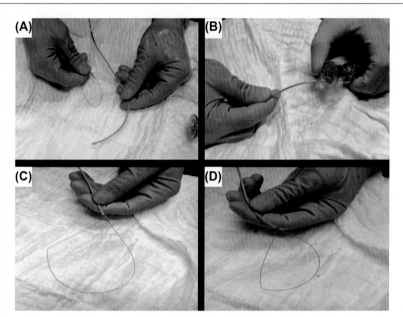

FIGURE 6.39 **How to make a KAM-snare** (see online video: "how to make a homemade snare"). The KAM-snare consists of a guidewire being inserted within the monorail lumen of a conventional angioplasty balloon. The distal end of the guide-wire is then shaped as a loop (panel A). The looped guidewire with the railed balloon are introduced into a guide catheter extension at the proximal entry site. Hence, the balloon entraps the distal returning end of the guidewire loop within the guide catheter extension (panel B). The KAM-snare is inserted into the proximal site of the Y-connector (panel B). By either pushing or pulling the exterior proximal end of the guidewire the diameter of the KAM-snare either increases or decreases (panels C, D). *Courtesy of Dr. Kambis Mashayekhi.*

6. Preparing for angioplasty and stenting.
   a. The microcatheter is retracted distal to the CTO (but continuing to cover the portion of the externalized guidewire that is coursing within the collateral vessel to prevent collateral injury). During withdrawal of the retrograde microcatheter, the donor guiding catheter should be disengaged and watched closely to avoid deep engagement and dissection of the donor vessel.
   b. A torquer or hemostat is attached at the proximal end of the externalized guidewire (to reduce the risk for inadvertent withdrawal of the wire into the microcatheter) (Fig. 6.38).
7. Alternatives to wire externalization.
   In some cases wire externalization may not be possible or should be avoided, such as when:
   a. Wire may not easily advance into the antegrade guide, usually due to tortuous and/or calcified retrograde course that hinders advancement of the retrograde microcatheter.
   b. Retrograde equipment is causing ischemia and needs to be removed.

Alternatives to externalization include antegrade wiring and retrograde stent delivery.

Antegrade Wiring

One way to complete the CTO PCI after retrograde wire crossing is to perform balloon angioplasty of the CTO segment using a balloon advanced over the retrograde guidewire to create a lumen through the CTO followed by antegrade wire crossing and PCI. Retrograde balloon angioplasty can be facilitated by advancing the retrograde wire far into the aorta, or, if possible, into the antegrade guide where it can be trapped with an antegrade balloon inflated at 10–15 atm. Other support techniques can improve the retrograde deliverability of equipment, such as the double-balloon anchoring technique, in which the retrograde wire is anchored into the antegrade guide, and the retrograde guide is anchored in the donor vessel ostium by inflating a balloon in a small vessel side branch.[66]

Other strategies used to enable antegrade wiring after retrograde crossing include:

1. The **antegrade microcatheter probing technique,** in which the retrograde microcatheter is advanced into the antegrade guide catheter, followed by removal of the retrograde guidewire and intubation of the microcatheter with an antegrade wire.[67]
2. The **bridge or rendezvous method,** in which the retrograde microcatheter is inserted into the antegrade guide and aligned with an antegrade microcatheter allowing insertion of an antegrade guidewire into the retrograde microcatheter, positioned within the antegrade guiding catheter[54,68] or in the CTO segment (see Online Cases 31 and 36).[54,68,69]
3. In the **tip-in technique** (Figs. 6.40–6.42),[70] an antegrade microcatheter is advanced over the retrograde guidewire, usually followed by insertion of an antegrade guidewire and antegrade delivery of balloons and stents (see Online Case 57).

    This places less strain on collaterals as compared with retrograde guidewire externalization, yet loss of wire position can occur.
4. The **reverse wire trapping technique,** which involves snaring the retrograde guidewire followed by withdrawing the retrograde guidewire, pulling the antegrade snare through the CTO into the distal true lumen.[63]
5. **Externalization of the wire (as described in step 8) followed by antegrade insertion of a microcatheter,** over which an antegrade wire is inserted.[71]

The advantage of all these techniques is that they minimize retrograde guidewire and balloon manipulations after retrograde crossing, but wiring a microcatheter can be challenging. Moreover, wire externalization can provide superior support for antegrade equipment delivery.

After antegrade wire crossing, antegrade crossing with balloons and stents can be challenging. Reverse anchoring (anchoring the antegrade wire by inflation of a retrograde balloon) can provide strong backup support for antegrade equipment delivery.[67,72]

If an antegrade wire cannot cross the CTO in spite of multiple retrograde balloon inflations, different strategies can be used, such as (1) externalization

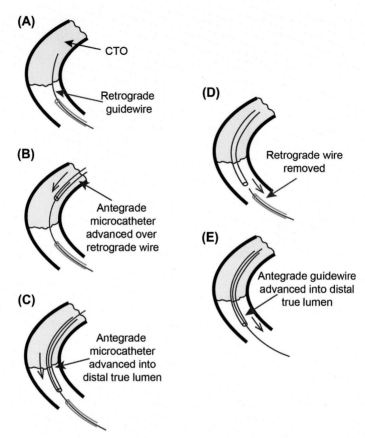

FIGURE 6.40   The tip-in technique. (A) Attempts for crossing with the retrograde guide-wire. (B) An antegrade microcatheter is advanced over the retrograde guidewire. (C) The antegrade microcatheter is advanced over the retrograde guidewire into the distal true lumen. (D) The retrograde wire is removed from the antegrade microcatheter. (E) A new wire is advanced through the antegrade microcatheter into the distal true lumen.

of the retrograde wire (step 8)[19] or (2) retrograde delivery of stents in selected cases.

Retrograde Delivery of Stents

This technique is almost never performed due to high risk for collateral channel injury, stent loss, and stent entrapment, but retrograde stent delivery has been reported through both septal[73] and epicardial[74] collaterals. It requires adequate predilatation of the collateral to minimize the risk of injury and stent entrapment or dislodgement.[75] After completion of the intervention the donor artery is imaged again to ascertain that no complication has occurred.

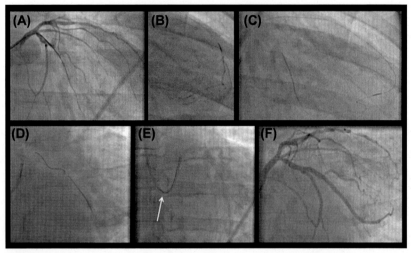

FIGURE 6.41 Percutaneous coronary intervention of a proximal circumflex chronic total occlusion (CTO), in which antegrade wire escalation using the parallel wire technique failed (panel A). An antegrade microcatheter was advanced within a diagonal branch of the left anterior descending artery, and super-selective antegrade contrast injection over this microcatheter showed a retrograde ipsilateral epi-myocardial collateral channel toward the distal circumflex (panel B). A retrograde Sion wire was advanced without distal protection of a microcatheter, because the retrograde microcatheter could not follow within the tiny retrograde collateral channel (panel C). An antegrade Sion marker wire supported correct retrograde advancement of an exchanged retrograde Gaia 1st wire (kissing wire technique) (panel D). The retrograde Gaia 1st wire crossed the CTO and was tipped-in the antegrade microcatheter that was located within the antegrade guide catheter (*white arrow*) (panel E). Panel F shows the final result after implantation of four drug-eluting stents (Xience Pro 3.5×8 mm, 3.0×23 mm, 2.5×18 mm, 2.5×12 mm). *Courtesy of Dr. Kambis Mashayekhi.*

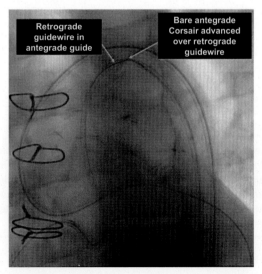

FIGURE 6.42 Illustration of the tip-in technique. *Courtesy of Dr. Michael Luna.*

Step 9 Treatment of the Chronic Total Occlusion

After the retrograde guidewire has been successfully externalized, balloon angioplasty and stenting can be performed using rapid-exchange equipment, followed by removal of the externalized guidewire.

Balloon Angioplasty and Stenting

1. The externalized guidewire provides outstanding support, facilitating device delivery.
2. The tip of the antegrade balloon/catheters should never be allowed to meet with the tip of the retrograde microcatheter/balloon on the same guidewire, because interlocking can occur, resulting in equipment entrapment that may require surgery for removal (see Section 12.1.1.3).

Step 10 Externalized Guidewire Removal (Fig. 6.43)

1. Once stenting of the CTO is completed (Fig. 6.43A), the retrograde wire should be removed in a safe manner.
2. The retrograde microcatheter is advanced back into the antegrade guide (through the recently deployed stents; Fig. 6.43B) unless resistance is encountered.
3. Both guide catheters are disengaged (Fig. 6.43C) (to minimize the risk for deep guide catheter advancement into the coronary ostia, potentially causing dissection).
    • The antegrade guide is disengaged by pushing the externalized guidewire.
    • The retrograde guide is disengaged by fixing the microcatheter and using it as rail for guide retraction.
4. The retrograde guidewire is withdrawn (Fig. 6.43D) paying careful attention to the position of the tip of the guide catheter (to prevent deep engagement and coronary artery injury).
5. The retrograde microcatheter is withdrawn into the donor vessel leaving the retrograde guidewire through the collateral (Fig. 6.43E).
6. Contrast is injected via the retrograde guide to ensure that no injury (perforation of rupture) of the collateral vessel has occurred. If injury is detected, the microcatheter (or a new microcatheter) can be readvanced over the retrograde guidewire to cover the collateral channel perforation and possibly deliver coils.
7. If no collateral vessel injury is detected, the guidewire is removed after readvancing the microcatheter over the guidewire to minimize the risk for collateral channel injury, especially in tortuous epicardial collaterals.

What Can Go Wrong?

1. Collateral dissection: Once a guidewire crosses a collateral, it should always be covered with a microcatheter or an over-the-wire balloon to minimize the risk for collateral vessel injury.
2. Dissection of the target vessel ostium (beware of guide catheter movement during externalization; the externalized guidewire should be pushed, rather than pulled).
3. Dissection of the donor vessel ostium. Care must be taken to disengage the retrograde guide when withdrawing the retrograde microcatheter and wire.

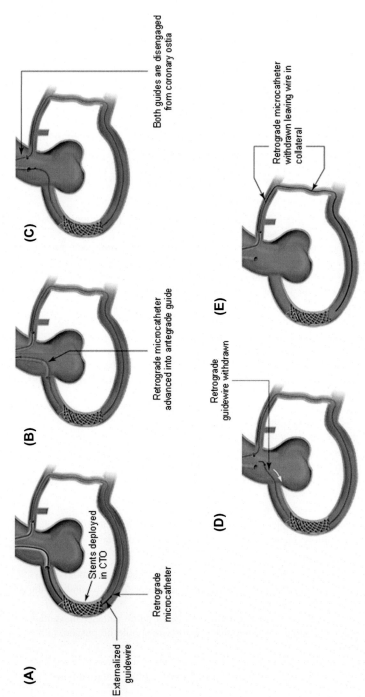

FIGURE 6.43   Illustration of the steps required for removal of the externalized guidewire.

4. Collateral perforation or rupture: Collateral injury can occur during attempts to cross the collateral, during snaring, or during equipment delivery.
5. Equipment entrapment: The antegrade balloon/stents should not meet the retrograde microcatheter or balloon in order to minimize the risk of catheter interlocking and entrapment. If the microcatheter and the balloon inadvertently become interlocked, the hub of the microcatheter can be cut after advancing its tip into the antegrade guide, then the antegrade balloon should be pulled out, pulling the microcatheter through the collateral. If the balloon and microcatheter became separated during the maneuver, the microcatheter should be trapped with a 3 mm semicompliant balloon in the antegrade guide catheter and the guide should be pulled out, thereby removing the microcatheter from the circulation.

6.5 CONTEMPORARY ROLE OF THE RETROGRADE APPROACH

6.5.1 Success

Use of the retrograde approach is critical for achieving high success rates,[14,18,20,35,37] although in some centers advanced antegrade techniques can result in fairly high success rates as well.

The retrograde approach is of particular importance for more complex CTOs, as shown in several contemporary CTO PCI registries (Fig. 6.44).[76,77] Some CTOs (such as flush aortocoronary CTOs; Section 9.3) can only be approached using retrograde crossing.

6.5.2 Complications

As described in detail in Chapter 12, the retrograde approach does carry a significantly higher risk for complications,[20] such as periprocedural myocardial infarction,[78–81] perforation and tamponade, and donor vessel injury, a potentially lethal complication. Use of the retrograde approach is independently associated with increased risk for periprocedural complications and is part of the PROGRESS-CTO Complications score.[82]

Therefore, the retrograde approach should be used cautiously and with meticulous attention to technique. Operators who perform retrograde CTO PCI should be equipped with the skills necessary to deal with the complications that might occur.

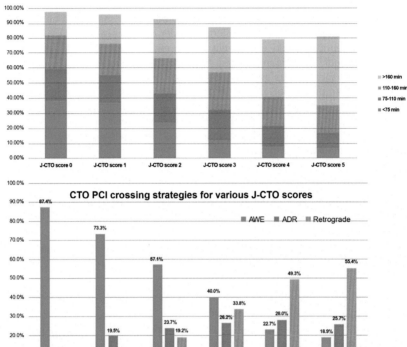

FIGURE 6.44 Successful crossing technique, technical success rates, and procedure time among 657 chronic total occlusion (CTO) percutaneous coronary interventions classified according to Japan-CTO score. AWE=antegrade wire escalation; ADR=antegrade dissection/re-entry. *Reproduced with permission from Christopoulos G, Wyman RM, Alaswad K, et al. Clinical utility of the Japan-Chronic total occlusion score in coronary chronic total occlusion interventions: results from a multicenter registry.* Circ Cardiovasc Interv *2015;8:e002171.*

REFERENCES

1. Brilakis ES, Grantham JA, Thompson CA, et al. The retrograde approach to coronary artery chronic total occlusions: a practical approach. *Catheter Cardiovasc Interv* 2012;**79**:3–19.
2. Joyal D, Thompson CA, Grantham JA, Buller CEH, Rinfret S. The retrograde technique for recanalization of chronic total occlusions: a step-by-step approach. *JACC Cardiovasc Interv* 2012;**5**:1–11.
3. Kahn JK, Hartzler GO. Retrograde coronary angioplasty of isolated arterial segments through saphenous vein bypass grafts. *Cathet Cardiovasc Diagn* 1990;**20**:88–93.
4. Silvestri M, Parikh P, Roquebert PO, Barragan P, Bouvier JL, Comet B. Retrograde left main stenting. *Cathet Cardiovasc Diagn* 1996;**39**:396–9.
5. Surmely JF, Tsuchikane E, Katoh O, et al. New concept for CTO recanalization using controlled antegrade and retrograde subintimal tracking: the CART technique. *J Invasive Cardiol* 2006;**18**:334–8.

6. Ozawa N. A new understanding of chronic total occlusion from a novel PCI technique that involves a retrograde approach to the right coronary artery via a septal branch and passing of the guidewire to a guiding catheter on the other side of the lesion. *Catheter Cardiovasc Interv* 2006;**68**:907–13.

7. Kumar SS, Kaplan B. Chronic total occlusion angioplasty through supplying collaterals. *Catheter Cardiovasc Interv* 2006;**68**:914–6.

8. Niccoli G, Ochiai M, Mazzari MA. A complex case of right coronary artery chronic total occlusion treated by a successful multi-step Japanese approach. *J Invasive Cardiol* 2006;**18**:E230–3.

9. Lin TH, Wu DK, Su HM, et al. Septum hematoma: a complication of retrograde wiring in chronic total occlusion. *Int J Cardiol* 2006;**113**:e64–6.

10. Rosenmann D, Meerkin D, Almagor Y. Retrograde dilatation of chronic total occlusions via collateral vessel in three patients. *Catheter Cardiovasc Interv* 2006;**67**:250–3.

11. Lane RE, Ilsley CD, Wallis W, Dalby MC. Percutaneous coronary intervention of a circumflex chronic total occlusion using an epicardial collateral retrograde approach. *Catheter Cardiovasc Interv* 2006;**69**:842–4.

12. Rathore S, Katoh O, Matsuo H, et al. Retrograde percutaneous recanalization of chronic total occlusion of the coronary arteries: procedural outcomes and predictors of success in contemporary practice. *Circ Cardiovasc Interv* 2009;**2**:124–32.

13. Rathore S, Katoh O, Tuschikane E, Oida A, Suzuki T, Takase S. A novel modification of the retrograde approach for the recanalization of chronic total occlusion of the coronary arteries intravascular ultrasound-guided reverse controlled antegrade and retrograde tracking. *JACC Cardiovasc Interv* 2010;**3**:155–64.

14. Okamura A, Yamane M, Muto M, et al. Complications during retrograde approach for chronic coronary total occlusion: sub-analysis of Japanese multicenter registry. *Catheter Cardiovasc Interv* 2016;**88**:7–14.

15. Sianos G, Barlis P, Di Mario C, et al. European experience with the retrograde approach for the recanalisation of coronary artery chronic total occlusions. A report on behalf of the euroCTO club. *EuroIntervention* 2008;**4**:84–92.

16. Biondi-Zoccai GG, Bollati M, Moretti C, et al. Retrograde percutaneous recanalization of coronary chronic total occlusions: outcomes from 17 patients. *Int J Cardiol* 2008;**130**:118–20.

17. Galassi AR, Tomasello SD, Reifart N, et al. In-hospital outcomes of percutaneous coronary intervention in patients with chronic total occlusion: insights from the ERCTO (European Registry of Chronic Total Occlusion) registry. *EuroIntervention* 2011;**7**:472–9.

18. Galassi AR, Sianos G, Werner GS, et al. Retrograde recanalization of chronic total occlusions in Europe: procedural, in-hospital, and long-term outcomes from the multicenter ERCTO registry. *J Am Coll Cardiol* 2015;**65**:2388–400.

19. Thompson CA, Jayne JE, Robb JF, et al. Retrograde techniques and the impact of operator volume on percutaneous intervention for coronary chronic total occlusions an early U.S. experience. *JACC Cardiovasc Interv* 2009;**2**:834–42.

20. Karmpaliotis D, Karatasakis A, Alaswad K, et al. Outcomes with the use of the retrograde approach for coronary chronic total occlusion interventions in a contemporary multicenter US registry. *Circ Cardiovasc Interv* 2016;9.

21. Sakakura K, Nakano M, Otsuka F, et al. Comparison of pathology of chronic total occlusion with and without coronary artery bypass graft. *Eur Heart J* 2014;**35**:1683–93.

22. Wu EB, Chan WW, Yu CM. Retrograde chronic total occlusion intervention: tips and tricks. *Catheter Cardiovasc Interv* 2008;**72**:806–14.

23. Fang HY, Wu CC, Wu CJ. Successful transradial antegrade coronary intervention of a rare right coronary artery high anterior downward takeoff anomalous chronic total occlusion by double-anchoring technique and retrograde guidance. *Int Heart J* 2009;**50**:531–8.

24. Nombela-Franco L, Werner GS. Retrograde recanalization of a chronic ostial occlusion of the left anterior descending artery: how to manage extreme takeoff angles. *J Invasive Cardiol* 2010;**22**:E7–12.

25. Kaneda H, Takahashi S, Saito S. Successful coronary intervention for chronic total occlusion in an anomalous right coronary artery using the retrograde approach via a collateral vessel. *J Invasive Cardiol* 2007;**19**:E1–4.

26. Dautov R, Manh Nguyen C, Altisent O, Gibrat C, Rinfret S. Recanalization of chronic total occlusions in patients with previous coronary bypass surgery and consideration of retrograde access via saphenous vein grafts. *Circ Cardiovasc Interv* 2016:9.

27. Aggarwal C, Varghese J, Uretsky BF. Left atrial inflow and outflow obstruction as a complication of retrograde approach for chronic total occlusion: report of a case and literature review of left atrial hematoma after percutaneous coronary intervention. *Catheter Cardiovasc Interv* 2013;**82**:770–5.

28. Karatasakis A, Akhtar YN, Brilakis ES. Distal coronary perforation in patients with prior coronary artery bypass graft surgery: the importance of early treatment. *Cardiovasc Revasc Med* 2016;**17**:412–7.

29. Wilson WM, Spratt JC, Lombardi WL. Cardiovascular collapse post chronic total occlusion percutaneous coronary intervention due to a compressive left atrial hematoma managed with percutaneous drainage. *Catheter Cardiovasc Interv* 2015;**86**:407–11.

30. Adusumalli S, Morris M, Pershad A. Pseudo-pericardial tamponade from right ventricular hematoma after chronic total occlusion percutaneous coronary intervention of the right coronary artery: successfully managed percutaneously with computerized tomographic guided drainage. *Catheter Cardiovasc Interv* 2016;**88**:86–8.

31. Marmagkiolis K, Brilakis ES, Hakeem A, Cilingiroglu M, Bilodeau L. Saphenous vein graft perforation during percutaneous coronary intervention: a case series. *J Invasive Cardiol* 2013;**25**:157–61.

32. Brilakis E, Banerjee S, Lombardi W. Retrograde recanalization of native coronary artery chronic occlusions via acutely occluded vein grafts. *Catheter Cardiovasc Interv* 2010;**75**:109–13.

33. Matsumi J, Adachi K, Saito S. A unique complication of the retrograde approach in angioplasty for chronic total occlusion of the coronary artery. *Catheter Cardiovasc Interv* 2008;**72**:371–8.

34. Werner GS, Ferrari M, Heinke S, et al. Angiographic assessment of collateral connections in comparison with invasively determined collateral function in chronic coronary occlusions. *Circulation* 2003;**107**:1972–7.

35. Dautov R, Urena M, Nguyen CM, Gibrat C, Rinfret S. Safety and effectiveness of the surfing technique to cross septal collateral channels during retrograde chronic total occlusion percutaneous coronary intervention. *EuroIntervention* 2017;**12**:e1859–67.

36. Brilakis ES, Badhey N, Banerjee S. "Bilateral knuckle" technique and Stingray re-entry system for retrograde chronic total occlusion intervention. *J Invasive Cardiol* 2011;**23**:E37–9.

37. Mashayekhi K, Behnes M, Akin I, Kaiser T, Neuser H. Novel retrograde approach for percutaneous treatment of chronic total occlusions of the right coronary artery using ipsilateral collateral connections: a European centre experience. *EuroIntervention* 2016;**11**:e1231–6.

38. Mashayekhi K, Behnes M, Valuckiene Z, et al. Comparison of the ipsi-lateral versus contralateral retrograde approach of percutaneous coronary interventions in chronic total occlusions. *Catheter Cardiovasc Interv* 2017;**89**:649–55.

39. Otsuji S, Terasoma K, Takiuchi S. Retrograde recanalization of a left anterior descending chronic total occlusion via an ipsilateral intraseptal collateral. *J Invasive Cardiol* 2008;**20**:312–6.

40. Utsunomiya M, Mukohara N, Hirami R, Nakamura S. Percutaneous coronary intervention for chronic total occlusive lesion of a left anterior descending artery using the retrograde approach via a septal-septal channel. *Cardiovasc Revasc Med* 2010;**11**:34–40.

41. Brilakis ES, Grantham JA, Banerjee S. "Ping-pong" guide catheter technique for retrograde intervention of a chronic total occlusion through an ipsilateral collateral. *Catheter Cardiovasc Interv* 2011;**78**:395–9.

42. Kawamura A, Jinzaki M, Kuribayashi S. Percutaneous revascularization of chronic total occlusion of left anterior descending artery using contralateral injection via isolated conus artery. *J Invasive Cardiol* 2009;**21**:E84–6.

43. Abdel-Karim AR, Vo M, Main ML, Grantham JA. Interventricular septal hematoma and coronary-ventricular Fistula: a complication of retrograde chronic total occlusion intervention. *Case Rep Cardiol* 2016;**2016**:8750603.

44. Araki M, Murai T, Kanaji Y, et al. Interventricular septal hematoma after retrograde intervention for a chronic total occlusion of a right coronary artery: echocardiographic and magnetic resonance imaging-diagnosis and follow-up. *Case Rep Med* 2016;**2016**:8514068.

45. Sianos G, Papafaklis MI. Septal wire entrapment during recanalisation of a chronic total occlusion with the retrograde approach. *Hellenic J Cardiol* 2011;**52**:79–83.

46. Franks RJ, de Souza A, Di Mario C. Left atrial intramural hematoma after percutaneous coronary intervention. *Catheter Cardiovasc Interv* 2015;**86**:E150–2.

47. Lichtenwalter C, Banerjee S, Brilakis ES. Dual guide catheter technique for treating native coronary artery lesions through tortuous internal mammary grafts: separating equipment delivery from target lesion visualization. *J Invasive Cardiol* 2010;**22**:E78–81.

48. Saeed B, Banerjee S, Brilakis ES. Percutaneous coronary intervention in tortuous coronary arteries: associated complications and strategies to improve success. *J Interv Cardiol* 2008;**21**:504–11.

49. Kawasaki T, Koga H, Serikawa T. New bifurcation guidewire technique: a reversed guidewire technique for extremely angulated bifurcation–a case report. *Catheter Cardiovasc Interv* 2008;**71**:73–6.

50. Shirai S, Doijiri T, Iwabuchi M. Treatment for LMCA ostial stenosis using a bifurcation technique with a retrograde approach. *Catheter Cardiovasc Interv* 2010;**75**:748–52.

51. Routledge H, Lefevre T, Ohanessian A, Louvard Y, Dumas P, Morice MC. Use of a deflectable tip catheter to facilitate complex interventions beyond insertion of coronary bypass grafts: three case reports. *Catheter Cardiovasc Interv* 2007;**70**:862–6.

52. Iturbe JM, Abdel-Karim AR, Raja VN, Rangan BV, Banerjee S, Brilakis ES. Use of the venture wire control catheter for the treatment of coronary artery chronic total occlusions. *Catheter Cardiovasc Interv* 2010;**76**:936–41.

53. Saito S. Different strategies of retrograde approach in coronary angioplasty for chronic total occlusion. *Catheter Cardiovasc Interv* 2008;**71**:8–19.

54. Furuichi S, Satoh T. Intravascular ultrasound-guided retrograde wiring for chronic total occlusion. *Catheter Cardiovasc Interv* 2010;**75**:214–21.

55. Wu EB, Chan WW, Yu CM. Antegrade balloon transit of retrograde wire to bail out dissected left main during retrograde chronic total occlusion intervention–a variant of the reverse CART technique. *J Invasive Cardiol* 2009;**21**:e113–8.

56. Wu EB, Chan WW, Yu CM. The confluent balloon technique–two cases illustrating a novel method to achieve rapid wire crossing of chronic total occlusion during retrograde approach percutaneous coronary intervention. *J Invasive Cardiol* 2009;**21**:539–42.

57. Tsuchikane E, Katoh O, Kimura M, Nasu K, Kinoshita Y, Suzuki T. The first clinical experience with a novel catheter for collateral channel tracking in retrograde approach for chronic coronary total occlusions. *JACC Cardiovasc Interv* 2010;**3**:165–71.

58. Alshamsi A, Bouhzam N, Boudou N. Cutting balloon in reverse CART technique for recanalization of chronic coronary total occlusion. *J Invasive Cardiol* 2014;**26**:E115–6.

59. Mozid AM, Davies JR, Spratt JC. The utility of a guideliner catheter in retrograde percutaneous coronary intervention of a chronic total occlusion with reverse cart-the "capture" technique. *Catheter Cardiovasc Interv* 2014;**83**:929–32.

60. Dai J, Katoh O, Kyo E, Tsuji T, Watanabe S, Ohya H. Approach for chronic total occlusion with intravascular ultrasound-guided reverse controlled antegrade and retrograde tracking technique: single center experience. *J Interv Cardiol* 2013;**26**:434–43.

61. Carlino M, Azzalini L, Colombo A. A novel maneuver to facilitate retrograde wire externalization during retrograde chronic total occlusion percutaneous coronary intervention. *Catheter Cardiovasc Interv* 2017;**89**:E7–12.

62. Michael TT, Papayannis AC, Banerjee S, Brilakis ES. Subintimal dissection/reentry strategies in coronary chronic total occlusion interventions. *Circ Cardiovasc Interv* 2012;**5**:729–38.

63. Ge J, Zhang F. Retrograde recanalization of chronic total coronary artery occlusion using a novel "reverse wire trapping" technique. *Catheter Cardiovasc Interv* 2009;**74**:855–60.

64. Ge JB, Zhang F, Ge L, Qian JY, Wang H. Wire trapping technique combined with retrograde approach for recanalization of chronic total occlusion. *Chin Med J (Engl)* 2008;**121**:1753–6.

65. Yokoi K, Sumitsuji S, Kaneda H, et al. A novel homemade snare, safe, economical and size-adjustable. *EuroIntervention* 2015;**10**:1307–10.

66. Lee NH, Suh J, Seo HS. Double anchoring balloon technique for recanalization of coronary chronic total occlusion by retrograde approach. *Catheter Cardiovasc Interv* 2009;**73**:791–4.

67. Christ G, Glogar D. Successful recanalization of a chronic occluded left anterior descending coronary artery with a modification of the retrograde proximal true lumen puncture technique: the antegrade microcatheter probing technique. *Catheter Cardiovasc Interv* 2009;**73**:272–5.

68. Muramatsu T, Tsukahara RI. "Rendezvous in coronary" technique with the retrograde approach for chronic total occlusion. *J Invasive Cardiol* 2010;**22**:E179–82.

69. Kim MH, Yu LH, Mitsudo K. A new retrograde wiring technique for chronic total occlusion. *Catheter Cardiovasc Interv* 2010;**75**:117–9.

70. Vo MN, Ravandi A, Brilakis ES. "Tip-in" technique for retrograde chronic total occlusion revascularization. *J Invasive Cardiol* 2015;**27**:E62–4.

71. Ng R, Hui PY, Beyer A, Ren X, Ochiai M. Successful retrograde recanalization of a left anterior descending artery chronic total occlusion through a previously placed left anterior descending-to-diagonal artery stent. *J Invasive Cardiol* 2010;**22**:E16–8.

72. Matsumi J, Saito S. Progress in the retrograde approach for chronic total coronary artery occlusion: a case with successful angioplasty using CART and reverse-anchoring techniques 3 years after failed PCI via a retrograde approach. *Catheter Cardiovasc Interv* 2008;**71**:810–4.

73. Utunomiya M, Katoh O, Nakamura S. Percutaneous coronary intervention for a right coronary artery stent occlusion using retrograde delivery of a sirolimus-eluting stent via a septal perforator. *Catheter Cardiovasc Interv* 2009;**73**:475–80.

74. Bansal D, Uretsky BF. Treatment of chronic total occlusion by retrograde passage of stents through an epicardial collateral vessel. *Catheter Cardiovasc Interv* 2008;**72**:365–9.

75. Utsunomiya M, Kobayashi T, Nakamura S. Case of dislodged stent lost in septal channel during stent delivery in complex chronic total occlusion of right coronary artery. *J Invasive Cardiol* 2009;**21**:E229–33.

76. Christopoulos G, Wyman RM, Alaswad K, et al. Clinical utility of the Japan-Chronic total occlusion score in coronary chronic total occlusion interventions: results from a multicenter registry. *Circ Cardiovasc Interv* 2015;**8**:e002171.

77. Wilson WM, Walsh SJ, Yan AT, et al. Hybrid approach improves success of chronic total occlusion angioplasty. *Heart* 2016;**102**:1486–93.

78. Kim SM, Gwon HC, Lee HJ, et al. Periprocedural myocardial infarction after retrograde approach for chronic total occlusion of coronary artery: demonstrated by cardiac magnetic resonance imaging. *Korean Circ J* 2011;**41**:747–9.
79. Lo N, Michael TT, Moin D, et al. Periprocedural myocardial injury in chronic total occlusion percutaneous interventions: a systematic cardiac biomarker evaluation study. *JACC Cardiovasc Interv* 2014;**7**:47–54.
80. Stetler J, Karatasakis A, Christakopoulos GE, et al. Impact of crossing technique on the incidence of periprocedural myocardial infarction during chronic total occlusion percutaneous coronary intervention. *Catheter Cardiovasc Interv* 2016;**88**:1–6.
81. Werner GS, Coenen A, Tischer KH. Periprocedural ischaemia during recanalisation of chronic total coronary occlusions: the influence of the transcollateral retrograde approach. *EuroIntervention* 2014;**10**:799–805.
82. Danek BA, Karatasakis A, Karmpaliotis D, et al. Development and validation of a scoring system for predicting periprocedural complications during percutaneous coronary interventions of chronic total occlusions: the prospective global registry for the study of chronic total occlusion intervention (PROGRESS CTO) complications score. *J Am Heart Assoc* 2016:5.

Chapter 7

Putting It All Together: The Hybrid Approach

The optimal approach to chronic total occlusion (CTO) percutaneous coronary intervention (PCI) continues to evolve. Although various CTO crossing techniques have been developed (antegrade wire escalation; antegrade dissection/reentry; and retrograde, as described in Chapters 4–6), there are different schools of thought about the relative merits and priority of each of those approaches.

In January 2011 several high-volume CTO operators convened in a workshop that took place in Bellingham, Washington and created a consensus algorithmic approach about how to optimally approach CTO crossing.[1] This approach, named the hybrid approach to CTO PCI (Fig. 7.1), focuses on opening the occluded vessel, using all available techniques (antegrade, retrograde, true-to-true lumen crossing, or reentry), tailored to the specific case in the safest and most effective and efficient way.

The main principle behind the hybrid approach is that operators should master all of the skill sets of CTO PCI and be able to alternate between these techniques during the same CTO PCI procedure to recanalize the CTO. The goal has been to demystify the procedure by breaking down its various components and gaining in-depth understanding of the principles underlying each technique, making it reproducible and teachable, and thus available to the broader interventional community. The hybrid approach has been used in a large number of cases in both the United States[2–8] and Europe[9,10] with high success rates, and is also useful in learning CTO PCI in a stepwise fashion.

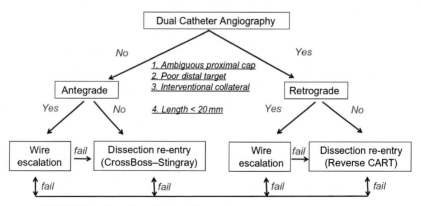

FIGURE 7.1   Overview of the hybrid chronic total occlusion crossing algorithm.

Manual of Chronic Total Occlusion Interventions. http://dx.doi.org/10.1016/B978-0-12-809929-2.00007-7

7.1 DESCRIPTION OF THE HYBRID ALGORITHM[11]

Step 1 Dual Injection

The first and arguably most important step of CTO PCI is to perform dual coronary injection, in nearly all cases, as described in detail in Section 3.2. Dual injection allows good visualization of both the proximal and distal vessel, as well as the collateral circulation, allowing selection of the most suitable initial crossing technique. It also clarifies the location of the guidewire(s) during crossing attempts. Routine performance of dual injection is the simplest and most important step to increase the success rate and safety of CTO PCI.

Step 2 Assessment of Chronic Total Occlusion Characteristics

In-depth review of diagnostic angiographic images prior to PCI is critical. Most experienced CTO operators recommend against performing ad hoc CTO PCI, instead performing CTO PCI only after a well thought out procedural plan has been developed. Time spent studying the diagnostic film is an investment toward a successful CTO PCI procedure. Moreover, radiation and contrast can be reduced during PCI, since the anatomical information has already been obtained during prior diagnostic angiography. Sometimes, however, dual injection images may not be performed until the time of PCI.

Four angiographic parameters are assessed (Chapter 3, Section 3.3): (1) the morphology of the proximal cap; (2) the length of the occlusion; (3) the vessel size and presence of bifurcations beyond the distal cap (i.e., landing zone); and (4) the location and suitability of collateral channels for retrograde access (Fig. 7.2).[1]

1. **Proximal cap location and morphology**. This characteristic refers to the ability to unambiguously localize the entry point to the CTO lesion by angiography or intravascular ultrasonography (see Chapter 9, Section 9.1) and to understand the course of the vessel in the CTO segment.

An ambiguous proximal cap increases the complexity of the procedure and decreases the likelihood of success.[12] A favorable proximal cap is one that is tapered, as opposed to blunt, and has no bridging collaterals or major side branches that would make engagement of the CTO segment difficult using traditional wire escalation techniques. A particularly challenging anatomic subset is that of flush aorto-ostial occlusions, which require use of a primary retrograde approach (see Section 9.3).

2. **Lesion Length**. Lesions are dichotomized into those that are <20mm and ≥20mm long.[13] As noted earlier, in most cases this characteristic can be accurately assessed only by using dual injections.

In CTOs in which antegrade crossing is attempted, short CTOs (<20mm) are usually best approached with antegrade wiring, whereas in long (≥20mm) CTOs, up-front use of a subintimal dissection/reentry technique is preferred, because there is high probability that wire-based crossing attempts will result in

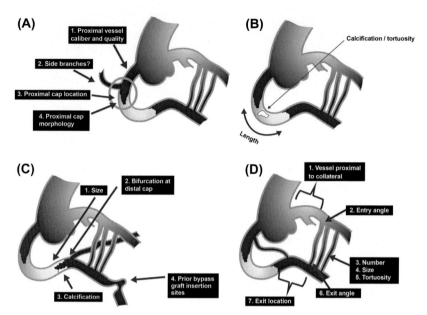

FIGURE 7.2 **Key anatomic characteristics for selecting a chronic total occlusion crossing strategy.** (A) Proximal cap and vessel; (B) lesion length and quality; (C) distal vessel; and (D) collateral assessment.

subintimal wire entry. A lesion length <20 mm has been identified as a predictor of rapid CTO crossing in the Multicenter CTO Registry in Japan (J-CTO; Chapter 1, Fig. 1.6).[13] With the wide adoption of dual injection, it has become evident that the length of the occlusion is frequently shorter than the one estimated by single injections.

3. **Target coronary vessel beyond the distal cap**. This refers to the size of the lumen, presence of significant side branches, vessel disease at the reconstitution point, and ability to adequately angiographically visualize this segment.

4. **Size and suitability of collateral circulation** for retrograde techniques. Optimal collateral vessels for retrograde CTO PCI:

    a. Are sourced from a healthy (or repaired) donor vessel.

    b. Can be easily accessed with wires and microcatheters.

    c. Have minimal tortuosity.

    d. Are not the only source of flow to the CTO segment (which places the patient at risk for intraprocedural ischemia during crossing of the collateral).

    e. Enter the CTO vessel well beyond the distal cap.

More favorable collateral circulation characteristics lower the barriers to utilizing retrograde techniques as an initial strategy or as an early crossover strategy.

What constitutes an interventional collateral (i.e., a collateral that can be wired during a retrograde approach) varies depending upon the experience and skills of the operator. In-depth understanding of the collateral circulation is also

Continued

of paramount importance during antegrade crossing attempts, since dissection reentry techniques and the formation of subintimal hematomas may compromise ipsilateral or bridging collaterals, leading to poor visualization of the distal vessel at the reentry zone and occasionally ischemia. Degenerated or even recently occluded bypass grafts anastomosed to the target vessel distal to the CTO can be also used as retrograde conduits to facilitate PCI.[14,15]

Assessment and utilization of these four angiographic characteristics is highly dependent on operator experience and skill set, and thus is constantly evolving.

Step 3 Antegrade Wiring (Chapter 4)

Antegrade wire escalation refers to the use of guidewires of increasing stiffness to cross a CTO. In the past, a gradual escalation was performed: starting with a workhorse guidewire and then increasing in stiffness to a Miracle 3, 6, 9, and eventually a Confianza Pro 12 guidewire. Currently, however, a more rapid escalation is favored from a tapered-tip polymer-jacketed guidewire (Fielder XT or Fighter) to either a stiff polymer-jacketed wire (Pilot 200) when the course of the CTO is uncertain, or a stiff-tapered tip guidewire (Gaia 2nd or Confianza Pro 12) in cases where the course of the CTO is well understood (Chapter 4, Fig. 4.5). Streamlined use of a relatively small array of guidewires can simplify clinical decisions and inventory management and also leads to in-depth understanding of the properties of the different wires.

Step 4 Antegrade Dissection and Reentry (Chapter 5)

For long lesions approached in the antegrade direction, up-front use of a dissection/reentry strategy is recommended (Fig. 7.3). Dissection can be achieved either by advancing a knuckle formed at the tip of a polymer-jacketed guidewire (such as the Fielder XT or the Pilot 200) or by using the CrossBoss catheter. Antegrade dissection minimizes the risk for perforation (by the blunt guidewire loop or by the CrossBoss catheter tip) and allows for rapid crossing of long occlusion segments. Reentry can be achieved using a stiff polymer-jacketed or tapered-tip guidewire with a sharp distal bend, or more consistently by using the Stingray system, as described in Chapter 5.

Once a particular strategy is selected based on the hybrid algorithm (e.g., Stingray-based reentry), the operator should resist the temptation of multiple attempts to reenter using different wires, as such attempts may expand the subintimal space and hinder reentry, ultimately leading to a failed procedure.

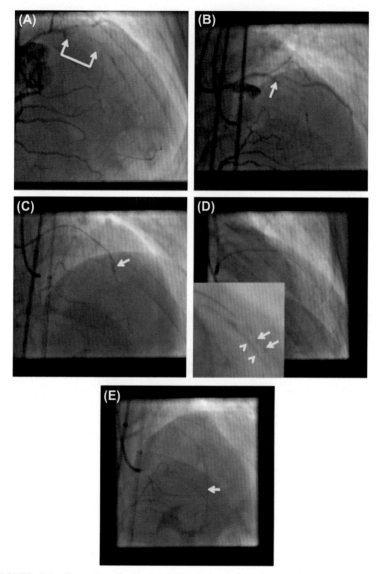

FIGURE 7.3 **Example of antegrade/dissection reentry.** Coronary angiography demonstrating a chronic total occlusion of the mid left-anterior descending artery (*arrows*, A). Angiographic characteristics included well-visualized proximal cap, good distal target, and lesion length >20 mm (H). A CrossBoss catheter was advanced to the proximal cap (*arrow*, B). Using the fast-spin technique the catheter crossed the occlusion subintimally into the mid left-anterior descending artery (*arrow*, C). The CrossBoss catheter was exchanged for a Stingray balloon (*arrows*, D), located adjacent to the true lumen (*arrowheads*, D). A Stingray guidewire (*arrow*, E) successfully crossed into the distal true lumen. Angiography after predilation showed expected dissection at the area of subintimal crossing (*arrows*, F). After implantation of multiple drug-eluting stents an excellent angiographic result was achieved (G). *Reproduced with permission from Brilakis ES, Grantham JA, Rinfret S, et al. A percutaneous treatment algorithm for crossing coronary chronic total occlusions.* JACC Cardiovasc Interv 2012;5:367–79.

Continued

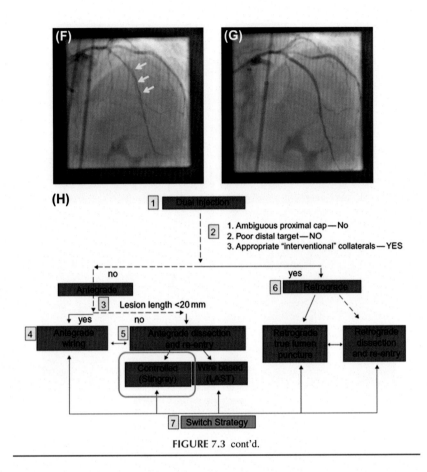

FIGURE 7.3 cont'd.

Step 5 Retrograde (Chapter 6)

The retrograde approach is a major component of a contemporary CTO PCI program. It is critical for achieving high success rates, especially in more complex occlusions,[9,16] however it also carries increased risk for complications such as myocardial infarction,[17–19] perforation, and donor vessel injury.[20,21] Hence, approaching a CTO in the antegrade direction first is preferred, if feasible.

The retrograde approach can be used either up front (primary retrograde, Fig. 7.4) or after a failed antegrade crossing attempt, and it enables high procedural success rates.[20–27] Factors that favor a primary retrograde approach include an ambiguous proximal cap (Section 9.1, see Online Cases 2, 5, 16, 30, 32, 34, 45, 47, 49, 51, 56, 69, and 80, 88, 93, and 104), distal cap at a bifurcation (see Online Cases 5, 45, 49, 64, 80, 100), poor distal target, good interventional collaterals, and heavy calcification. Retrograde wire crossing can occur by advancing the retrograde guidewire into the proximal true lumen (retrograde true lumen puncture, conceptually similar to antegrade wire escalation), by antegrade wiring toward a retrograde-placed guidewire into the distal true lumen (just-marker technique), or by using one of the dissection/reentry techniques, such as controlled antegrade and retrograde tracking (CART), or, more commonly, the reverse CART technique, as described in Chapter 6.

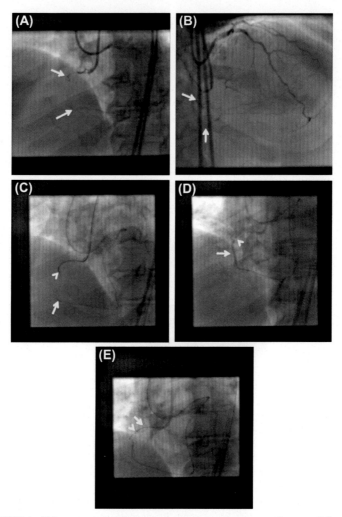

FIGURE 7.4 **Primary retrograde percutaneous coronary intervention to a right coronary artery chronic total occlusion (CTO).** Dual coronary angiography demonstrating a CTO of the proximal right coronary artery (*arrows*, A, B) with distal filling via collaterals from the left anterior descending artery. Angiography demonstrated a poorly defined proximal cap, long lesion length, diffusely diseased distal target vessel, and good interventional collaterals (H). A Corsair microcatheter (*arrow*, C) was advanced retrogradely via a septal collateral over a soft, polymer-jacketed, nontapered guidewire, to the distal right coronary artery. A second Corsair microcatheter (*arrowhead*, C) was advanced antegradely to the mid right coronary artery. After antegrade insertion of a Guideliner catheter (*arrowhead*, D), a 2.5×20 mm balloon was inflated over the antegrade guidewire in the mid right coronary artery (*arrow*, D) and reverse controlled antegrade and retrograde tracking and dissection (reverse CART) was performed, followed by successful advancement of a retrograde Pilot 200 guidewire (E) into the antegrade Guideliner. A ViperWire Advance 335 cm guidewire was externalized followed by predilation of the right coronary artery (F) and restoration of antegrade flow was achieved after implantation of multiple drug-eluting stents (G). *Reproduced with permission from Brilakis ES, Grantham JA, Rinfret S, et al. A percutaneous treatment algorithm for crossing coronary chronic total occlusions. JACC Cardiovasc Interv 2012;5:367–79. 79.*

Continued

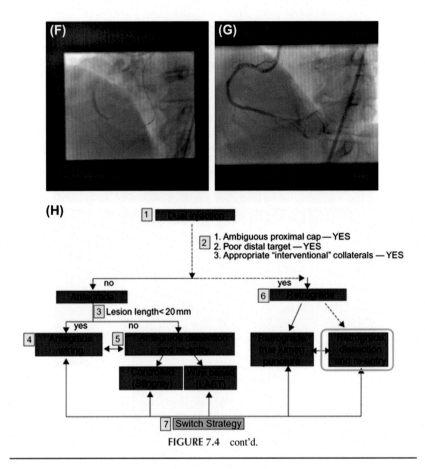

(F)

(G)

(H)

1. Ambiguous proximal cap — YES
2. Poor distal target — YES
3. Appropriate "interventional" collaterals — YES

no

yes

Antegrade

Retrograde

3 Lesion length < 20 mm

yes

no

4 Antegrade wiring

5 Antegrade dissection and re-entry

Retrograde true lumen puncture

Retrograde dissection and re-entry

Controlled (Stingray)

Wire-based (LAST)

7 Switch Strategy

FIGURE 7.4 cont'd.

Step 6 Change

Alternating between different CTO PCI techniques is at the heart of the hybrid algorithm (Fig. 7.5). When one approach fails, something different should be attempted ("don't get stuck in a failure mode"). For example, if antegrade wire crossing fails, then antegrade dissection/reentry should be tried, and if this fails too, retrograde crossing should be attempted (if, of course, appropriate collaterals exist and the risk is acceptable).

Every CTO is different and as a result may require different strategies for success. **Excessive persistence** in the face of minimal progress increases the chances for procedural failure due to utilization of limited resources (radiation, contrast, time). However, the operator should not **change too early,** but instead invest enough effort in the utilized strategy to maximize its chance for success. What constitutes an adequate effort varies from lesion to lesion and operator to operator, and is best determined with increased CTO PCI experience. **Generally, no more than 5–10 min should be spent in a stagnant mode without minor (such as reshaping the tip of wire, or changing to a wire with significantly different properties), or major (such as switching from an antegrade to a retrograde approach) technique adjustments being made.** Efficient change of strategy can result in shorter procedure time and lower patient and staff radiation exposure and contrast utilization. The hybrid approach message is not to let the case stall! The hybrid approach requires a high level of familiarity and comfort with all crossing strategies, so that there are no impediments to making a change.

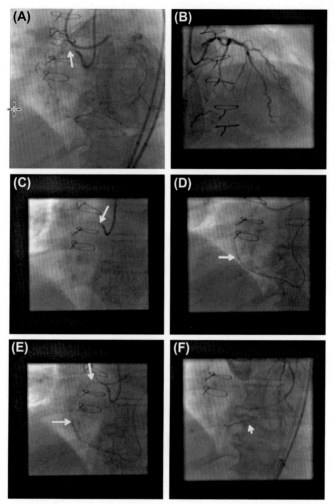

FIGURE 7.5 **Example of intraprocedural strategy changes: complex antegrade and retrograde intervention of a right coronary artery chronic total occlusion (CTO).** Bilateral coronary angiography demonstrating a proximal right coronary artery CTO due to in-stent restenosis (*arrow*, A). The right posterior descending artery filled by septal collaterals from the left anterior descending artery (B). Angiographic characteristics included a well-defined proximal cap, good distal target vessel, long lesion length, and appropriate retrograde interventional collaterals. Antegrade crossing attempts using a CrossBoss catheter (*arrow*, C) and antegrade wire escalation failed. A Corsair microcatheter (*arrow*, D) was advanced retrogradely via a septal collateral to the mid right coronary artery over a nontapered polymer-jacketed guidewire, but attempts to cross the occlusion retrogradely failed (E). Repeat antegrade crossing attempts with a moderately stiff, nontapered polymer-jacketed wire (Pilot 200 wire) using the limited subintimal tracking technique were successful in subintimal advancement of the wire (*arrow*, F). After a Guideliner catheter (*arrowhead*, G) was advanced into the proximal right coronary artery, a Confianza Pro 12 wire (*arrow*, G) successfully crossed the occlusion into the distal true lumen. After implantation of multiple drug-eluting stents an excellent angiographic result was achieved (H). This case highlights the importance of multiple modifications in the procedural plan (I), and switching strategies to be adaptive during CTO percutaneous coronary interventions. *Reproduced with permission from Brilakis ES, Grantham JA, Rinfret S, et al. A percutaneous treatment algorithm for crossing coronary chronic total occlusions. JACC Cardiovasc Interv 2012;5:367–79.*

Continued

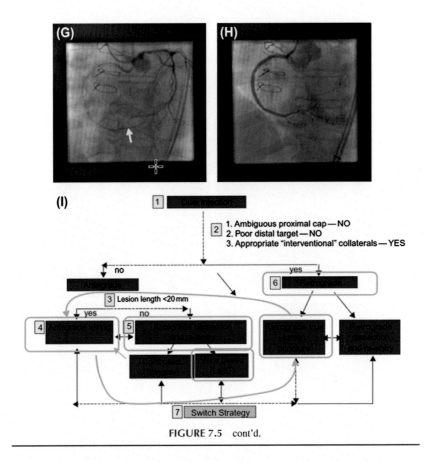

FIGURE 7.5 cont'd.

7.2 IMPACT OF THE HYBRID APPROACH

The hybrid approach algorithm has had a major impact on the dissemination and application of CTO PCI techniques by a wide population of interventional cardiologists, in different practice settings (from small private practice groups, to academic institutions, to nonacademic tertiary centers).[2–10] It has allowed for the first time an approach to CTO PCI that is simple (but not simplistic), systematic, reproducible, and teachable.

Procedural success at experienced centers is approximately 85%-90% with approximately 3% risk for a major complication (Fig. 7.6).[3]

Full application of the CTO hybrid algorithm requires long-term commitment, ongoing training, and experience with all types of CTO PCI techniques, and hence may not be fully applicable to all operators at different stages of their learning curve. It does provide, however, a framework for effective communication between operators that can facilitate acquiring and building a comprehensive CTO PCI skill set.

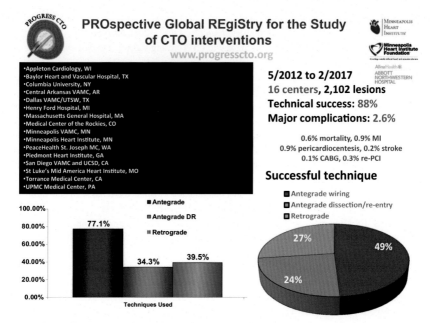

FIGURE 7.6   Outcomes with the hybrid approach from the PROGRESS-CTO registry.

REFERENCES

1. Brilakis ES, Grantham JA, Rinfret S, et al. A percutaneous treatment algorithm for crossing coronary chronic total occlusions. *JACC Cardiovasc Interv* 2012;**5**:367–79.
2. Vo MN, McCabe JM, Lombardi WL, Ducas J, Ravandi A, Brilakis ES. Adoption of the hybrid CTO approach by a single non-CTO operator: procedural and clinical outcomes. *J Invasive Cardiol* 2015;**27**:139–44.
3. Christopoulos G, Karmpaliotis D, Alaswad K, et al. Application and outcomes of a hybrid approach to chronic total occlusion percutaneous coronary intervention in a contemporary multicenter US registry. *Int J Cardiol* 2015;**198**:222–8.
4. Michael TT, Mogabgab O, Fuh E, et al. Application of the "hybrid approach" to chronic total occlusion interventions: a detailed procedural analysis. *J Interv Cardiol* 2014;**27**:36–43.
5. Christopoulos G, Menon RV, Karmpaliotis D, et al. Application of the "hybrid approach" to chronic total occlusions in patients with previous coronary artery bypass graft surgery (from a Contemporary Multicenter US registry). *Am J Cardiol* 2014;**113**:1990–4.
6. Christopoulos G, Menon RV, Karmpaliotis D, et al. The efficacy and safety of the "hybrid" approach to coronary chronic total occlusions: insights from a contemporary multicenter US registry and comparison with prior studies. *J Invasive Cardiol* 2014;**26**:427–32.
7. Shammas NW, Shammas GA, Robken J, et al. The learning curve in treating coronary chronic total occlusion early in the experience of an operator at a tertiary medical center: the role of the hybrid approach. *Cardiovasc Revasc Med* 2016;**17**:15–8.
8. Pershad A, Eddin M, Girotra S, Cotugno R, Daniels D, Lombardi W. Validation and incremental value of the hybrid algorithm for CTO PCI. *Catheter Cardiovasc Interv* 2014;**84**:654–9.
9. Wilson WM, Walsh SJ, Yan AT, et al. Hybrid approach improves success of chronic total occlusion angioplasty. *Heart* 2016;**102**:1486–93.

10. Maeremans J, Walsh S, Knaapen P, et al. The hybrid algorithm for treating chronic total occlusions in Europe: the RECHARGE registry. *J Am Coll Cardiol* 2016;**68**:1958–70.

11. Brilakis ES. The "hybrid" approach: the key to CTO crosing success. *Cardiol Today's Interv* November/December 2012.

12. Christopoulos G, Kandzari DE, Yeh RW, et al. Development and validation of a novel scoring system for predicting technical success of chronic total occlusion percutaneous coronary interventions: the progress CTO (prospective global registry for the study of chronic total occlusion intervention) score. *JACC Cardiovasc Interv* 2016;**9**:1–9.

13. Morino Y, Abe M, Morimoto T, et al. Predicting successful guidewire crossing through chronic total occlusion of native coronary lesions within 30 minutes: the J-CTO (Multicenter CTO Registry in Japan) score as a difficulty grading and time assessment tool. *JACC Cardiovasc Interv* 2011;**4**:213–21.

14. Kahn JK, Hartzler GO. Retrograde coronary angioplasty of isolated arterial segments through saphenous vein bypass grafts. *Catheter Cardiovasc Diagn* 1990;**20**:88–93.

15. Brilakis ES, Banerjee S, Lombardi WL. Retrograde recanalization of native coronary artery chronic occlusions via acutely occluded vein grafts. *Catheter Cardiovasc Interv* 2010;**75**:109–13.

16. Christopoulos G, Wyman RM, Alaswad K, et al. Clinical utility of the Japan-Chronic total occlusion score in coronary chronic total occlusion interventions: results from a multicenter registry. *Circ Cardiovasc Interv* 2015;**8**:e002171.

17. Werner GS, Coenen A, Tischer KH. Periprocedural ischaemia during recanalisation of chronic total coronary occlusions: the influence of the transcollateral retrograde approach. *EuroIntervention* 2014;**10**:799–805.

18. Lo N, Michael TT, Moin D, et al. Periprocedural myocardial injury in chronic total occlusion percutaneous interventions: a systematic cardiac biomarker evaluation study. *JACC Cardiovasc Interv* 2014;**7**:47–54.

19. Stetler J, Karatasakis A, Christakopoulos GE, et al. Impact of crossing technique on the incidence of periprocedural myocardial infarction during chronic total occlusion percutaneous coronary intervention. *Catheter Cardiovasc Interv* 2016;**88**:1–6.

20. Karmpaliotis D, Karatasakis A, Alaswad K, et al. Outcomes with the use of the retrograde approach for coronary chronic total occlusion interventions in a contemporary multicenter US registry. *Circ Cardiovasc Interv* 2016;**9**.

21. El Sabbagh A, Patel VG, Jeroudi OM, et al. Angiographic success and procedural complications in patients undergoing retrograde percutaneous coronary chronic total occlusion interventions: a weighted meta-analysis of 3,482 patients from 26 studies. *Int J Cardiol* 2014;**174**:243–8.

22. Rathore S, Katoh O, Matsuo H, et al. Retrograde percutaneous recanalization of chronic total occlusion of the coronary arteries: procedural outcomes and predictors of success in contemporary practice. *Circ Cardiovasc Interv* 2009;**2**:124–32.

23. Karmpaliotis D, Michael TT, Brilakis ES, et al. Retrograde coronary chronic total occlusion revascularization: procedural and in-hospital outcomes from a multicenter registry in the United States. *JACC Cardiovasc Interv* 2012;**5**:1273–9.

24. Tsuchikane E, Yamane M, Mutoh M, et al. Japanese multicenter registry evaluating the retrograde approach for chronic coronary total occlusion. *Catheter Cardiovasc Interv* 2013;**82**:E654–61.

25. Mashayekhi K, Behnes M, Akin I, Kaiser T, Neuser H. Novel retrograde approach for percutaneous treatment of chronic total occlusions of the right coronary artery using ipsi-lateral collateral connections – a European Centre experience. *EuroIntervention* 2016;**11**:e1231–6.

26. Galassi AR, Sianos G, Werner GS, et al. Retrograde recanalization of chronic total occlusions in Europe: procedural, in-hospital, and long-term outcomes from the multicenter ERCTO registry. *J Am Coll Cardiol* 2015;**65**:2388–400.
27. Yamane M, Muto M, Matsubara T, et al. Contemporary retrograde approach for the recanalisation of coronary chronic total occlusion: on behalf of the Japanese Retrograde Summit Group. *EuroIntervention* 2013;**9**:102–9.

Chapter 8

"Balloon-Uncrossable" and "Balloon-Undilatable" Chronic Total Occlusions

8.1 BALLOON-UNCROSSABLE LESIONS

See Online Cases 1, 5, 15, 18, 27, 30, 31, 47, 49, 52, 53, 57, 73, 82.

GOAL

Cross the chronic total occlusion (CTO) with a balloon (or microcatheter) after successful guidewire crossing.

The main reason for failure of CTO interventions is inability to cross the occlusion with a guidewire. However, in some cases, a balloon cannot cross the lesion after successful guidewire crossing and confirmation of guidewire placement into the distal true lumen. Such lesions are called balloon-uncrossable CTOs, and represent 6%–9% of CTO lesions.[1–3]

Fig. 8.1 outlines a step-by-step algorithm for approaching such lesions.

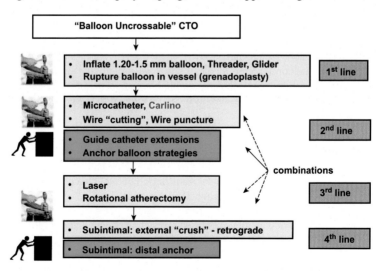

FIGURE 8.1   Algorithm for crossing a balloon-uncrossable chronic total occlusion.

Manual of Chronic Total Occlusion Interventions. http://dx.doi.org/10.1016/B978-0-12-809929-2.00008-9

Step 1 Advance and Inflate a Small Balloon, Grenadoplasty
How?

Step 1a Small Balloons
a. Use single-marker, rapid-exchange compliant balloons with low crossing profile (1.20, 1.25, and 1.5 mm in diameter) and long length (20–30 mm). The balloon profile is highest at the marker segment, hence longer balloons may allow for deeper lesion penetration before the balloon marker reaches the uncrossable segment of the CTO.
b. If the balloon stops advancing, it can be inflated while maintaining forward pressure. This may dilate the proximal cap and allow lesion crossing, sometimes even with the same balloon (balloon-wedge technique).
c. If the balloon fails to advance after inflation, one can try with a new small balloon (since balloons do not return to their original profile after inflation), or one manufactured by another company, as different crossing profile and tip characteristics may assist in crossing. Rapid-exchange balloon catheters allow more pushability into the lesion.
d. Alternatively, one can attempt crossing with a larger 2.5–3.0 mm diameter rapid-exchange balloon. Sometimes inflation with a larger diameter balloon just proximal to the CTO lesion will disrupt the architecture of the proximal CTO cap enough to allow subsequent passage of a small profile balloon or microcatheter.

Step 1b Threader and Glider Balloons
a. The Threader (Section 2.8.1, Fig. 2.53, Boston Scientific) is a combined balloon and microcatheter with excellent penetrating capacity and is easy to use (same as a small balloon as described in Step 1a). It has a hydrophilic coating and 0.017″ lesion entry profile.
b. The Threader is available in both rapid-exchange and over-the-wire versions. The rapid-exchange Threader is preferred to the over-the-wire version for balloon-uncrossable lesions, as it has more penetrating capacity (likely due to stiffer shaft). On the other hand, the over-the-wire Threader allows guidewire changes and contrast injection.
c. The Glider balloon (Trireme Medical) has a beveled tip and was developed to cross through the struts of stents during bifurcation stenting, but can also be useful in hard-to-cross lesions, as it can be torqued to present different tip configurations to the lesion.

What Can Go Wrong?
a. Guide catheter and guidewire position can be lost during attempts to advance the balloon or Threader. Carefully monitor the guide catheter position and stop advancing if the guide catheter starts backing out of the coronary ostium or if the distal wire position is being compromised.
b. Injury of the distal target vessel can occur (dissection or perforation) due to significant distal guidewire movement (see-saw action of wire with forward push of balloon and retraction of force), especially when stiff (such as Confianza Pro 12) or polymer-jacketed (such as the Pilot 200) guidewires are used.
c. Balloon entrapment can occur inside the occlusion, although this is highly unlikely.

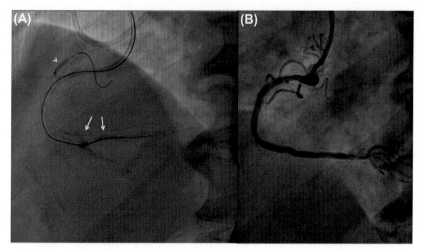

FIGURE 8.2 **Illustration of the grenadoplasty technique to cross a distal right coronary artery balloon-uncrossable chronic total occlusion.** The lesion could not be crossed despite using an 8 Fr Amplatz guide and an anchor balloon (*arrowhead*, panel A). A 1.2 mm balloon was ruptured with contrast spreading proximally and distally (*arrows*, panel A). A Finecross microcatheter could then be advanced through the lesion followed by wire exchange and a 2.0 mm balloon with an excellent final result (panel B). *Courtesy of Dr. Gabriele Gasparini.*

Step 1c Grenadoplasty (Intentional Balloon Rupture; also Called Balloon-Assisted Microdissection)

This is a simple, safe, and often effective technique, which is increasingly being used in the treatment algorithm for balloon-uncrossable lesions.[4]

How?

A small (1.20–1.50 mm) balloon is advanced as far as possible into the lesion and inflated at high pressure until it ruptures (Fig. 8.2). When the balloon ruptures, suction should be immediately applied through the inflating device. The balloon rupture can often modify the plaque, resulting in successful crossing with a new balloon.

What Can Go Wrong?

a. Proximal vessel dissection and perforation. This is extremely unlikely when small (1.20–1.50 mm) balloons are used. **Larger balloons should *not* be used for grenadoplasty.**

b. The balloon should be meticulously prepared to empty all air and hence minimize the risk for air embolism.

c. Watching the indeflator rather than the screen allows more rapid deflation of the balloon as soon as rupture has occurred. This will reduce the chance of pinhole contrast injury from the rupture site of the balloon.

d. One may encounter difficulty removing the ruptured balloon. In some cases the ruptured balloon becomes entangled with the guidewire, requiring removal of both, hence losing guidewire position. This is more likely to happen with the 1.25 mm balloons, hence utilizing 1.5 mm diameter balloons is recommended.[4]

Step 2 Microcatheter Advancement, Carlino, Wire-Cutting, Wire Puncture, and Increasing Guide Catheter Support

Treating the balloon-uncrossable lesion can be achieved using a combination of plaque modification (e.g., using a microcatheter) and increasing guide catheter support.

Step 2a Microcatheter Advancement

How?

a. The concept behind use of a microcatheter is that advancement of a microcatheter through the CTO can modify the occlusion, enabling subsequent crossing with a balloon.

b. There are several microcatheters that can be utilized as described in Chapter 2.

c. The following microcatheters are especially designed for balloon-uncrossable lesions:

- The Tornus catheter (Section 2.8.2) was designed for advancing through calcified and difficult-to-penetrate lesions and should be advanced using counterclockwise rotation and withdrawn using clockwise rotation.[5]
- The Turnpike Spiral and Turnpike Gold catheters (Section 2.8.2) were also designed with threads to screw into the lesion and modify it. In contrast to the Tornus catheter, they are advanced by turning clockwise and withdrawn by turning counterclockwise.

d. Standard microcatheters can also be used:

- The Corsair and Corsair Pro catheter (Section 2.4.2) can be advanced by rotating in either direction (in contrast to the Tornus catheter).
- The Turnpike and Turnpike LP catheter (Section 2.4.6) can also be rotated in either direction.
- Similarly, the Finecross (Section 2.4.4) or Micro 14 (Section 2.4.5) can be rotated in either direction, although rotation may be challenging to achieve.
- The Caravel (Section 2.4.3) is a low profile microcatheter, but is not designed for aggressive torqueing as the Corsair and Turnpike.

e. If successful advancement of a microcatheter is achieved, a balloon can often subsequently cross the lesion. Alternatively, the guidewire could be exchanged for a more supportive guidewire or an atherectomy wire, if the latter is planned as the next lesion preparation step.

What Can Go Wrong?

a. Guide catheter and guidewire position may be lost with aggressive pushing of the microcatheters.

b. Distal vessel injury (dissection and/or perforation) can occur from uncontrolled guidewire movement during microcatheter advancement attempts.

c. The microcatheter can get damaged if overtorqued, leading to catheter entrapment or tip/shaft breakage. Rotation should not exceed 10 turns before allowing the catheter to unwind. A guidewire should always be kept within the microcatheter lumen to prevent kinking and possible entrapment. If the tip of the microcatheter breaks off it can become entrapped in the lesion (Online Case 87).

d. Rarely excessive manipulation of the microcatheter can disrupt the device and/or the guidewire and lock both devices together, requiring withdrawal of both (See Chapter 2, Fig. 2.21). A polymer-jacketed guidewire can sometimes be advanced through the track that has been established, allowing the crossing attempts to restart.

Step 2b Carlino
The Carlino technique is described in detail in Section 5.6.3.

How?
a. A microcatheter is advanced as close to the proximal cap as possible.
b. A small amount of contrast (0.5–1.0 mL) is injected through the microcatheter under cineangiography.
c. Contrast injection can cause microdissection and facilitate subsequent advancement of a balloon or a microcatheter.

What Can Go Wrong?
a. Perforation, if a large amount of contrast is injected and if the catheter is inserted into a small side branch. The risk is low with injection of a small amount of contrast.

Step 2c Wire-Cutting (Fig. 8.3)[6]

How?
a. A second guidewire is advanced through the occlusion (which may be challenging to achieve).
b. A balloon is advanced over the first guidewire, as far as possible into the proximal cap and inflated.

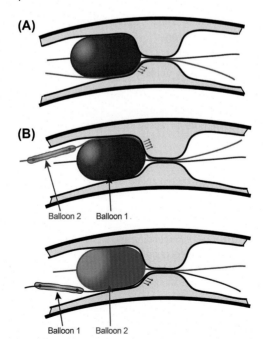

(A)

(B)

Balloon 2   Balloon 1

Balloon 1   Balloon 2

FIGURE 8.3 Illustration of the wire-cutting (A) and see-saw wire-cutting (B) techniques.

c. The second guidewire is withdrawn with the balloon inflated, effectively cutting the proximal cap and modifying it.

d. After deflation and removal of the original balloon, a new balloon is advanced over the first guidewire, often successfully crossing the modified lesion.

e. A modified version of this technique is the **see-saw wire-cutting technique**. In this technique two balloons are advanced over the two guidewires (one balloon over each wire). One of the balloons is first advanced as distally as possible and inflated, pressing the other wire against the proximal cap. Then the other balloon is advanced distally and inflated, producing a similar cutting effect to modify the cap on the other side. This process is repeated multiple times until one of the balloons crosses the lesion. A retrospective study of 80 patients found this technique to be associated with higher device and procedural success rates and shorter procedure time as compared with use of the Tornus catheter.[7,8]

What Can Go Wrong?

a. Withdrawal of the second guidewire may result in deep intubation of the guide catheter, potentially leading to proximal vessel dissection. This complication could be prevented by careful visualization of the guide catheter during withdrawal of the second guidewire. If, however, a dissection occurs it could be used for subintimal wire advancement and subintimal lesion modification (or subintimal distal anchor) as described in step 4.

Step 2d Wire Puncture

This technique uses a stiff guidewire to modify the proximal cap.

How?

a. A second stiff guidewire (such as Gaia 2nd or 3rd or Confianza Pro 12) is advanced to the proximal cap.

b. Several punctures of the proximal cap are performed modifying the cap and facilitating equipment crossing.

What Can Go Wrong?

a. Perforation, although this is unlikely even if the guidewire exits the vessel architecture (unless it is followed by a balloon or microcatheter).

Step 2e Increase Guide Catheter Support

Better guide catheter support increases the likelihood of successful balloon or microcatheter crossing.

How?

Guide catheter support can be enhanced by using a side branch anchor technique,[9] or a guide catheter extension,[10] as described in detail in Section 3.6.

a. Side branch anchor technique. A workhorse guidewire is advanced into a side branch (usually a conus or acute marginal branch for the right coronary artery or a diagonal for the left anterior descending artery), followed by a small balloon (usually 1.5–2.0 mm in diameter depending on the side branch vessel size) (Fig. 8.4). The balloon is inflated, usually at 6–8 atm, anchoring the guide into the vessel and enhancing advancement of balloons or microcatheters. Sometimes, patients may develop chest pain during inflation of the balloon in the side branch.

b. Guide catheter extension. A GuideLiner or Guidezilla guide catheter extension is advanced into the vessel, enhancing guide catheter support and the pushability of balloons/microcatheters. In a randomized trial, use of a 5-in-6 guide catheter extension was more effective and efficient in facilitating the success of transradial percutaneous coronary intervention for complex coronary lesions, as compared with buddy wire or balloon anchoring.[11]

c. Buddy wire stent anchor (Fig. 8.5). If the proximal vessel needs stenting, a buddy wire can be inserted and a stent deployed over this guidewire, effectively trapping the buddy wire, which then provides strong guide catheter support.

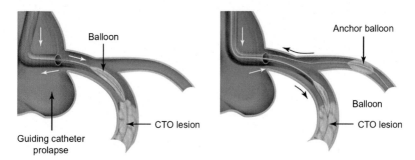

FIGURE 8.4   Example of side branch anchor to facilitate delivery of a balloon across a chronic total occlusion.

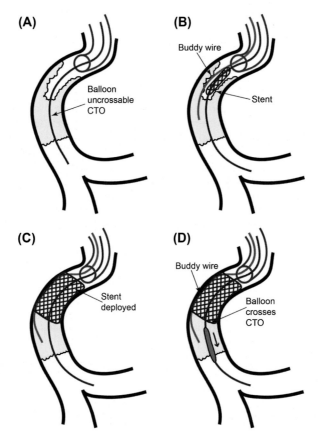

FIGURE 8.5   Illustration of the buddy-wire stent anchor technique.

What Can Go Wrong?
a. Guidewire and guide catheter position loss or distal vessel injury.
b. Side branch anchor can cause injury or dissection of the side branch, but this is infrequent and usually does not have adverse consequences.
c. Perforation of the side branch may rarely occur. Oversizing of the anchor balloon should be avoided to minimize the risk for side branch perforation and dissection.
d. Guide catheter extensions can cause ostial or mid-target vessel dissection.[10] Other potential complications include dislodgement of a stent during attempts to advance it through the guide catheter extension metal collar (Chapter 2, Figure 2.13) and dislodgement of the guide catheter extension distal marker.[12]
e. In the buddy-wire stent anchor technique, potential risks include inability to remove the buddy wire or failure to advance the subsequent distal stent through the proximal stent.

Step 3 Laser, Atherectomy

See online video: "Impact of contrast on laser activation".
 Online Cases 5, 18, 27, 47, 52, 73, 86.

Step 3a Laser
Laser is increasingly being used for balloon-uncrossable and balloon-undilatable lesions, in part because in contrast to rotational and orbital atherectomy, which require specialized guidewires, it can be advanced over any standard 0.014 in. guidewire.

How?

a. The 0.9 mm excimer laser atherectomy catheter should be used, usually at maximum repetition and fluence levels (repetition rate 80 Hz and fluence of 80 mJ/mm²).[13] The peripheral turbo catheter (Spectranetics) is generally preferred over the coronary catheter, since in contrast to the coronary catheter the peripheral catheter does not automatically stop after 10 s. The coronary catheter is programmed to minimize the risk for coronary ischemia, as laser activation is designed to work with a slow, continuous saline infusion but this is usually not of concern in CTOs. The laser can modify the uncrossable segment and facilitate balloon entry into the lesion.[14] Flushing with saline is recommended during passes of the laser catheter.[15]
b. The laser catheter requires a 5-min warming period as well as calibration prior to use. Anticipation that use of a laser may be required and early set-up of the system can minimize delays and improve the efficiency of the procedure.
c. A variation of the previous technique is to perform laser while injecting contrast through the guide catheter or via the over-the-wire side port of the laser catheter itself (if an over-the-wire laser is used). The latter is accomplished by injecting pure or diluted contrast using a Y-connector and an inflating device inflated up to 20 atm. Laser activation in contrast can cause profound plaque modification through the acousticomechanical effect of the rapidly exploding bubbles. However, it also carries risk for vessel injury and/or perforation.

What Can Go Wrong?

a. Vessel perforation, which is why the location of the guidewire into the distal true lumen should be confirmed before attempting to advance the laser catheter.

b. Some operators advocate avoiding use of laser if CTO crossing was achieved using dissection/reentry to minimize the risk for perforation, however the laser has been used several times in such lesions without complications.

c. Laser activation over polymer-jacketed guidewires could result in polymer damage, potentially making the wire sticky or gummy.

Step 3b Atherectomy

How?

a. Rotational or orbital atherectomy can greatly facilitate lesion crossing with a balloon. However, these forms of atherectomy require wire exchange for a thinner (0.009 in. for rotational and 0.012 in. for orbital atherectomy) dedicated guidewire, which may not always be feasible through the CTO.[16]

b. If no other maneuver is successful in crossing the CTO, it is sometimes possible to "burry" a microcatheter as far as possible into the CTO lesion, pull the CTO wire, and attempt to rewire the CTO with a dedicated atherectomy guidewire. If rewiring is successful, then atherectomy can be performed.

c. Rotational atherectomy may have advantages over the current version of orbital atherectomy, as it provides forward cutting in contrast to orbital atherectomy, which provides sideways cutting.

d. Atherectomy will differentially cut calcific tissue but will not cut through elastic adventitia.

e. In some challenging cases rotational atherectomy has been performed in the subintimal space for balloon-uncrossable lesions (see Online Case 53).

What Can Go Wrong?

a. Loss of guidewire position across the CTO (if the guidewire has to be removed and replaced with a dedicated atherectomy guidewire) that may fail to recross the lesion.

b. Vessel perforation, which is why a small-diameter rotational atherectomy burr (1.25–1.50 mm) is usually used. The risk of perforation may be higher with atherectomy in the subintimal space.

c. Burr entrapment upon forceful forward advancement of the rotational atherectomy burr. The burr should be advanced in a repetitive and gentle manner, avoiding forceful wedging into the occlusion. Prevention and management of this complication is discussed in Chapter 12.

Step 4 Subintimal Crossing Techniques (Fig. 8.6)

See Online Case 1.

Subintimal techniques can significantly facilitate treatment of balloon-uncrossable lesions, but require experience in the use of these techniques, as described in Chapter 5.

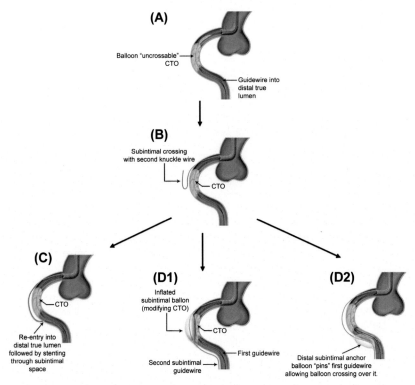

FIGURE 8.6 **Illustration of subintimal crossing techniques for crossing a balloon-uncrossable chronic total occlusion (CTO).** If a balloon cannot cross the CTO after successful guidewire crossing (A), a second guidewire is used to subintimally cross the CTO (B) and reenter into the distal true lumen (C). A balloon can be inflated over the subintimal guidewire next to the CTO to crush and modify it (D1), or distal to the CTO to anchor the true lumen guidewire and allow balloon crossing over the true lumen guidewire (D2).

How?

Step 4a External Crush

See Online Case 15.

a. A second guidewire (antegrade or retrograde) is advanced subintimally around the balloon-uncrossable segment of the CTO (or the CrossBoss catheter is used to achieve subintimal position and exchanged for the guidewire, as explained in Chapter 5) (Fig. 8.6A–C). A balloon is advanced over the subintimal guidewire next to the CTO and inflated (usually at 8–10 atm), crushing the CTO plaque from the outside (Fig. 8.6D1). This can modify the plaque enough to allow passage of a balloon over the guidewire that had previously entered the distal true lumen.[17,18]

Step 4b Subintimal Distal Anchor

a. In a variation of this technique entitled subintimal distal anchor, a second guidewire is advanced subintimally distal to the CTO. A balloon is advanced over the subintimal guidewire distal to the CTO. The balloon is inflated distally, anchoring the true lumen guidewire, and enabling antegrade delivery of a balloon over the true lumen guidewire (Fig. 8.6D2).[19]

What Can Go Wrong?

a. The subintimal techniques require subintimal wire crossing, which may not always be feasible (e.g., the guidewire may track side branches).

b. Positioning of the subintimal guidewire within the vessel architecture should be confirmed to prevent inadvertent perforation; for example, if the guidewire enters side branches or if it exits the vessel and enters the pericardium. Confirmation of wire position is best done usually with contralateral injection in two orthogonal projections.

c. Subintimal crossing and balloon inflation in the subintimal space can cause a large subintimal hematoma that may compress the distal true lumen.

Step 5 Simultaneous Use of Strategies

Simultaneous application of lesion modification strategies and techniques that increase guide catheter support can enhance the likelihood of successful crossing. For example the Anchor-Tornus,[20] Proxis-Tornus,[21] and Anchor-Laser (Fig. 8.7)[13] techniques have been described for crossing balloon-uncrossable

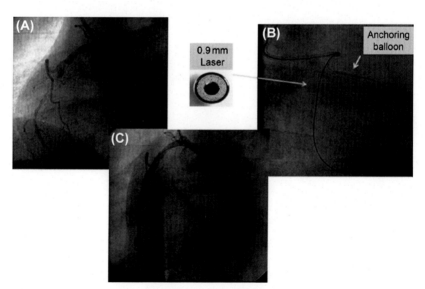

FIGURE 8.7   Example of combined use of side branch anchoring and laser to cross a balloon-uncrossable chronic total occlusion (CTO). A right coronary artery CTO (panel A) was successful crossed with a guidewire but no balloon would cross the lesion. A laser catheter could not be delivered, but did cross with use of a side branch anchor balloon (panel B) enabling an excellent final angiographic result after stenting (panel C). *Reproduced with permission from Ben-Dor I, Maluenda G, Pichard AD, et al. The use of excimer laser for complex coronary artery lesions. Cardiovasc Revasc Med 2011;12:69e1–8.*

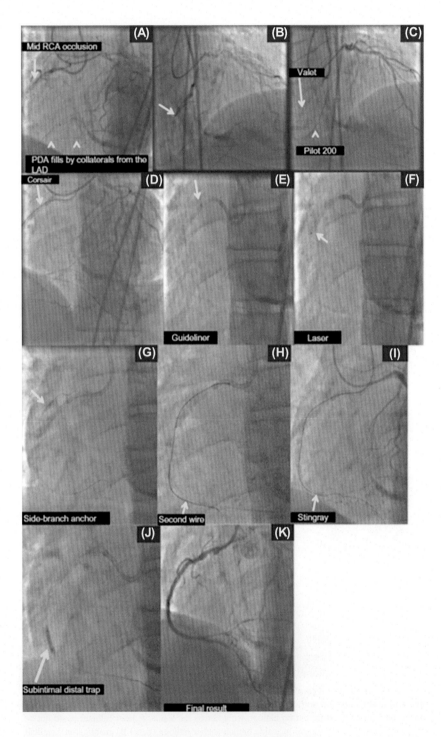

FIGURE 8.8 **Example of a balloon-uncrossable chronic total occlusion (CTO) case, in which multiple strategies were used as part of the algorithm presented in Fig. 8.1.** Coronary angiography using dual injection demonstrating a CTO of the mid right coronary artery (RCA; *arrow*, panel A).

lesions. In addition to simultaneous application, sequential application of various techniques is used until a final successful outcome is achieved (Fig. 8.8).[19] One must be creative in using these strategies in various combinations to achieve the desired goal. The process can be difficult but also very rewarding.

8.2 BALLOON-UNDILATABLE CHRONIC TOTAL OCCLUSIONS

Online Case 26, 86

Goal

Dilate the lesion that does not expand in spite of multiple balloon inflations.

Prevention

Balloon-undilatable lesions are more frequent in CTO than non-CTO lesions. It is important to avoid implanting a stent in a balloon-undilatable lesion. Adequate predilation with a balloon sized according to the vessel reference diameter is critical before stenting a CTO (or any coronary) lesion to ensure proper stent expansion, especially when the lesion is severely calcified. Intravascular imaging can be useful in determining the appropriate predilation balloon size and plaque characteristics that may benefit from particular forms of atherectomy.[22,23] If the lesion is resistant, additional predilatation and/or atherectomy should be pursued prior to stent deployment.

←————————————————————————————

The posterior descending artery (PDA; *arrowheads*, panel A) filled via collaterals from the left anterior descending artery (LAD). Dual injection coronary angiography showing a long mid RCA occlusion (*arrow*, panel B). The RCA CTO was successfully crossed using a Pilot 200 guidewire (*arrowhead*, panel C) over a Valet microcatheter (*arrow*, panel C). A Corsair catheter (*arrow*, panel D) failed to cross the lesion, despite using a GuideLiner (*arrow*, panel E) and a 0.9 mm laser catheter (*arrow*, panel F). Crossing the CTO with a balloon failed after use of a side branch anchor balloon (*arrow*, panel G). The CTO was crossed subintimally with a second guidewire that was knuckled to the distal RCA (panel H), but reentry attempts using the Stingray balloon and guidewire failed (*arrow*, panel I). A 3.0 mm balloon was inflated over the subintimal guidewire in the distal RCA (*arrow*, panel J) that anchored the true lumen guidewire allowing advancement of a 1.5 mm balloon through the CTO, enabling lesion dilation. After stenting an excellent final result was achieved (panel K). *Reproduced from Michael TT, Banerjee S, Brilakis ES. Subintimal distal anchor technique for "balloon-uncrossable" chronic total occlusions. J Invasive Cardiol 2013;25:552–4, with permission from HMP Communications.*

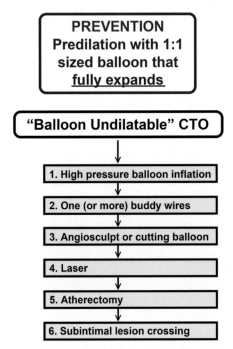

FIGURE 8.9 Algorithm for dilating a balloon-undilatable chronic total occlusion.

Treatment

Fig. 8.9 outlines an algorithm for approaching balloon-undilatable CTOs.

Step 1 High Pressure Balloon Inflation

How?

a. Using noncompliant balloons inflated at high pressures (often up to 30 atm).[24] Specialized noncompliant balloons have been developed to deliver very high (35 atm) pressures (Schwager OPN balloon, SIS Medical, Switzerland), but are not currently available in the United States.[24] It is important to be familiar with the rated burst pressure and percentage diameter growth of the available noncompliant balloons.

b. Performing prolonged inflations (30–60 s or more). Alternatively, once at high pressure, reducing and increasing the pressure quickly multiple times may help break the resistant lesion.

c. Using slightly undersized balloons (usually by 0.5 mm).

d. Avoiding balloon injury of nonstented vessel segments. All ballooned segments should subsequently be covered with stent(s) to minimize the risk for restenosis (geographic miss).

e. Using shorter balloons might help to concentrate the force of the dilatation and achieve the desired balloon expansion inside the lesion.

f. Simultaneous inflation of two undersized balloons positioned side by side within the lesion, which creates asymmetric pressure, sometimes resulting in lesion expansion.

What Can Go Wrong?
a. Vessel perforation: Best avoided by conservative balloon sizing (in nearly all cases 1:1 or lower balloon-to-vessel ratio should be used).
b. The buddy wire may enter the subintimal space during advancement; however, this can be prevented by using a dual lumen microcatheter for wire delivery.
c. Balloon rupture leading to vessel dissection and/or perforation. If balloon rupture occurs, immediate angiography should be performed to determine if perforation has occurred.

Step 2 Buddy Wire(s)

How?

a. Advancing one[25–27] or more[28] buddy wires through the undilatable CTO, followed by high-pressure balloon inflation. The wires can modify the balloon forces exerted on the vessel wall, leading to plaque modification and expansion (this is a poor man's version of a cutting or scoring balloon).

What Can Go Wrong?
a. Vessel injury from advancement of the buddy wire(s), Using soft workhorse guidewires and careful wiring technique is advised.

Step 3 Angiosculpt or Cutting Balloon

How?

a. Similar to the buddy wire technique, the vessel wall is modified by application of focused pressure through the blades of the cutting balloon[29] or the nitinol wire of the Angiosculpt balloon (described in Section 2.8.3).

What Can Go Wrong?
a. Loss of guidewire and guide catheter position, as the Angiosculpt or cutting balloon may not deliver easily through the balloon-undilatable lesion. Use of enhanced guide catheter support techniques (such as anchor techniques and guide catheter extensions) may be needed to deliver the Angiosculpt/cutting balloon through the lesion.
b. Vessel rupture or perforation.
c. Balloon entrapment (especially if the balloon ruptures). Excessive inflation forces or balloon rupture within the lesion may make equipment withdrawal challenging.[30] The cutting balloon should be inflated slowly and >14 atm inflation pressures should be avoided.

Step 4 Laser

How?

a. The 0.9 mm excimer laser atherectomy catheter should be used, usually at maximum repetition and fluence levels (repetition rate 80 Hz and fluence of 80 mJ/mm^2).[13,31–34]
b. Laser is activated during saline flushing (or with simultaneous contrast injection if performed within an underexpanded stent Online Case 86).

Continued

c. If saline flush is used, several activations are performed over several seconds.

d. If contrast injection is used, two to three brief (3–4 s) activations are performed.[13,31,34]

e. High-pressure balloon inflation is subsequently performed to determine whether sufficient lesion modification has occurred, to allow expansion of the underexpanded lesion.

What Can Go Wrong?

a. Vessel dissection, which is the reason why laser activation with contrast is typically reserved for use within an underexpanded stent.

b. Perforation.

Step 5 Atherectomy

Rotational and/or orbital atherectomy can be used to expand a balloon-undilatable lesion, as discussed in step 3 of the balloon-uncrossable lesion treatment. The combination of laser and rotablation has been described for highly resistant lesions[35] and can offer a solution to lesions resistant to either one approach applied in isolation.

Step 6 Subintimal Lesion Crossing

This may also allow expansion of the lesion, as discussed in step 4 of the balloon-uncrossable lesion treatment.

In the future additional treatment modalities, such as lithoplasty (Shockwave Medical, Fremont, California) may help expand balloon undilatable lesions.

8.3 SPECIAL CASE SCENARIO: STENTED BALLOON-UNDILATABLE LESIONS

Discovering that a lesion is balloon undilatable *after* stenting poses significant challenges, as stent underexpansion can predispose to stent thrombosis and restenosis. Prevention is key and can be achieved by performing careful balloon predilation with a 1:1 sized balloon that fully expands, and ensuring complete balloon expansion before stent implantation. If the balloon fails to fully expand, additional vessel preparation (often with atherectomy) is needed before stent implantation.

All steps just described can be used to expand the lesion and the deployed stent, however rotational atherectomy (stentablation)[36,37] is rarely performed due to the risk of stent material or plaque embolization, burr entrapment, and stent damage necessitating implantation of an additional stent.[37] Orbital atherectomy is contraindicated in in-stent restenotic lesions. To avoid burr entrapment, larger sized burrs (≥1.75 mm) should be selectively used for undilatable stents. Moreover, in-stent atherectomy carries higher risk for restenosis as compared with use of laser.[38,39] Similarly, use of cutting balloons or Angiosculpt balloons may be complicated by fracture of the cutting balloon blade,[40] stent strut avulsion,[41,42] and balloon entrapment.[43–45]

Laser with simultaneous contrast activation is the preferred treatment strategy if high-pressure noncompliant balloon inflations (with and without buddy wires) fail to expand the stent,[13,31] with atherectomy reserved for cases where the laser fails (Fig. 8.10).

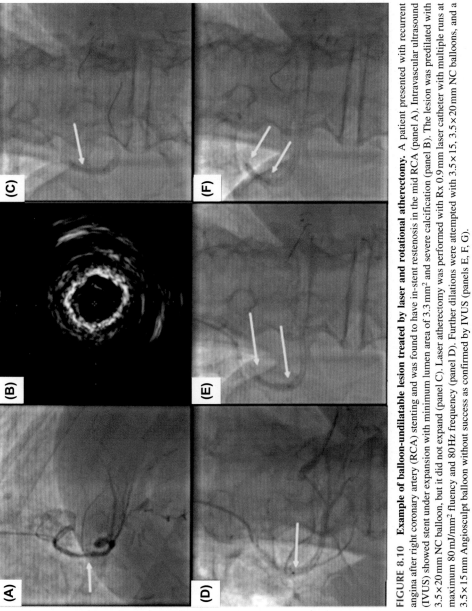

FIGURE 8.10 **Example of balloon-undilatable lesion treated by laser and rotational atherectomy.** A patient presented with recurrent angina after right coronary artery (RCA) stenting and was found to have in-stent restenosis in the mid RCA (panel A). Intravascular ultrasound (IVUS) showed stent under expansion with minimum lumen area of 3.3 mm^2 and severe calcification (panel B). The lesion was predilated with 3.5×20 mm NC balloon, but it did not expand (panel C). Laser atherectomy was performed with Rx 0.9 mm laser catheter with multiple runs at maximum 80 ml/mm^2 fluency and 80 Hz frequency (panel D). Further dilations were attempted with 3.5×15, 3.5×20 mm NC balloons, and a 3.5×15 mm Angiosculpt balloon without success as confirmed by IVUS (panels E, F, G).

Continued

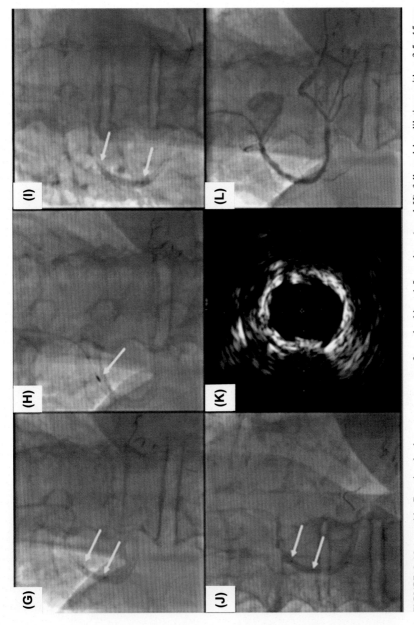

FIGURE 8.10 cont'd—Rotational atherectomy was performed with a 1.5 mm burr (panel H) followed by dilations with a 3.5×15 mm Angiosculpt balloon and a 4.0×15 mm NC balloon (panels I, J) achieving stent expansion as confirmed by IVUS (panel K) and angiography (panel L).

REFERENCES

1. Kovacic JC, Sharma AB, Roy S, et al. GuideLiner mother-and-child guide catheter extension: a simple adjunctive tool in PCI for balloon uncrossable chronic total occlusions. *J Interv Cardiol* 2013;**26**:343–50.
2. Patel SM, Pokala NR, Menon RV, et al. Prevalence and treatment of "balloon-uncrossable" coronary chronic total occlusions. *J Invasive Cardiol* 2015;**27**:78–84.
3. Karacsonyi J, Karmpaliotis D, Alaswad K, et al. Prevalence, indications and management of balloon uncrossable chronic total occlusions: Insights from a contemporary multicenter US registry. *Catheter Cardiovasc Interv* 2017;**90**:12–20.
4. Vo MN, Christopoulos G, Karmpaliotis D, Lombardi WL, Grantham JA, Brilakis ES. Balloon-assisted microdissection "BAM" technique for balloon-uncrossable chronic total occlusions. *J Invasive Cardiol* 2016;**28**:E37–41.
5. Fang HY, Lee CH, Fang CY, et al. Application of penetration device (Tornus) for percutaneous coronary intervention in balloon uncrossable chronic total occlusion-procedure outcomes, complications, and predictors of device success. *Catheter Cardiovasc Interv* 2011;**78**:356–62.
6. Hu XQ, Tang L, Zhou SH, Fang ZF, Shen XQ. A novel approach to facilitating balloon crossing chronic total occlusions: the "wire-cutting" technique. *J Interv Cardiol* 2012;**25**:297–303.
7. Xue J, Li J, Wang H, et al. "Seesaw balloon-wire cutting" technique is superior to Tornus catheter in balloon uncrossable chronic total occlusions. *Int J Cardiol* 2017;**228**:523–7.
8. Li Y, Li J, Sheng L, et al. "Seesaw balloon-wire cutting" technique as a novel approach to "balloon-uncrossable" chronic total occlusions. *J Invasive Cardiol* 2014;**26**:167–70.
9. Di Mario C, Ramasami N. Techniques to enhance guide catheter support. *Catheter Cardiovasc Interv* 2008;**72**:505–12.
10. Luna M, Papayannis A, Holper EM, Banerjee S, Brilakis ES. Transfemoral use of the GuideLiner catheter in complex coronary and bypass graft interventions. *Catheter Cardiovasc Interv* 2012;**80**:437–46.
11. Zhang Q, Zhang RY, Kirtane AJ, et al. The utility of a 5-in-6 double catheter technique in treating complex coronary lesions via transradial approach: the DOCA-TRI study. *EuroIntervention* 2012;**8**:848–54.
12. Papayannis AC, Michael TT, Brilakis ES. Challenges associated with use of the GuideLiner catheter in percutaneous coronary interventions. *J Invasive Cardiol* 2012;**24**:370–1.
13. Ben-Dor I, Maluenda G, Pichard AD, et al. The use of excimer laser for complex coronary artery lesions. *Cardiovasc Revasc Med* 2011;**12**(69):e1–8.
14. Niccoli G, Giubilato S, Conte M, et al. Laser for complex coronary lesions: impact of excimer lasers and technical advancements. *Int J Cardiol* 2011;**146**:296–9.
15. Shen ZJ, Garcia-Garcia HM, Schultz C, van der Ent M, Serruys PW. Crossing of a calcified "balloon uncrossable" coronary chronic total occlusion facilitated by a laser catheter: a case report and review recent four years' experience at the Thoraxcenter. *Int J Cardiol* 2010;**145**:251–4.
16. Pagnotta P, Briguori C, Mango R, et al. Rotational atherectomy in resistant chronic total occlusions. *Catheter Cardiovasc Interv* 2010;**76**:366–71.
17. Vo MN, Ravandi A, Grantham JA. Subintimal space plaque modification for "balloon-uncrossable" chronic total occlusions. *J Invasive Cardiol* 2014;**26**:E133–6.
18. Christopoulos G, Kotsia AP, Rangan BV, et al. "Subintimal external crush" technique for a "balloon uncrossable" chronic total occlusion. *Cardiovasc Revasc Med* 2017;**18**:63–5.
19. Michael TT, Banerjee S, Brilakis ES. Subintimal distal anchor technique for "balloon-uncrossable" chronic total occlusions. *J Invasive Cardiol* 2013;**25**:552–4.
20. Kirtane AJ, Stone GW. The Anchor-Tornus technique: a novel approach to "uncrossable" chronic total occlusions. *Catheter Cardiovasc Interv* 2007;**70**:554–7.

21. Brilakis ES, Banerjee S. The "Proxis-Tornus" technique for a difficult-to-cross calcified saphenous vein graft lesion. *J Invasive Cardiol* 2008;**20**:E258–61.
22. Kim BK, Shin DH, Hong MK, et al. Clinical impact of intravascular ultrasound-guided chronic total occlusion intervention with Zotarolimus-eluting versus Biolimus-eluting stent implantation: randomized study. *Circ Cardiovasc Interv* 2015;**8**:e002592.
23. Tian NL, Gami SK, Ye F, et al. Angiographic and clinical comparisons of intravascular ultrasound- versus angiography-guided drug-eluting stent implantation for patients with chronic total occlusion lesions: two-year results from a randomised AIR-CTO study. *EuroIntervention* 2015;**10**:1409–17.
24. Raja Y, Routledge HC, Doshi SN. A noncompliant, high pressure balloon to manage undilatable coronary lesions. *Catheter Cardiovasc Interv* 2010;**75**:1067–73.
25. Yazdanfar S, Ledley GS, Alfieri A, Strauss C, Kotler MN. Parallel angioplasty dilatation catheter and guide wire: a new technique for the dilatation of calcified coronary arteries. *Catheter Cardiovasc Diagn* 1993;**28**:72–5.
26. Stillabower ME. Longitudinal force focused coronary angioplasty: a technique for resistant lesions. *Catheter Cardiovasc Diagn* 1994;**32**:196–8.
27. Meerkin D. My buddy, my friend: focused force angioplasty using the buddy wire technique in an inadequately expanded stent. *Catheter Cardiovasc Interv* 2005;**65**:513–5.
28. Lindsey JB, Banerjee S, Brilakis ES. Two "buddies" may be better than one: use of two buddy wires to expand an underexpanded left main coronary stent. *J Invasive Cardiol* 2007;**19**:E355–8.
29. Wilson A, Ardehali R, Brinton TJ, Yeung AC, Lee DP. Cutting balloon inflation for drug-eluting stent underexpansion due to unrecognized coronary arterial calcification. *Cardiovasc Revasc Med* 2006;**7**:185–8.
30. Pappy R, Gautam A, Abu-Fadel MS. AngioSculpt PTCA Balloon Catheter entrapment and detachment managed with stent jailing. *J Invasive Cardiol* 2010;**22**:E208–10.
31. Karacsonyi J, Danek BA, Karatasakis A, Ungi I, Banerjee S, Brilakis ES. Laser coronary atherectomy during contrast injection for treating an underexpanded stent. *JACC Cardiovasc Interv* 2016;**9**:e147–8.
32. Fernandez JP, Hobson AR, McKenzie D, et al. Beyond the balloon: excimer coronary laser atherectomy used alone or in combination with rotational atherectomy in the treatment of chronic total occlusions, non-crossable and non-expansible coronary lesions. *EuroIntervention* 2013;**9**:243–50.
33. Badr S, Ben-Dor I, Dvir D, et al. The state of the excimer laser for coronary intervention in the drug-eluting stent era. *Cardiovasc Revasc Med* 2013;**14**:93–8.
34. Sunew J, Chandwaney RH, Stein DW, Meyers S, Davidson CJ. Excimer laser facilitated percutaneous coronary intervention of a nondilatable coronary stent. *Catheter Cardiovasc Interv* 2001;**53**:513–7.
35. Egred M. RASER angioplasty. *Catheter Cardiovasc Interv* 2012;**79**:1009–12.
36. Kobayashi Y, Teirstein P, Linnemeier T, Stone G, Leon M, Moses J. Rotational atherectomy (stentablation) in a lesion with stent underexpansion due to heavily calcified plaque. *Catheter Cardiovasc Interv* 2001;**52**:208–11.
37. Medina A, de Lezo JS, Melian F, Hernandez E, Pan M, Romero M. Successful stent ablation with rotational atherectomy. *Catheter Cardiovasc Interv* 2003;**60**:501–4.
38. Ferri LA, Jabbour RJ, Giannini F, et al. Safety and efficacy of rotational atherectomy for the treatment of undilatable underexpanded stents implanted in calcific lesions. *Catheter Cardiovasc Interv* 2017;**90**:E19–E24.

39. Latib A, Takagi K, Chizzola G, et al. Excimer Laser LEsion modification to expand non-dilatable stents: the ELLEMENT registry. *Cardiovasc Revasc Med* 2014;**15**:8–12.
40. Haridas KK, Vijayakumar M, Viveka K, Rajesh T, Mahesh NK. Fracture of cutting balloon microsurgical blade inside coronary artery during angioplasty of tough restenotic lesion: a case report. *Catheter Cardiovasc Interv* 2003;**58**:199–201.
41. Harb TS, Ling FS. Inadvertent stent extraction six months after implantation by an entrapped cutting balloon. *Catheter Cardiovasc Interv* 2001;**53**:415–9.
42. Wang HJ, Kao HL, Liau CS, Lee YT. Coronary stent strut avulsion in aorto-ostial in-stent restenosis: potential complication after cutting balloon angioplasty. *Catheter Cardiovasc Interv* 2002;**56**:215–9.
43. Kawamura A, Asakura Y, Ishikawa S, et al. Extraction of previously deployed stent by an entrapped cutting balloon due to the blade fracture. *Catheter Cardiovasc Interv* 2002;**57**:239–43.
44. Sanchez-Recalde A, Galeote G, Martin-Reyes R, Moreno R. AngioSculpt PTCA balloon entrapment during dilatation of a heavily calcified lesion. *Rev Esp Cardiol* 2008;**61**:1361–3.
45. Giugliano GR, Cox N, Popma J. Cutting balloon entrapment during treatment of in-stent restenosis: an unusual complication and its management. *J Invasive Cardiol* 2005;**17**:168–70.

Chapter 9

Complex Lesion Subsets

9.1 PROXIMAL CAP AMBIGUITY

Online Cases 2, 5, 16, 30, 32, 33, 34, 45, 47, 49, 51, 55, 56, 69, 80, 88, 93, 104.

Proximal cap ambiguity refers to the inability to determine the exact location of the proximal cap of the occlusion, due to the presence of obscuring side branches or overlapping branches that cannot be resolved despite multiple angiographic projections. It is encountered in approximately 31% of chronic total occlusion (CTO) percutaneous coronary interventions (PCIs) and is independently associated with technical failure.[1] In the hybrid algorithm, proximal cap ambiguity is an indication for a primary retrograde approach (Fig. 9.1[2]); however, several antegrade options also exist (Fig. 9.2).

9.1.1 Better Angiography

See Online Cases 2 and 47.

High-quality angiography, including dual injections and multiple, possibly steep, angiographic projections, may help resolve proximal cap ambiguity. Here are some examples:

1. A Vieussens collateral (which is a collateral from the conus branch of the right coronary artery [RCA] to the left anterior descending artery [LAD]) may not fill with contrast if the RCA catheter is deeply engaged, or it may have a separate ostium instead of originating from the aorta. Contrast injections with the catheter less deeply engaged in the RCA can allow filling of the Vieussens collateral and help clarify proximal cap ambiguity (Fig. 9.3, Online Case 2).
2. The origin of the CTO may overlap with the origin of a side branch. Various angiographic views with different angulation may help to separate the branches.

9.1.2 Computed Tomography Angiography

Computed tomography angiography (CTA) can help clarify the course of the occluded vessel (Fig. 9.4), as well as provide information on the presence of calcification and tortuosity.[3,4] CTA is particularly useful when the proximal lesion anatomy is unclear (Section 3.3.6). There are ongoing efforts for coregistration of the coronary angiography and computed tomography images to facilitate crossing.[5,6]

Manual of Chronic Total Occlusion Interventions. https://doi.org/10.1016/B978-0-12-809929-2.00009-0

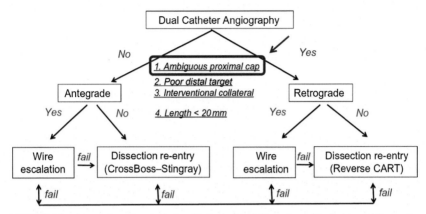

FIGURE 9.1   Approach to chronic total occlusions with ambiguous proximal cap according to the hybrid algorithm.

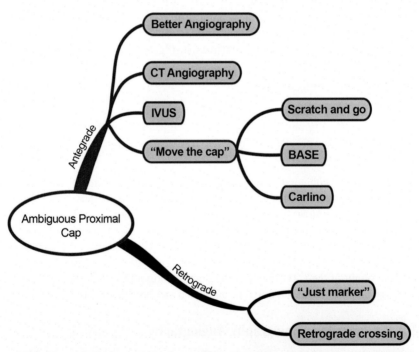

FIGURE 9.2   Antegrade and retrograde approaches to chronic total occlusions with ambiguous proximal cap.

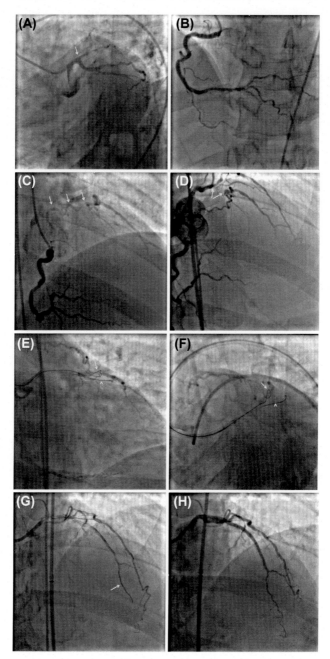

FIGURE 9.3 Example of proximal cap ambiguity clarified by use of multiple angiographic projections and use of the scratch-and-go technique (see Online Case 2). (A) Flush occlusion of the proximal left anterior descending artery (LAD) with unclear proximal cap (*arrow*). The LAD was not filling with contrast. (B) Right coronary artery injection did not provide any filling of the LAD. (C) Repeat right coronary artery injection with the catheter less deeply engaged demonstrated a Vieussens collateral (conus branch of the right coronary artery to the LAD, *arrows*). (D) Dual injection demonstrated that the LAD chronic total occlusion was relatively short (*arrows*). (E and F) Antegrade wire escalation (*arrowhead*) was unsuccessful. A subintimal dissection was created, with a knuckled guidewire (*arrow*) following the vessel course (dancing with the vessel). (G) Successful reentry into distal true lumen was achieved using the Stingray system and a stick-and-swap technique. (H) Successful final result with recanalization of the LAD.

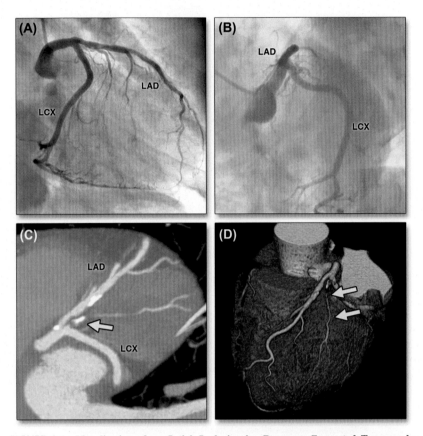

FIGURE 9.4 **Visualization of an Ostial Occlusion by Coronary Computed Tomography Angiography (CTA).** (A and B) Invasive angiography in a 53-year-old woman fails to visualize the large intermediate branch that is occluded at its origin. (C and D) Visualization of the occluded intermediate branch by coronary CTA (*arrows*). *LAD*, left anterior descending artery; *LCX*, left circumflex coronary artery. *(Reproduced with permission from Opolski MP, Achenbach S. CT angiography for revascularization of CTO: crossing the borders of diagnosis and treatment. JACC Cardiovasc Imaging 2015;8:846–58 (Fig. 4 in that paper)).*

9.1.3 Intravascular Ultrasonography

Intravascular ultrasonography (IVUS) can help clarify the location of the proximal cap, especially when there is a side branch close to the occlusion (Fig. 9.5).[7,8] A short-tip, solid-state IVUS catheter is preferred as it can reach further down the vessel, allowing enhanced visualization. However, sometimes a smaller diameter rotational IVUS catheter can be used for smaller side branches that have a longer landing zone.

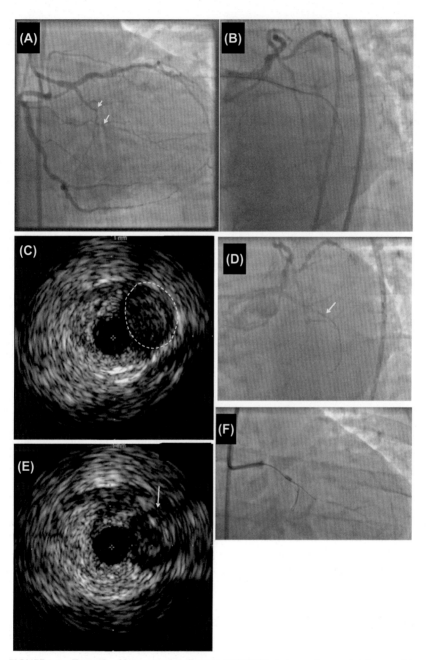

FIGURE 9.5   **Example of intravascular ultrasound (IVUS) use to resolve proximal cap ambiguity (Online Case 34).** Ostial chronic total occlusion (CTO) of the first obtuse marginal branch (*arrows*, A). Repeat antegrade crossing attempts were unsuccessful and the guidewire frequently entered the distal circumflex (B). IVUS demonstrated that the CTO (*yellow circle*, C) actually originated proximal (*arrow*, D) to the distal circumflex's apparent origin. During repeat antegrade crossing attempts a Confianza Pro 12 guidewire was utilized and its location within the CTO was confirmed by IVUS (*arrow*, E) before advancing it through the occlusion (F).

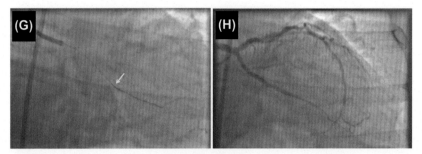

FIGURE 9.5 cont'd—Due to subintimal crossing the Stingray wire and balloon were used for reentry (*arrow*, G) with a successful final outcome (H).

How?

- In ostial branch occlusions (e.g., ostial obtuse marginal CTO), inserting an IVUS in the main vessel can demonstrate the entry point into the occlusion (see Online Case 34) (Fig. 9.5).
- In main vessel occlusions, if there is a side branch adjacent to the occlusion (classic example is LAD CTO at the takeoff of a large diagonal branch, see Online Case 93), the IVUS is inserted into the side branch to identify the beginning of the CTO.
- IVUS guidance can be either (a) real-time (i.e., crossing attempts with simultaneous IVUS visualization), or (b) intermittent (i.e., imaging, followed by crossing attempts, followed by reimaging, etc.).[7]
- Real-time guidance has several limitations:
  1. It requires 8 Fr guide catheters if the Corsair or Turnpike microcatheters are used, or 7 Fr guide catheters if lower profile microcatheters (such as the FineCross or Micro 14) are being used. Alternatively, a ping-pong guide catheter technique can be used.
  2. Intravascular position of the IVUS during crossing attempts may interfere with guidewire manipulation and hinder simultaneous contrast injection.
  3. The IVUS catheter may require constant repositioning due to movement during guidewire manipulation. Hence, live IVUS guidance is infrequently used for wiring through ambiguous proximal caps, and intermittent (serial imaging) is preferred instead.
- In both scenarios, IVUS can be used to demonstrate the wire position during antegrade wire-crossing attempts. If the wire is in the intima or the subintimal space, but within the occlusion, a microcatheter can be advanced over the wire to provide additional support and facilitate crossing. If not, the wire is withdrawn and redirected.
- If a suitable sized artery is available, the short-tip solid-state IVUS catheter is preferred (Eagle Eye short tip, Volcano); it minimizes the extent of distal advancement, which is needed for distal imaging, and is also more deliverable.

- Increasing the diameter of the field of view can be useful in visualizing the occluded vessel, particularly as it traverses away from the IVUS catheter.

What Can Go Wrong?

1. Injury (including perforation) of the side branch from advancing the IVUS catheter, hence the IVUS catheter should not be advanced through very small or tortuous vessels.
2. Thrombosis, due to insertion of multiple equipment in the coronary artery.

9.1.4 Move-the-Cap

See Online Cases 32, 45, 82, 83, and 104.

The move-the-cap techniques use antegrade dissection/reentry to clarify the course of the occluded vessel and achieve crossing.[9] There are three variations of this technique: the balloon-assisted subintimal entry (BASE) technique, the scratch-and-go technique, and the Carlino technique.[10,11] Each allows the operator to decide on the site of proximal subintimal entry.

9.1.4.1 Balloon-Assisted Subintimal Entry (Fig. 9.6)

Step 1 Wire the Vessel Proximal to the Chronic Total Occlusion

How?
- Use a workhorse guidewire.
- Confirm in orthogonal projections that the wire is actually within the intended segment.

What Can Go Wrong?
- Wire may advance into a side branch. This should be appreciated using orthogonal projections and corrected before advancing the balloon.

Step 2 Advance Balloon Proximal to Proximal Cap

How?
- Use a slightly oversized compliant balloon (1.1:1 or 1.2:1 balloon:vessel diameter ratio).
- Check in orthogonal projections that the wire and balloon are actually within the intended segment.

What Can Go Wrong?
- Balloon may not advance due to severely diseased proximal vessel and/ or severe calcification. In such cases predilation with a smaller balloon or other lesion preparation (e.g., with atherectomy or laser) may need to be performed first. Alternatively, a larger or more supportive guide catheter (such as Amplatz 1 for the RCA) or other guide supporting techniques, such as side branch anchor or guide catheter extension (Section 3.6) may need to be used.

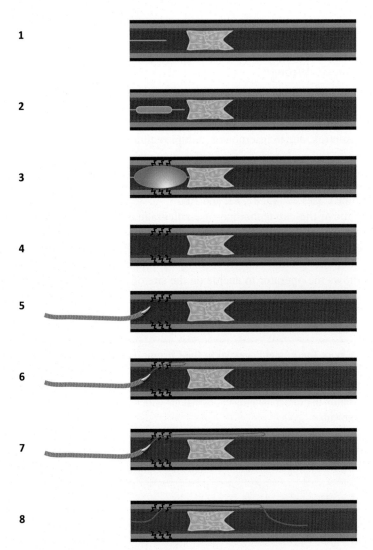

FIGURE 9.6 Illustration of the balloon-assisted subintimal entry technique.

Step 3 Balloon Inflation

How?

- At 10–15 atm

What Can Go Wrong?

- Perforation of the proximal vessel, given balloon oversizing, hence high inflation pressures (20 atm or more) should be avoided.

Step 4 Contrast Injection

How?

- Through the guide catheter to verify that proximal vessel dissection has indeed occurred.

What Can Go Wrong?

- Propagation of the dissection either downstream (potentially compressing the distal true lumen) or upstream (causing aortocoronary dissection). This can be prevented by gentle injection under fluoroscopic or cineangiographic imaging. Side-hole guides, or partially disengaging the guide, can also reduce the contrast injection force.

Step 5 Delivery of Microcatheter Proximal to Proximal Cap

How?

- Over the workhorse guidewire that was used to deliver the angioplasty balloon.

What Can Go Wrong?

- Inability to deliver microcatheter due to tortuosity or calcification (unlikely given prior balloon inflation). If it occurs, additional balloon dilations or increased guide catheter support may be needed.

Step 6 Insert Polymer-Jacketed Guidewire and Create Knuckle Into Dissection Plane

How?

- Advance a polymer-jacketed guidewire (such as Fielder XT, Fighter, or Pilot 200) through the microcatheter. Wire is advanced by pushing, not turning, to minimize the risk for fracture

What Can Go Wrong?

- Perforation. Can be prevented by checking the wire course in orthogonal projections.
- Inability to form a knuckle. The wire tip may be reshaped into an umbrella-handle or other configuration, ensuring that the wire is folding back on itself rather than dissecting forward, before reinserting.
- Guidewire entrapment in the vessel wall (Fig. 9.7).[12] This is a very infrequent complication. Potential solutions include advancing a second guidewire next to the entrapped wire and performing balloon inflations in an attempt to free the wire. If the guidewire fractures, IVUS can help ascertain that there is no wire unraveling into the proximal part of the vessel or into the aorta.

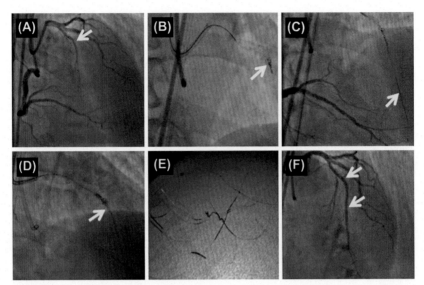

FIGURE 9.7 **Guidewire entrapment during antegrade dissection reentry.** Bilateral coronary angiography demonstrating a chronic total occlusion of the mid-left anterior descending artery (*arrow*, A). Entrapment of a knuckled Fielder XT guidewire (*arrow*, B). The chronic total occlusion was successfully crossed with a Pilot 200 guidewire (*arrow*, C) advanced parallel to the entrapped guidewire. After balloon angioplasty was performed around the entrapped guidewire (*arrow*, D), the entrapped guidewire was successfully retrieved (E) with an excellent final angiographic result (*arrow*, F). *(Reproduced with permission from Danek BA, Karatasakis A, Brilakis ES. Consequences and treatment of guidewire entrapment and fracture during percutaneous coronary intervention.* Cardiovasc Revasc Med *2016;17:129–33 (see* Online Case 24*).)*

Step 7 Chronic Total Occlusion Crossing (as Described in Section 5.4)

How?

- Once a wire knuckle enters the subintimal space it can be advanced through the occluded segment with very low risk of causing perforation due to the distensibility of the subintimal space. Reentry can now be set up beyond the distal cap of the CTO, ideally proximal to the origin of any large branches.

Step 8 Reentry (as Described in Section 5.4)

How?

- In most cases reentry is achieved as close to the distal cap as possible, using the Stingray system, as described in Chapter 5. High-end reentry approaches, such as the "double Stingray" (see Online Case 80) or use of the retrograde approach, may be needed if the distal cap is at the bifurcation of a large branch (e.g., right posterior descending and right posterolateral vessel).

9.1.4.2 Scratch-and-Go (Fig. 9.8)

See Online Cases 2, 16, 32, 45, 59, and 61.

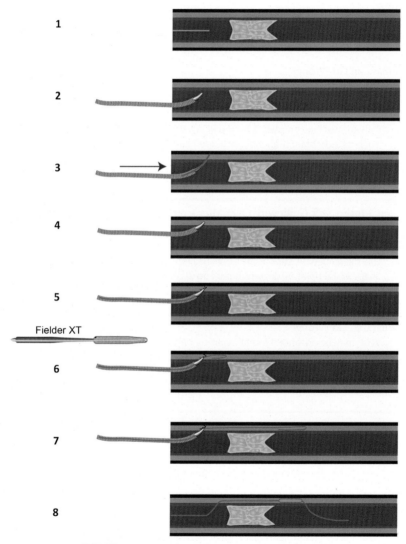

FIGURE 9.8   Illustration of the scratch-and-go technique.

Step 1 Wire the Vessel Proximal to the Chronic Total Occlusion

Step 2 Advance Microcatheter Proximal to Proximal Cap

How?
- Use any standard microcatheter, such as the Corsair and Turnpike.
- Check in orthogonal projections that the wire and microcatheter are actually within the intended segment.

What Can Go Wrong?
- Microcatheter may not advance due to severely diseased proximal vessel and/ or severe calcification. In such cases predilation with a small balloon may need to be performed first. Alternatively, a larger or more supportive guide catheter (such as Amplatz 1 for the RCA) or other guide-supporting techniques (such as side branch anchor, or guide catheter extensions; Section 3.6) may need to be used.

Step 3 Insert Stiff Guidewire Over Microcatheter and Advance Toward Vessel Wall

How?
- Use stiff guidewire, such as the Confianza Pro 12, Gaia 2nd or 3rd, or Hornet 14.
- Shape distal tip at 90 degrees bend and 2–3 mm length.
- Advance guidewire into vessel wall proximal to the proximal cap. Only advance 1–2 mm into the wall to avoid perforation.

What Can Go Wrong?
- Wire may be advanced too far, causing perforation. Wire advancement alone is extremely rare to cause perforation, but if the microcatheter follows, perforation is possible.

Step 4 Advance Microcatheter Over Stiff Guidewire Inside Vessel Wall

How?
- Advance the microcatheter tip over the stiff guidewire toward the vessel wall (only a minimal distance, usually 1 mm or less).

What Can Go Wrong?
- Perforation, if the guidewire or microcatheter has exited the vessel wall, hence confirmation that the guidewire has not perforated (by antegrade contrast injection) is needed before advancing the microcatheter.

Step 5 Insert Polymer-Jacketed Guidewire Through Microcatheter

How?

- Insert a polymer-jacketed guidewire (such as the Fielder XT, Fighter, or Pilot 200) through the microcatheter.
- Advance (pushing without turning) to form a knuckle.

What Can Go Wrong?

- Perforation, inability to form a knuckle, and guidewire entrapment in the vessel wall, as described in Step 6 above.

Step 6 Creation of Knuckle Into Dissection Plane (as described in Section 5.4)

Step 7 Chronic Total Occlusion Crossing (as described in Section 5.4)

Step 8 Reentry (as described in Section 5.4)

9.1.4.3 Carlino Technique for Resolving Proximal Cap Ambiguity (Figs. 9.9 and 9.10)

The Carlino microdissection technique is described in detail in Chapter 5, Section 5.6.3. The Carlino microdissection technique has multiple other uses, such as to facilitate crossing of wire-uncrossable lesions and forward advancement during antegrade subintimal crossing attempts (Figs. 9.9 and 9.10).

How?

- The Carlino technique can be used after a microcatheter tip has entered the subintimal space before, during, or after guidewire advancement attempts, to facilitate subintimal dissection, and also to confirm that the wire tip is indeed located within the subintimal space.

Step 1 Wire the Vessel Proximal to the Chronic Total Occlusion

---

Step 2 Advance Microcatheter Proximal to Proximal Cap

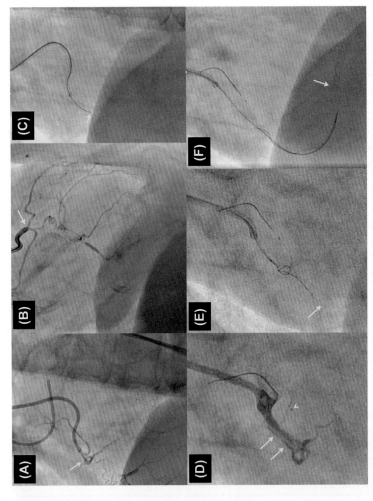

FIGURE 9.9 **Chronic total occlusion (CTO) with ambiguous proximal cap and in-stent occluded segment, successfully recanalized using the scratch-and-go technique with crushing of the occluded stent (see Online Case 32).** (A) Proximal right coronary artery CTO with ambiguous proximal cap (*arrow*). (B) Lateral view showing the ambiguous proximal cap of the right coronary artery CTO (*arrow*). (C) Scratch-and-go technique: creation of a dissection proximal to the proximal cap with a Confianza Pro 12 guidewire (*arrow*) through a Corsair microcatheter. (D) Proximal vessel dissection (*arrows*). A side branch anchor balloon has been placed in an acute marginal branch (*arrowhead*). (E) A knuckled guidewire (*arrow*) is advanced subintimally around the proximal cap.

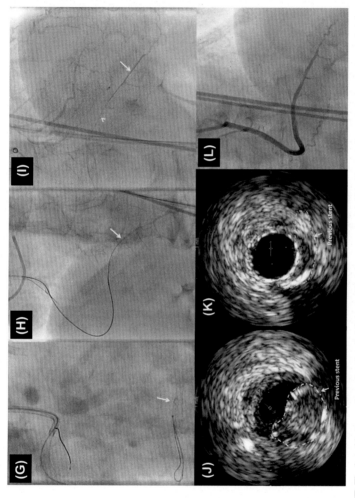

FIGURE 9.9 cont'd—(F) The knuckle (*arrow*) reaches the distal stent. (G) Advancement of a CrossBoss catheter and the knuckled wire (*arrow*) around the previously placed (now occluded) stent. (H) Subintimal guidewire advancement into the right posterior descending artery. (I) Successful reentry into the distal true lumen with a Pilot guidewire (*arrow*) advanced through a Stingray balloon (*arrowhead*). (J) Intravascular ultrasound demonstrating subintimal crossing around the previously placed stent. (K) Intravascular ultrasound after stent implantation showing crushing of the previously placed stents. (L) Successful recanalization of the right coronary artery CTO.

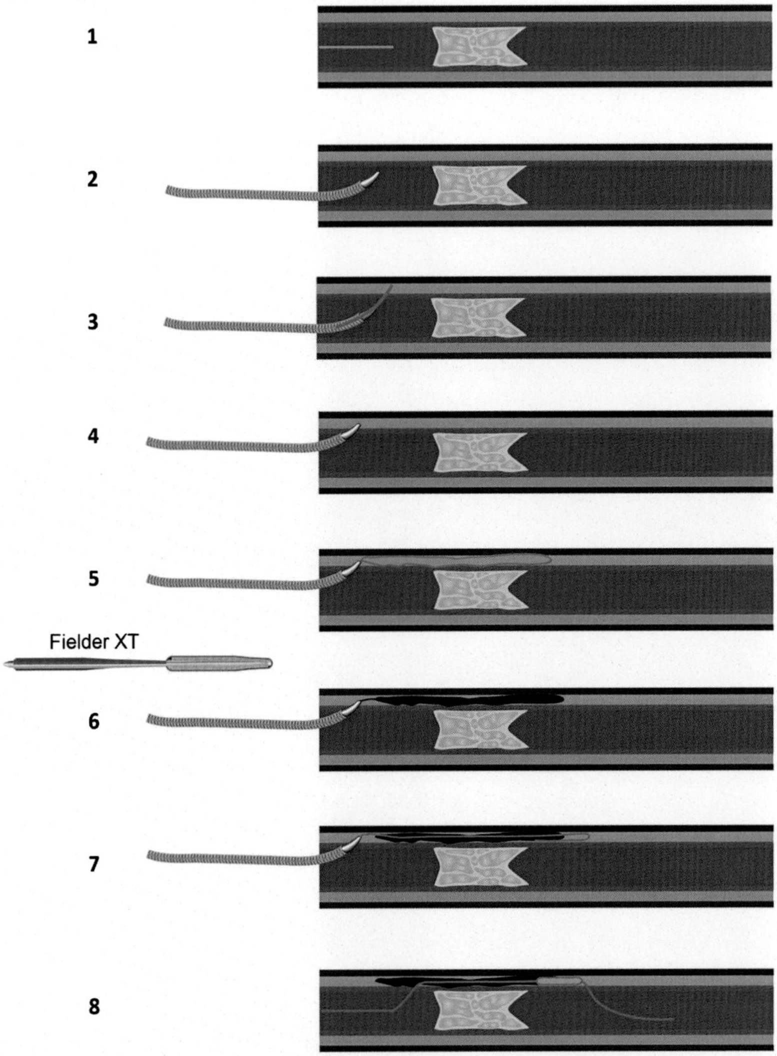

FIGURE 9.10 Illustration of the Carlino technique for resolving proximal cap ambiguity.

Step 3 Insert Stiff Guidewire Over Microcatheter and Advance Toward Vessel Wall

Step 4 Advance Microcatheter Over Stiff Guidewire Inside Vessel Wall

Step 5 Inject Small Amount of Contrast Through Microcatheter

How?
- Use a small (usually 3 cc) luer lock syringe.
- Inject a small amount (0.5–1.0 mL) of contrast gently under cineangiographic guidance.[10]

What Can Go Wrong?
- Perforation. The risk can be minimized by injecting a small amount of contrast and meticulous fluoroscopic visualization of the injection.
- Retrograde contrast propagation causing side branch or proximal branch occlusion.

Step 6 Advance Guidewire Into Dissection Plane

(Sometimes the contrast might dissect into the distal true lumen.)

Step 7 Chronic Total Occlusion Crossing (as Described in Section 5.4)

Step 8 Reentry (as Described in Section 5.4)

(In case of subintimal crossing.)

9.1.5 Retrograde Crossing

See Online Cases 5, 31, 33, 36, 51, 56, 69, 70, 88, and 104.

Retrograde crossing can provide an excellent solution to proximal cap ambiguity, as the retrograde guidewire can be advanced either in the true lumen or in the subintimal space to the proximal cap, clarifying the vessel course and resolving the proximal cap ambiguity. Moreover, the retrograde guidewire can modify the proximal cap, facilitating antegrade crossing (see Online Case 102). The Carlino technique can be used through the retrograde microcatheter (see Online Case 62) and retrograde balloons can be used to perform the controlled antegrade and retrograde tracking and dissection (CART) technique.

9.2 IMPENETRABLE PROXIMAL (OR DISTAL) CAP

Heavily calcified or fibrotic proximal (or distal) caps may be challenging or impossible to penetrate with a guidewire. Distal caps may be particularly challenging

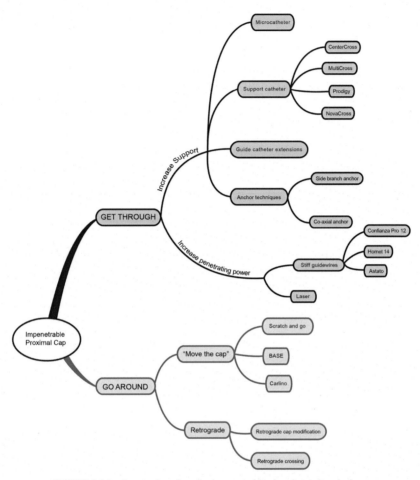

FIGURE 9.11   Approach to the wire-impenetrable chronic total occlusions.

to cross in patients with prior coronary bypass graft surgery, presumably because the distal cap was exposed to systemic arterial pressure when the bypass graft was patent and also because bypass graft surgery can lead to severe calcification. There are several possible options for crossing such caps, which can be grouped into two major categories: (1) get-through and (2) go-around (Fig. 9.11).

9.2.1 Get-Through

Advancing through a wire impenetrable lesion can be achieved via (1) strong guidewire support and/or (2) high penetrating power guidewires or devices. Novel technologies are currently in development and could facilitate crossing of such lesions, such as the Soundbite system[13] and administration of collagenase.[14]

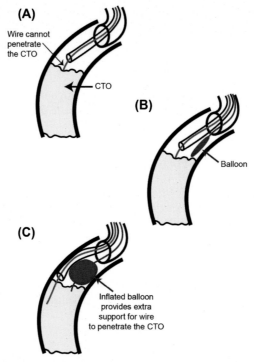

FIGURE 9.12 **Illustration of the power microcatheter technique.** When the support of the microcatheter and guide catheter is not enough for crossing a chronic total occlusion (A), a balloon (sized 1:1 to the vessel) can be advanced next to the microcatheter (B) and inflated (C). Balloon inflation anchors the microcatheter against the vessel wall, providing extra support (power) for guidewire advancement.

Strong guidewire support can be achieved via various microcatheters (especially the more supportive Corsair and Turnpike Spiral), support catheters, guide catheter extensions, and anchoring techniques, as described in Section 3.6. A variation of the coaxial anchoring technique is inflating a balloon next to a microcatheter, hence providing support by pressing against the vessel wall (power microcatheter; Fig. 9.12 and Online Case 82). This approach can also be used to power anchor a guide extension catheter for even more support, but requires a second guide catheter system (e.g., ping-pong guide).

High penetration power can be provided via stiff guidewires (the classic guidewire for this purpose is the Confianza Pro 12, although the Hornet 14, the Stingray, and the Astato 20 guidewire can also be used). Some operators recommend short laser activation (if the proximal cap anatomy is very clear), however this approach may carry increased risk for perforation and is not favored by most CTO operators. Use of the Carlino technique (contrast injection through microcatheter; Chapter 5, Section 5.6.3) can help create a dissection plane and allow subsequent wire crossing (see Online Case 37).

9.2.2 Go-Around

See Online Cases 12, 13, 46, 49, 57, 58, 60, and 62.

Going around instead of through a wire-impenetrable CTO using subintimal techniques can be a very effective and safe strategy for crossing wire-impenetrable CTOs.

As described in the approach to CTOs with ambiguous proximal cap, subintimal crossing can be achieved either in the antegrade direction (move the cap techniques) or in the retrograde direction.

9.3 FLUSH AORTOOSTIAL CHRONIC TOTAL OCCLUSIONS

Flush aortoostial CTOs are aortoostial occlusions without a stump (RCA, left main, or bypass graft). Such lesions can be challenging to cross, as guide catheters cannot be seated, antegrade crossing is not feasible until possibly after a retrograde guidewire is advanced to the coronary ostium (see Online Case 90), and therefore, a primary retrograde approach is required. Similarly challenging can be CTOs in vessels that are too diffusely diseased or of anomalous origin that cannot be engaged with a guide catheter.

There are two important considerations/challenges in patients with flush aortoostial CTOs, as described next.

9.3.1 Confirm That the Chronic Total Occlusion Is Truly a Flush Aortoostial Chronic Total Occlusion

1. **Angiography in multiple projections, including aortography**
   Sometimes the vessel is patent, but has anomalous origin or cannot be engaged with a guide catheter. Various angiographic projections, cusp angiography (injection in the coronary cusp), and ascending aortography (typically $20\,\text{mL/s} \times 3\,\text{s}$ injection for a total of $60\,\text{mL}$ of contrast in the left anterior oblique projection) can in some cases demonstrate the origin of a vessel.
2. **Computed tomography angiography**
   CTA (Section 3.3.6) provides a definitive answer as to whether the CTO vessel indeed has an aortoostial occlusion or not, and provides additional information on the morphology of the occlusion (such as length, calcification, tortuosity), facilitating subsequent CTO recanalization attempts.

9.3.2 Retrograde Crossing Into the Aorta

Since flush aortoostial CTOs by definition cannot be engaged with a guide catheter (as they do not have a stump), retrograde wire crossing into the aorta is necessary (see Online Cases 18, 37, 58 and 77). Such crossing can be challenging due to calcification of the ostial occlusion and occasional tendency of the guidewire to advance into the subaortic subintimal space.

How?

Retrograde crossing of a flush aortoostial CTO can be facilitated by:

1. **Use of stiff guidewires**
   Stiff guidewires supported by the retrograde microcatheter can facilitate penetration into the aorta. Typical guidewires used in this manner include the Gaia series (usually 2nd and 3rd), the Confianza Pro 12, Astato 20, Hornet 14, and the Pilot 200 (see Online Case 18).

2. **The Carlino technique**
   Injection of a small amount of contrast through the tip of the retrograde microcatheter can assist with crossing,[10,11] similar to crossing the antegrade cap, as described in Section 9.1.4 (Fig. 9.13 and Online Case 37).

3. **The e-CART (ElectroCautery-Assisted Re-enTry) technique**
   This is a last-resort technique, to be used when all other techniques fail. It uses the power of electrocautery to penetrate through very challenging occlusions (see Online Case 58).[15] The following steps are involved (Fig. 9.14):
   a. The cautery pad is placed on the patient's hip.
   b. The retrograde guidewire is advanced retrogradely pointing at the ostium.
   c. The guidewire should be covered by a microcatheter in its entire length, except for the tip.
   d. Typically a Confianza Pro 12 guidewire is used with the distal 3 mm of the tip amputated (although wire amputation is not necessary).
   e. A hemostat is used to clamp the back of the wire to the cautery needle.
   f. The cautery power is set at 50 Watts on Cut.
   g. The cautery is activated on the guidewire for 1 s under cineangiography to observe the wire pop through.
   h. Guidance with transesophageal echocardiography may minimize the risk for guidewire perforation.

The e-CART technique should be used with caution, as it can cause a significant perforation or entry in nearby cardiac structures, such as the pulmonary artery. The retrograde electrocautery can facilitate antegrade wire advancement, providing enough wire overlap to perform CART or reverse CART.

9.4 OSTIAL CIRCUMFLEX AND OSTIAL LEFT ANTERIOR DESCENDING ARTERY CHRONIC TOTAL OCCLUSIONS

Both ostial circumflex and ostial LAD CTOs can have ambiguous proximal caps, requiring use of the techniques described in Section 9.1 to resolve the ambiguity.

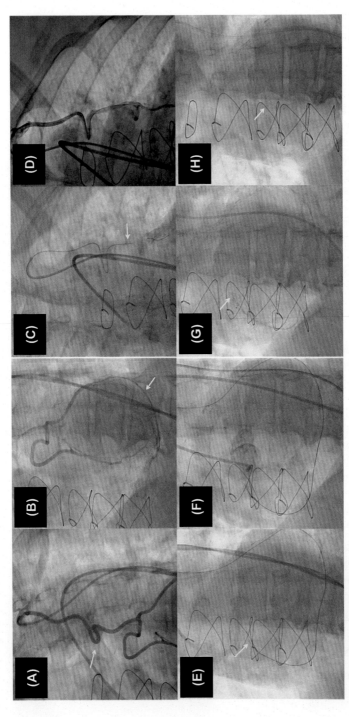

FIGURE 9.13   Left internal mammary artery with significant mid-portion tortuosity (*arrow*, A), which supplies the distal right coronary artery via epicardial collaterals (*arrow*, B) from the circumflex. After advancement of a Corsair catheter (*arrow*, C) over a Sion guidewire (Asahi Intecc) through the left internal mammary artery (LIMA), antegrade flow was preserved (D). The guidewire and Corsair catheter were advanced to the distal chronic total occlusion (CTO) cap (*arrow*, E), but retrograde wiring was extremely challenging due to LIMA tortuosity. Only Gaia guidewires (Asahi Intecc) could be advanced to the tip of the Corsair catheter but still could not penetrate the occlusion (F). The Corsair catheter was exchanged for a Finecross catheter and 0.5 cc of contrast was injected, advancing slightly more into the occlusion (*arrow*, G). Several additional Gaia 2nd guidewires (Asahi Intecc) could not be advanced through the lesion. Intralesion contrast injection was repeated, achieving contrast jet entry into the aorta (*arrow*, H).

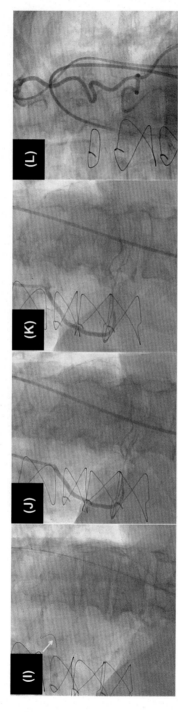

FIGURE 9.13 cont'd—An RG3 guidewire (Asahi Intecc) was advanced into the aorta (*arrow*, I), snared, and externalized (J), enabling stenting of the right coronary artery CTO with an excellent final result (K). LIMA flow was not affected during the procedure and no injury was seen on angiography performed after equipment removal (L) (see Online Case 37). (*Reproduced with permission from Amsavelu S, Carlino M, Brilakis ES. Carlino to the rescue: use of intralesion contrast injection for bailout antegrade and retrograde crossing of complex chronic total occlusions. Catheter Cardiovasc Interv 2016;87:1118–23 (Fig. 1.3 in that paper).*)

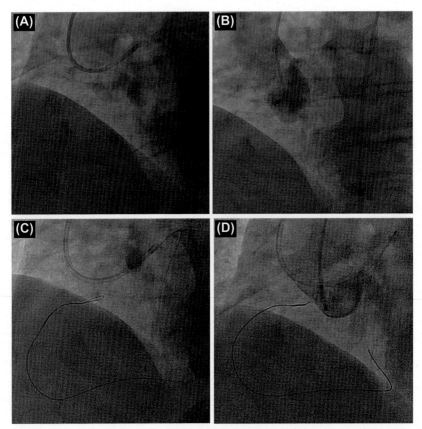

FIGURE 9.14 Illustration of the e-CART (ElectroCautery-Assisted Re-enTry) technique. Initial coronary angiograms showing (A) the retrograde filling of the distal right coronary artery from the left system and (B) true ostial occlusion of the right coronary artery. Subintimal passage of a knuckled guidewire and microcatheter was achieved retrograde to the ostium of the right coronary artery occlusion (C). A pigtail catheter was placed at the site of the right coronary artery aortoostial occlusion (D).

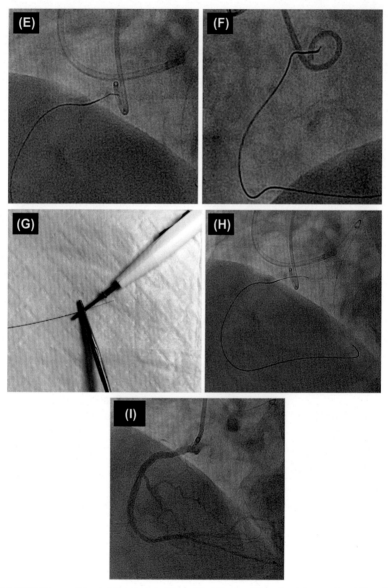

FIGURE 9.14   cont'd—A left anterior oblique 40 projection (E) and a right anterior oblique 30
projection (F) were utilized to serve as a target for our retrograde wire to ensure accurate directing
of the energized retrograde Confianza Pro 12 (Asahi Intecc). The back end of the guidewire was
connected to a unipolar electrosurgery pencil using forceps (G). Distal crossing tip of the guidewire
was energized in cutting mode at 50 W for a 1-s burst, with immediate unimpeded crossing into the
lumen of the aorta (H). Final angiographic result after snaring and externalizing the retrograde wire
utilizing standard chronic total occlusion percutaneous coronary intervention techniques in creating a neo-ostium of the right coronary artery (I). *(Reproduced with permission from Nicholson W,
Harvey J, Dhawan R. E-CART (ElectroCautery-Assisted Re-entry) of an aorto-ostial right coronary
artery chronic total occlusion: first-in-man. JACC Cardiovasc Interv 2016;9:2356–8, Elsevier.)*

If the retrograde approach is being used for such lesions, it may be best to:

1. Use a guide catheter extension to perform reverse CART to minimize the risk for left main dissection (see Online Case 36).
2. Insert a safety guidewire into the nonoccluded vessel (circumflex for LAD CTOs and LAD for circumflex CTOs) to maintain access to the vessel in case of dissection or plaque shift during recanalization of the CTO (see Online Case 102).

Ostial circumflex or LAD CTOs may be challenging to cross due to angulation: use of the Venture catheter (see Online Case 48; Section 2.4.10) (or the angled SuperCross microcatheters, see Online Case 91; Section 2.4.12) can help direct and support the wire to cross the lesion (Fig. 9.15). Due to tortuosity and frequent lack of interventional collaterals circumflex CTOs often have lower success rates than CTOs or the RCA or the LAD.[16]

9.5 BIFURCATION AT PROXIMAL CAP

There are two potential challenges associated with bifurcations at the proximal or distal cap (Fig. 9.16):

1. **Inability to wire the occlusion** due to preferential guidewire entry into the patent branch and failure to engage the CTO (Fig. 9.16A) (see Online Case 55).
2. **Occlusion of the side branch** (Fig. 9.16B). This may be inconsequential for small branches, but can lead to periprocedural myocardial infarction[17–19] or arrhythmias. As described in Section 9.4, inserting a safety guidewire into the patent branch can help maintain access to the vessel in case of dissection or plaque shift during recanalization of the CTO (see Online Case 49).

How to cross CTOs with bifurcation at the proximal cap (Fig. 9.17) is explained next.

There are several techniques to successfully cross such lesions:

1. **Insertion of a guidewire into the side branch**, which can act as a marker of the side branch origin, facilitating antegrade crossing attempts with a second guidewire (Fig. 9.17, 1). In addition, this wire can protect the side branch if it becomes compromised/dissected during aggressive CTO wiring of the proximal cap.
2. **Balloon inflation in the side branch** can induce a geometrical shift of the hard plaque in the proximal cap and enable guidewire entry into the CTO (this has been called the **open-sesame** technique) (Fig. 9.17, 2).[20]
3. **Deflecting balloon or blocking balloon**: A balloon is inflated at the ostium of the side branch, blocking entry of the guidewire into the side branch, which can then engage the CTO proximal cap (Fig. 9.17, 3).

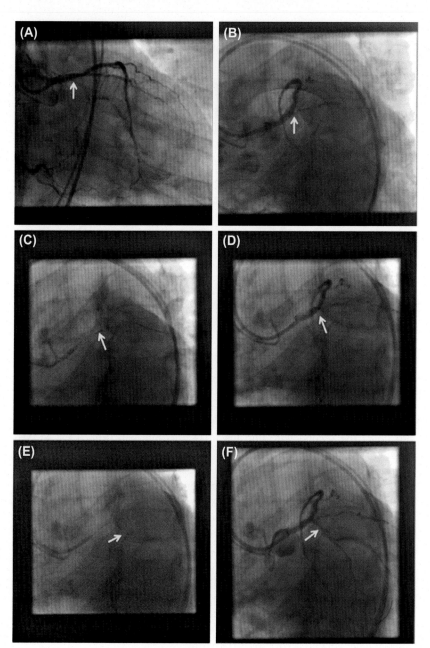

FIGURE 9.15 Use of the Venture catheter for treating an ostial circumflex occlusion. The case was complicated by stent loss. Bilateral coronary angiography demonstrating an ostial chronic total occlusion (CTO) of the circumflex (*arrows*, A and B). Intravascular ultrasonography was performed to identify the ostium of the circumflex (*arrow*, C), followed by insertion of a Venture catheter (*arrow*, D), through which a Pilot 200 wire easily crossed the occlusion (*arrow*, E). After predilation antegrade flow in the circumflex was restored (*arrow*, F). In spite of multiple balloon predilations and use of a GuideLiner catheter, a 2.5 × 23 mm stent could not be delivered and during attempts to retrieve it into the guide catheter it was dislodged from the balloon into the left main coronary artery. Attempts to retrieve the stent using a 4 mm Gooseneck snare were unsuccessful. During attempts to deliver a stent to the circumflex, the stent became dislodged and was crushed with another stent in the left main coronary artery followed by rewiring and balloon angioplasty of the circumflex with a satisfactory final angiographic result (G). This case highlights the value of the Venture catheter in highly angulated lesions, especially in ostial circumflex CTOs (H). The Venture catheter enabled rapid lesion crossing of a very challenging CTO in a tortuous and calcified vessel. In such vessels stent delivery may also be very difficult and may be complicated by stent loss, as in this case. Aggressive lesion preparation and use of guide catheter extensions may minimize the risk for stent loss.

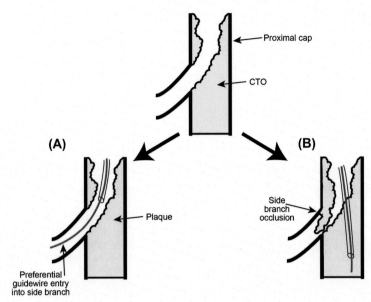

(A) (B)

FIGURE 9.16 Challenges associated with chronic total occlusions (CTOs) that have a bifurcation at the proximal cap.

4. **Use of a dual lumen microcatheter** (such as the Twin Pass; Section 2.4.11) to direct the second wire into the CTO proximal cap (Fig. 9.17, 4) (see Online Case 55).[21]

   A dual lumen microcatheter helps (1) enhance the wire penetrating capacity, (2) straighten the proximal vessel bending, and (3) change the puncture position by altering the catheter position. When the CTO wire is deeply advanced into the occlusion, the catheter should be replaced by a general microcatheter by using the trapping technique to facilitate subsequent wire manipulations. In terms of wire penetrating capacity, this strategy provides the most powerful force in combination with the following additional method when the size of side branch is large enough to advance a balloon: (1) two floppy wires are advanced into the same side branch, (2) an adequate size rapid exchange balloon is inserted and inflated in the branch to trap one of the wires as anchoring, and (3) a dual lumen microcatheter is advanced through the anchored wire for CTO wiring.

5. **Use of a wire directing catheter**, such as the Venture or the SuperCross microcatheter (Fig. 9.17, 5; Fig. 9.18).[22]

6. **Use of a support catheter**, such as the CenterCross (Section 2.4.13, Fig. 9.17, 6), Prodigy (Section 2.4.14), and Novacross (Section 2.4.15) to point more directly to the proximal cap.

7. **Use of antegrade dissection reentry techniques** (if the side branch is small and its loss is acceptable) (move-the-cap techniques; Section 9.1.4) (Fig. 9.17, 7).

8. **Use of retrograde crossing** (Fig. 9.17, 8).

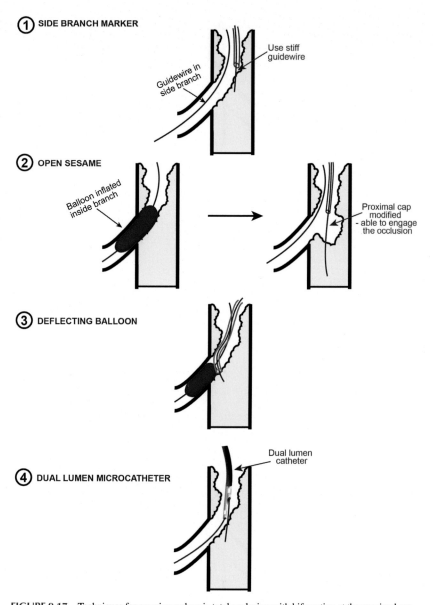

FIGURE 9.17 Techniques for crossing a chronic total occlusion with bifurcation at the proximal cap.

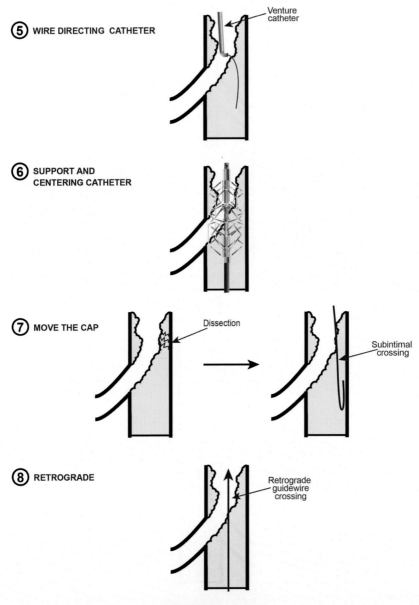

FIGURE 9.17 cont'd.

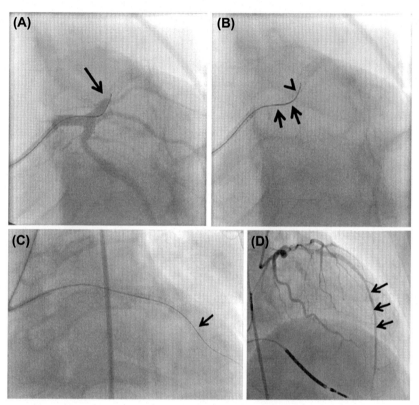

FIGURE 9.18 Example of use of the Venture catheter to direct the guidewire away from a side branch arising at the chronic total occlusion proximal cap. *(Modified with permission from Iturbe JM, Abdel-Karim AR, Raja VN, Rangan BV, Banerjee S, Brilakis ES. Use of the venture wire control catheter for the treatment of coronary artery chronic total occlusions. Catheter Cardiovasc Interv 2010;**76**:936–41.)*

9.6 BIFURCATION AT DISTAL CAP

The same challenges associated with bifurcations at the proximal cap exist for CTOs with bifurcation at the distal cap, namely being able to cross the CTO (retrograde wiring can be challenging due to preferential entry into the branch) and maintaining patency of both branches (Fig. 9.19).

If antegrade dissection/reentry is used for crossing a CTO with bifurcation at the distal cap, reentering immediately proximal or at the bifurcation is ideal for maintaining patency of both branches (see Online Case 45). Otherwise separate reentry into both branches may be needed, followed by kissing angioplasty and most commonly a two-stent bifurcation stenting technique (see Online Case 80).

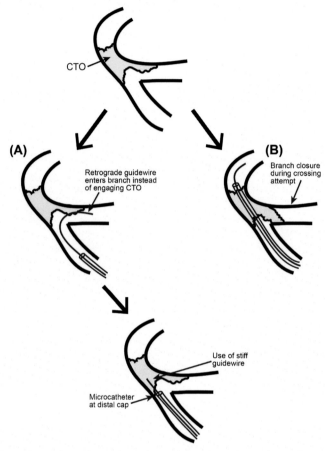

FIGURE 9.19 Challenges associated with percutaneous coronary intervention of a chronic total occlusion with bifurcation at the distal cap.

CTOs with a bifurcation at the distal cap can be approached in a primary retrograde or antegrade direction, as shown next.

9.6.1 Retrograde Techniques (Fig. 9.20)

See Online Cases 5, 12, 23, and 46.

A primary retrograde approach allows engagement of the CTO from a favorable angle and can increase the likelihood of maintaining patency of both branches. However, entering the CTO can be challenging, requiring advancement of the retrograde microcatheter all the way to the distal cap and use of stiff guidewires, such as Gaia, Confianza Pro 12, and Pilot 200.

A. If retrograde crossing is achieved without compromising the origin of the branch vessel, the lesion can be stented while maintaining patency of both branches (Fig. 9.20A).

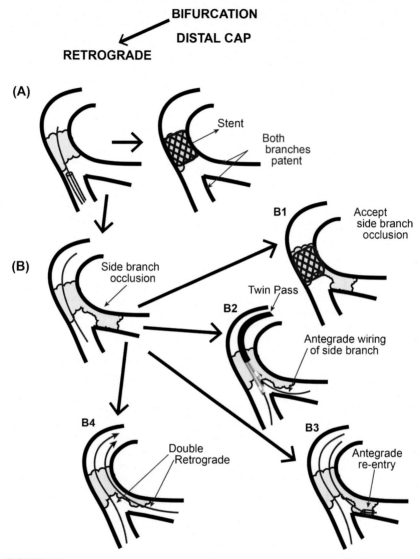

FIGURE 9.20   Retrograde techniques for crossing a chronic total occlusion with bifurcation at the distal cap.

B. If retrograde crossing succeeds but causes occlusion of an (important) branch (Fig. 9.20B), the following options exist:

   B1. Accept occlusion of the side branch without further recanalization attempts. This may be the best course of action, if the branch is relatively small or in long procedures requiring large amounts of radiation or contrast (see Online Case 42).

   B2. Attempt antegrade wiring of the side branch (see Online Case 100), which can be facilitated by use of a dual lumen microcatheter.

B3. Attempt reentry into the occluded branch, usually using the Stingray system.

B4. Double retrograde: attempt second retrograde crossing of the occluded side branch (see Online Case 64).

9.6.2 Antegrade Techniques (Fig. 9.21)

See Online Cases 11, 25, 70, and 80.

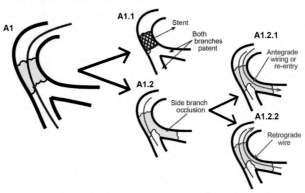

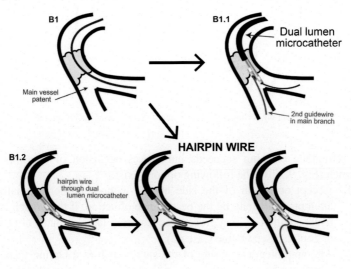

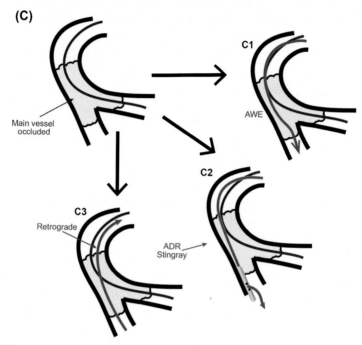

FIGURE 9.21 Antegrade techniques for crossing a chronic total occlusion with bifurcation at the distal cap.

Antegrade crossing of a CTO with bifurcation at the distal cap carries the risk of compromising one of the branches, especially if dissection/reentry is required. Antegrade crossing can be achieved into the main branch or the distal side branch.

A. Antegrade crossing into the main branch.

    A.1.1. In the best-case scenario distal true lumen crossing is achieved without affecting the patency of the side branch, even after stenting.

    A.1.2. If the side branch is compromised after crossing into the distal main branch, the side branch is rewired, either in the antegrade direction (Fig. 9.21; A.1.2.1) or the retrograde direction (Fig. 9.21; A.1.2.2). If the side branch is small or the procedure is challenging, recanalization of the side branch may be deferred or not performed at all (see Online Cases 8 and 67).

B. Antegrade crossing into the side branch maintaining main branch patency.

    B.1.1. A second guidewire is then advanced into the main branch, ideally using a dual lumen microcatheter to maintain access to the side branch.

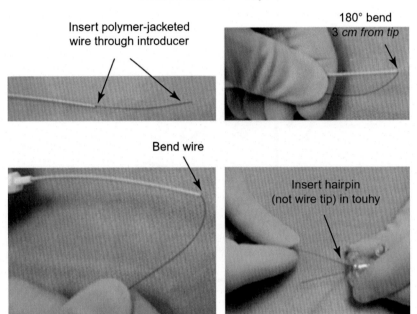

FIGURE 9.22 Illustration of hairpin wire creation.

B.1.2. Prolonged antegrade wiring attempts with a second guidewire may cause dissection of the other branch ostium and side branch occlusion. In such cases, the hairpin wire (also called reversed guidewire[23,24]) technique can be useful in directing a guidewire from the side branch into the main vessel (see Online Case 71).[25] In this technique a polymer-jacketed wire is bent approximately 3 cm from the wire tip (Fig. 9.22), advanced into the side branch (Fig. 9.23C), and pulled back (Fig. 9.23D) entering the main branch (Fig. 9.23E). Alternatively, the hairpin wire can be advanced through a dual lumen microcatheter (Fig. 9.24).

What Can Go Wrong?

Use of the "hairpin wire" technique may cause vessel dissection. Also after the "hairpin" enters the main vessel, further advancement may be challenging due to the bend in the wire.

C. Antegrade crossing into the side branch occluding the main branch.

Crossing into the main vessel is critical in such cases and can be achieved in several ways:

C.1 Antegrade wire escalation, potentially using a dual lumen microcatheter.

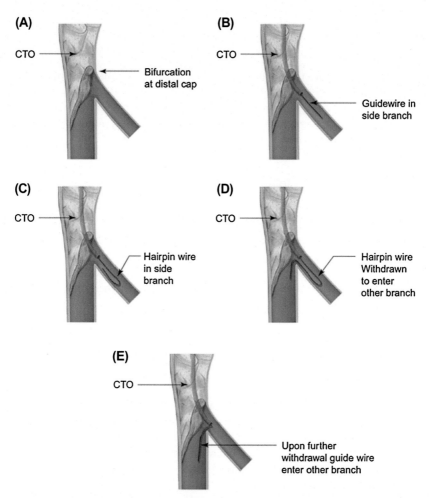

FIGURE 9.23 Technique for approaching chronic total occlusions (CTOs) after crossing into a side branch at the distal cap. A CTO with a bifurcation at the distal cap (A) is successfully crossed with a guidewire that enters the side branch (B). A hairpin guidewire is advanced into the side branch (C) and withdrawn, entering the main branch (D and E).

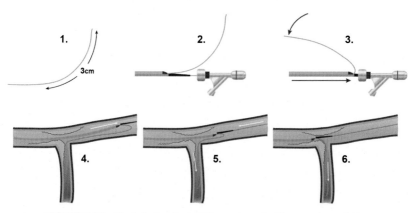

FIGURE 9.24 The hairpin wire technique using a dual lumen microcatheter.

C.2 Reentry into the main branch distal true lumen, usually by using the Stingray system (see Online Cases 21 and 25).

C.3 By retrograde crossing through the main branch (see Online Case 7). In most cases a two-stent bifurcation stenting technique is needed to maintain patency of both branches.

9.7 IN-STENT CHRONIC TOTAL OCCLUSIONS

In-stent CTOs are usually due to occlusive in-stent restenosis, represent approximately 11%–12% of the total CTO intervention case volume,[26,27] and can be challenging to recanalize,[28–32] especially in the presence of a large side branch in LAD occlusions or in the presence of tortuosity in RCA lesions.[33,34] Although similar procedural success can be achieved for in-stent and de novo CTOs,[26,27,35] the risk of subsequent restenosis is likely higher among in-stent CTOs as well as non-CTO lesions.[36]

9.7.1 Approach to In-Stent Chronic Total Occlusions (Fig. 9.25)

The major differences between approaching in-stent CTOs and de novo CTOs are the following:

1. **Subintimal techniques are best avoided for in-stent CTOs**, as wire exit behind the stent would require crushing of the restenosed stent after additional stent implantation. However, crushing of the occluded stents is an option if other approaches fail with encouraging short and mid-term outcomes in case reports (see Online Cases 32, 54, 72, and 100).[31,37–41]

2. **Use of the CrossBoss catheter (Section 2.6.1) is preferred in in-stent CTOs** (see Online Cases 8 and 19).[42,43] The CrossBoss catheter is well suited for crossing in-stent CTOs, because the stent struts may act as

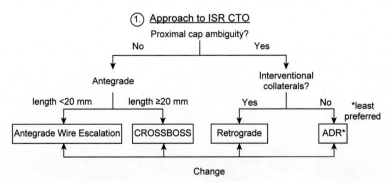

FIGURE 9.25 Approach to chronic total occlusions (CTOs) due to in-stent restenosis.

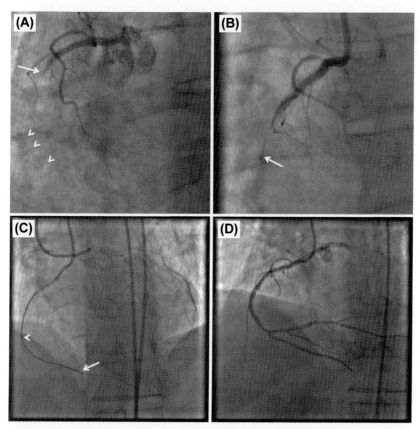

FIGURE 9.26   **Example of treating an in-stent restenotic chronic total occlusion (CTO) using the CrossBoss catheter.** Coronary angiography demonstrating a CTO of the right coronary artery (*arrow*, A) due to in-stent restenosis (*arrowheads*, A). The CrossBoss catheter (*arrow*, B) was inserted into the lesion and advanced using the fast-spin technique. The CrossBoss catheter could not be advanced through the stent (*arrowhead*, C), but a Confianza Pro 12 wire (*arrow*, C) crossed into the distal true lumen, as confirmed by contralateral injection. After stent implantation an excellent final result was achieved (D). (*Reproduced with permission from Papayannis A, Banerjee S, Brilakis ES. Use of the Crossboss catheter in coronary chronic total occlusion due to in-stent restenosis.* Catheter Cardiovasc Interv *2012;80:E30–6.*)

barrier, preventing advancement of the CrossBoss catheter behind the restenosed stent (Fig. 9.26). Rarely, the CrossBoss catheter can advance behind stent struts, leading to stent crushing (Fig. 9.27).[44] Occasionally the CrossBoss advancement may stop, requiring redirection with a guidewire (usually a polymer-jacketed guidewire, such as the Pilot 200).[42,45]

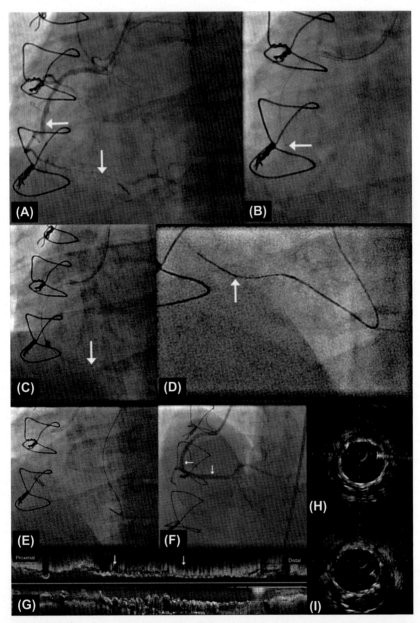

FIGURE 9.27 Example of CrossBoss catheter exit between stent struts. (A) Diagnostic coronary angiography demonstrated occlusive in-stent restenosis with ipsilateral and contralateral collaterals (*arrows* show proximal and distal edges of the previously implanted overlapping stents). (B) The CrossBoss catheter entered the chronic total occlusion, but stopped advancing after 5–10 mm. (C) Following further rapid rotation, it progressed with relative ease toward the distal vessel. (D) Rendezvous in the distal right coronary artery (RCA). Favorable alignment and advancement of the retrograde wire into the CrossBoss (the *arrow* is at the level of the CrossBoss tip). (E) Inflation of a high-pressure balloon at the point of stent exit. (F) Final angiographic result. (G) Intravascular ultrasound (IVUS) longitudinal view after implantation of three overlapping Xience stents. (H) Double stent strut layer in the middle RCA. (I) IVUS image poststent implantation in the distal RCA, also showing the old crushed stent. *(Reproduced with permission from Ntatsios A, Smith WHT. Exit of CrossBoss between stent struts within chronic total occlusion to subintimal space: completion of case via retrograde approach with rendezvous in coronary. J Cardiol Cases 2014;9:183–6, Elsevier.)*

9.7.2 Challenges Associated With Percutaneous Coronary Intervention of In-Stent Chronic Total Occlusions (Fig. 9.28)

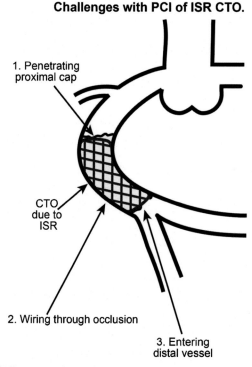

Challenges with PCI of ISR CTO.

1. Penetrating proximal cap

CTO due to ISR

2. Wiring through occlusion

3. Entering distal vessel

FIGURE 9.28 Challenges associated with percutaneous coronary intervention of in-stent chronic total occlusions (CTOs).

9.7.2.1 Penetrating the Proximal Cap (Fig. 9.29)

In-stent restenotic CTOs can be calcified and hard to penetrate. As outlined in Section 9.2, penetrating a tough proximal cap requires use of support techniques and penetrating guidewires, often in combination (see Online Case 7).

9.7.2.2 Crossing the Occlusion (Fig. 9.30)

- Similar to de novo lesions, several guidewires can be used to cross in-stent CTOs, usually starting from soft, tapered polymer-jacketed guidewires and escalating to stiff polymer-jacketed guidewires or stiff composite core guidewires. Laser could also be useful.[46]
- Frequent visualization in orthogonal projections ensures that the guidewire is advancing within the stent struts.

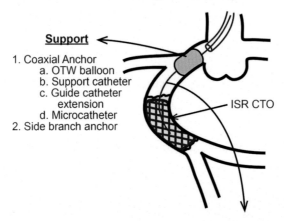

Penetrating Proximal Gap

Support
1. Coaxial Anchor
   a. OTW balloon
   b. Support catheter
   c. Guide catheter
      extension
   d. Microcatheter
2. Side branch anchor

ISR CTO

Penetrating Guidewire
1. Confianza Pro 12
2. Astato 20
3. Gaia 2nd, 3rd
4. Hornet 14

FIGURE 9.29   Penetrating the proximal cap of in-stent chronic total occlusions.

2.2. Wiring Through ISR Occlusion

A. Wires

1. Soft tapered polymer jacketed guidewires
   -Fielder XT, Fielder XT-A, Fighter

2. Stiff polymer jacketed guidewires
   Pilot 200

3.Stiff composite core guidewires
   Non-tapered: Ultimate 3
   Tapered: Gaia 1st, 2nd, 3rd

B. Wire goes outside stent

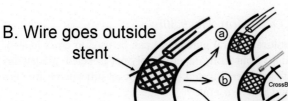

Solutions

a. Change wire (polymer wires more likely to go subintimal)

b. Use CrossBoss

CrossBoss

c. Subintimal crossing + re-entry

FIGURE 9.30   How to cross in-stent chronic total occlusions.

- If the guidewire exits the stent struts, there are three options:
  1. Change the guidewire (polymer-jacketed wires may be more likely to enter into the subintimal space).
  2. Change for a CrossBoss catheter (although occasionally the CrossBoss catheter may track the course of a previously inserted guidewire.
  3. Complete subintimal crossing and reentry following by crushing the stent with a new stent (Fig. 9.9; see Online Cases 32, 54, and 100).[37–39]

Conservative sizing of the new stent, ideally informed by IVUS, is important to minimize the risk for perforation.

9.7.2.3 Entering the Distal Vessel (Fig. 9.31)

Entry into distal lumen

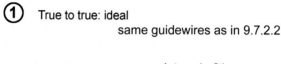

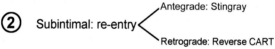

FIGURE 9.31   How to enter the distal vessel during percutaneous coronary intervention of in-stent chronic total occlusions.

9.8 SAPHENOUS VEIN GRAFT CHRONIC TOTAL OCCLUSIONS

See Online Case 44.

Treatment of a saphenous vein graft (SVG) CTO is given a class III (level of evidence C) recommendation in the 2011 ACC/AHA PCI guidelines, due to high restenosis and repeat revascularization rates.[47] In patients with prior coronary bypass graft surgery treatment of a native coronary artery CTO is preferable to treatment of an SVG CTO supplying the same territory.[48,49] Moreover, in patients who develop an SVG aneurysm recanalization of the native coronary artery followed by coiling of the SVG can be performed to treat the aneurysm (see Online Cases 16, 61, and 102).[49] The SVGs (patent or occluded) can serve as retrograde conduits for recanalization of the native coronary CTO (see Online Cases 4, 16, 61, 81, 87, 96, 102, and 103).[48,50,51] However, if native CTO PCI is not possible, PCI of the SVG-CTO can provide a treatment option (Online Case 44).[48,52–57]

Crossing of occluded SVGs is usually attempted with a Corsair microcatheter and a stiff, polymer-jacketed guidewire, such as the Pilot 200. Retrograde crossing into the native vessel proximal to the SVG touchdown may be challenging due to acute angulation and may require use of a Venture catheter (Section 2.4.10) or the SuperCross angulated microcatheter (Section 2.4.12), or the hairpin technique. Coronary CTA may be helpful for understanding the angle of SVG insertion into the target vessel.

9.9 SEVERE CALCIFICATION

See Online Cases 5, 18, 49, 65, 85, and 92.

Calcification can hinder all stages of CTO PCI—crossing, equipment delivery, and stent expansion. It is common (especially in older CTOs) and is part of the J-CTO score for determining the complexity of the procedure.[58] Calcification can have worse adverse consequences when combined with severe tortuosity.

How Should Calcified Chronic Total Occlusions Be Treated?

1. **Calcification can be deceiving**
   Although calcification can provide an outline of the vessel, it can also be deceiving: for example, the calcification may not be located within the occluded lumen, but may be present in the vessel wall.

2. **Wire advancement**
   Wire advancement and penetration can be challenging through calcified lesions. Use of stiff, tapered-tip guidewires (such as Confianza Pro 12) can help penetrate through the proximal cap. Alternative solutions include use of dissection/reentry techniques (both antegrade and retrograde) and use of the more torquable, composite-core guidewires, such as the Gaia family of wires (see Section 2.5.2).

3. **Reentry**
   Severe calcification may prevent reentry after subintimal guidewire advancement. Selecting a different reentry area (more proximal or distal to calcification—bobsled technique, see Fig. 3.25 in Chapter 3) or using the retrograde approach are potential solutions. Successful puncture of a calcified reentry zone into the true lumen may require a stiff, tapered guidewire such as a Confianza Pro 12 or Astato 20 rather than the standard Stingray wire, followed by swapping to a Pilot 200 to wire the distal vessel.

4. **Equipment delivery**
   Severe calcification may prevent equipment advancement. Careful lesion preparation (with high-pressure balloon inflations and rotational/orbital atherectomy) can assist with delivery, but additional techniques to increase support (such as use of guide catheter extensions) may be required. Atherectomy of the subintimal space should generally be avoided due to increased risk for perforation, but can be useful in highly selected cases (see Online Case 53).

5. **Stent expansion**
   Severe calcification may prevent stent expansion. Moreover, aggressive postdilation of severely calcified lesions may lead to perforation (Fig. 9.32).
   See Online Cases 17 and 83.

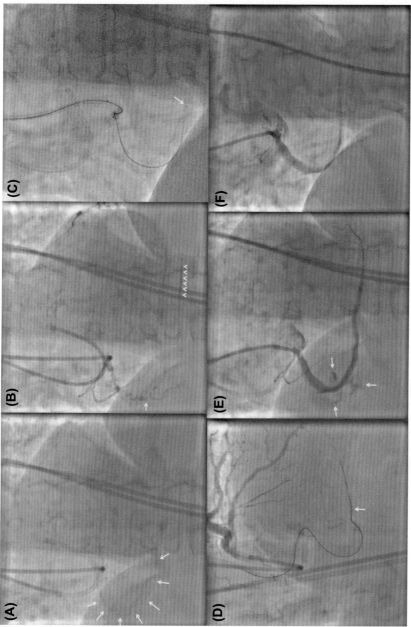

FIGURE 9.32   Percutaneous coronary intervention of a heavily calcified right coronary artery complicated by perforation (Online Case 17). (A) Severely calcified right coronary artery (*arrows*), cineangiography obtained without contrast injection). (B) Chronic total occlusion (CTO) of the mid-right coronary artery (*arrow*) with the right posterior descending artery filling via collaterals (*arrowheads*). (C) Subintimal crossing of the CTO with a CrossBoss catheter. (D) Successful reentry into the distal true lumen using a Stingray balloon (*arrow*). (E) Mid-right coronary artery perforation (*arrow*). (F) Sealing of the perforation after prolonged balloon inflations.

Prevention

1. Ensure that calcified lesions are predilated with noncompliant balloons, sized one to one with the vessel before implanting a stent.
2. If balloons fail to expand, additional lesion preparation (using Angiosculpt, cutting balloon, one or more buddy wires, and rotational or orbital atherectomy) should be performed. Stents should not be implanted until after adequate lesion expansion is achieved.

Treatment

If stent underexpansion is detected after stent placement, several maneuvers can be performed, as outlined in Section 8.2.

9.10 HEART FAILURE

Patients with heart failure who require CTO PCI may benefit from preprocedural optimization, intraprocedural monitoring, and use of hemodynamic support.

9.10.1 Preprocedural Optimization and Planning

Optimizing the patient's hemodynamic status before CTO PCI (e.g., with aggressive diuresis) can help reduce the risk for intraprocedural hemodynamic deterioration.

Working together with the advanced heart failure team can be invaluable for coordinating care, selecting the need and type of hemodynamic support, and determining bailout options (such as ventricular assist devices) if a complication or hemodynamic decompensation occurs during the procedure.

9.10.2 Intraprocedural Monitoring

Performing right heart catheterization before (and during) CTO PCI can help manage the patient's hemodynamic status (e.g., help determining if the patient needs intravenous fluid administration or diuresis) and allow early detection of adverse hemodynamic changes (see Online Cases 11 and 85).

9.10.3 Hemodynamic Support

See Online Cases 11, 14, 20, 29, 31, 46, 51, 67, and 89. Use of hemodynamic support can reduce the risk for hemodynamic collapse and improve the safety of the procedure in high-risk patients. The decision on when to use hemodynamic support focus around coronary anatomy and the hemodynamic status of the patients should consider:

1. **Coronary anatomy**
   Support is generally recommended for challenging anatomy, especially with large areas of myocardium at risk, such as PCI through the last remaining vessel, PCI of the left main in a left-dominant system (see Online Case 67) or

in patients with occluded RCAs, use of the retrograde approach (especially when retrograde crossing of a left internal mammary artery is performed, see Online Case 46), and use of atherectomy.

2. **Hemodynamic status**

Factors favoring use of hemodynamic support include low ejection fraction, increased pulmonary capillary wedge pressure, severe mitral regurgitation, and severe pulmonary hypertension.

Use of hemodynamic support during CTO PCI can be associated with several unique challenges:

1. **Arterial access**

All hemodynamic support devices require a large (13–18 Fr) arterial access (except for the intraaortic balloon pump, which requires 8 Fr access). Such access is commonly obtained from the femoral artery. In patients with severe peripheral arterial disease, peripheral intervention can be performed to facilitate insertion of a hemodynamic support device, subclavian access can be obtained (percutaneously or surgically), or transcaval access can be used. CTO PCI can then be performed using femoral and radial access, biradial access, or by doing a double stick of the contralateral femoral artery.

2. **Overcrowding in the aorta**

When the Impella device is used for hemodynamic support its location in the aortic root may interfere with guide engagement of the coronary arteries. In general the Impella device is inserted first, followed by the RCA guide catheter, leaving insertion of the left coronary artery guide catheter for the end. If the Impella is affecting the guide catheter position, it may be better to reposition the guide catheter before CTO PCI is started.

9.11 PRIOR FAILURE

Prior failure is one of the adverse predictors for being able to cross a CTO within 30 min in the J-CTO score.[58] However, the importance of the prior failure depends on the experience of the operator who attempted the initial procedure and on whether a complication occurred necessitating stopping the procedure. In the PROGRESS-CTO registry technical and procedural successes were similar for never-attempted and for previously-failed cases.[59]

Important considerations for previously-failed cases include:

a. Radiation dose. If large (>5–7 Gy) air kerma radiation dose was used, the skin should be examined for radiation-related rash and repeat CTO PCI attempts should be delayed for 2–3 months to minimize the risk for radiation skin injury.

b. Mechanism of failure. Understanding the mechanism of failure (for example, formation of subintimal hematoma) can help optimally select the subsequent crossing strategy (e.g., primary retrograde crossing in patient in whom antegrade crossing attempts failed).

REFERENCES

1. Karatasakis A, Danek BA, Karmpaliotis D, et al. Impact of proximal cap ambiguity on outcomes of chronic total occlusion percutaneous coronary intervention: insights from a multicenter US registry. *J Invasive Cardiol* 2016;**28**:391–6.
2. Brilakis ES, Grantham JA, Rinfret S, et al. A percutaneous treatment algorithm for crossing coronary chronic total occlusions. *JACC Cardiovasc Interv* 2012;**5**:367–79.
3. Opolski MP, Achenbach S. CT angiography for revascularization of CTO: crossing the borders of diagnosis and treatment. *JACC Cardiovasc Imaging* 2015;**8**:846–58.
4. Dautov R, Abdul Jawad Altisent O, Rinfret S. Stumpless chronic total occlusion with no retrograde option: multidetector computed tomography-guided intervention via bi-radial approach utilizing bioresorbable vascular scaffold. *Catheter Cardiovasc Interv* 2015;**86**:E258–62.
5. Opolski MP, Debski A, Borucki BA, et al. First-in-Man computed tomography-guided percutaneous revascularization of coronary chronic total occlusion using a wearable computer: proof of concept. *Can J Cardiol* 2016;**32**:e11–3.
6. Ghoshhajra BB, Takx RA, Stone LL, et al. Real-time fusion of coronary CT angiography with x-ray fluoroscopy during chronic total occlusion PCI. *Eur Radiol* 2017;**27**:2464–73.
7. Galassi AR, Sumitsuji S, Boukhris M, et al. Utility of intravascular ultrasound in percutaneous revascularization of chronic total occlusions: an overview. *JACC Cardiovasc Interv* 2016;**9**:1979–91.
8. Karacsonyi J, Alaswad K, Jaffer FA, et al. Use of intravascular imaging during chronic total occlusion percutaneous coronary intervention: insights from a contemporary multicenter registry. *J Am Heart Assoc* 2016:5, pii:e003890.
9. Vo MN, Karmpaliotis D, Brilakis ES. "Move the cap" technique for ambiguous or impenetrable proximal cap of coronary total occlusion. *Catheter Cardiovasc Interv* 2016;**87**:742–8.
10. Carlino M, Ruparelia N, Thomas G, et al. Modified contrast microinjection technique to facilitate chronic total occlusion recanalization. *Catheter Cardiovasc Interv* 2016;**87**:1036–41.
11. Amsavelu S, Carlino M, Brilakis ES. Carlino to the rescue: use of intralesion contrast injection for bailout antegrade and retrograde crossing of complex chronic total occlusions. *Catheter Cardiovasc Interv* 2016;**87**:1118–23.
12. Danek BA, Karatasakis A, Brilakis ES. Consequences and treatment of guidewire entrapment and fracture during percutaneous coronary intervention. *Cardiovasc Revasc Med* 2016;**17**:129–33.
13. Benko A, Berube S, Buller CE, et al. Novel crossing system for chronic total occlusion recanalization: first-in-man experience with the SoundBite crossing system. *J Invasive Cardiol* 2017;**29**:E17–20.
14. Strauss BH, Osherov AB, Radhakrishnan S, et al. Collagenase Total Occlusion-1 (CTO-1) trial: a phase I, dose-escalation, safety study. *Circulation* 2012;**125**:522–8.
15. Nicholson W, Harvey J, Dhawan R. E-CART (ElectroCautery-Assisted Re-entry) of an aorto-ostial right coronary artery chronic total occlusion: first-in-man. *JACC Cardiovasc Interv* 2016;**9**:2356–8.
16. Christopoulos G, Karmpaliotis D, Wyman MR, et al. Percutaneous intervention of circumflex chronic total occlusions is associated with worse procedural outcomes: insights from a multicentre US registry. *Can J Cardiol* 2014;**30**:1588–94.
17. Nguyen-Trong PK, Rangan BV, Karatasakis A, et al. Predictors and outcomes of side-branch occlusion in coronary chronic total occlusion interventions. *J Invasive Cardiol* 2016;**28**:168–73.
18. Jang WJ, Yang JH, Choi SH, et al. Association of periprocedural myocardial infarction with long-term survival in patients treated with coronary revascularization therapy of chronic total occlusion. *Catheter Cardiovasc Interv* 2016;**87**:1042–9.

19. Lo N, Michael TT, Moin D, et al. Periprocedural myocardial injury in chronic total occlusion percutaneous interventions: a systematic cardiac biomarker evaluation study. *JACC Cardiovasc Interv* 2014;**7**:47–54.

20. Saito S. Open Sesame Technique for chronic total occlusion. *Catheter Cardiovasc Interv* 2010;**75**:690–4.

21. Chiu CA. Recanalization of difficult bifurcation lesions using adjunctive double-lumen micro-catheter support: two case reports. *J Invasive Cardiol* 2010;**22**:E99–103.

22. Iturbe JM, Abdel-Karim AR, Raja VN, Rangan BV, Banerjee S, Brilakis ES. Use of the venture wire control catheter for the treatment of coronary artery chronic total occlusions. *Catheter Cardiovasc Interv* 2010;**76**:936–41.

23. Kawasaki T, Koga H, Serikawa T. New bifurcation guidewire technique: a reversed guide-wire technique for extremely angulated bifurcation–a case report. *Catheter Cardiovasc Interv* 2008;**71**:73–6.

24. Suzuki G, Nozaki Y, Sakurai M. A novel guidewire approach for handling acute-angle bifurca-tions: reversed guidewire technique with adjunctive use of a double-lumen microcatheter. *J Invasive Cardiol* 2013;**25**:48–54.

25. Michael T, Banerjee S, Brilakis ES. Distal open sesame and hairpin wire techniques to facilitate a chronic total occlusion intervention. *J Invasive Cardiol* 2012;**24**:E57–9.

26. Christopoulos G, Karmpaliotis D, Alaswad K, et al. The efficacy of "hybrid" percutaneous coronary intervention in chronic total occlusions caused by in-stent restenosis: insights from a US multicenter registry. *Catheter Cardiovasc Interv* 2014;**84**:646–51.

27. Azzalini L, Dautov R, Ojeda S, et al. Procedural and long-term outcomes of percutaneous coronary intervention for in-stent chronic total occlusion. *JACC Cardiovascular Interventions* 2017;**10**:892–902.

28. Abbas AE, Brewington SD, Dixon SR, Boura J, Grines CL, O'Neill WW. Success, safety, and mechanisms of failure of percutaneous coronary intervention for occlusive non-drug-eluting in-stent restenosis versus native artery total occlusion. *Am J Cardiol* 2005;**95**:1462–6.

29. Yang YM, Mehran R, Dangas G, et al. Successful use of the frontrunner catheter in the treat-ment of in-stent coronary chronic total occlusions. *Catheter Cardiovasc Interv* 2004;**63**:462–8.

30. Ho PC. Treatment of in-stent chronic total occlusions with blunt microdissection. *J Invasive Cardiol* 2005;**17**:E37–9.

31. Lee NH, Cho YH, Seo HS. Successful recanalization of in-stent coronary chronic total occlu-sion by subintimal tracking. *J Invasive Cardiol* 2008;**20**:E129–32.

32. Werner GS, Moehlis H, Tischer K. Management of total restenotic occlusions. *EuroIntervention* 2009;**5**(Suppl. D):D79–83.

33. Brilakis ES, Lombardi WB, Banerjee S. Use of the Stingray guidewire and the Venture cath-eter for crossing flush coronary chronic total occlusions due to in-stent restenosis. *Catheter Cardiovasc Interv* 2010;**76**:391–4.

34. Abdel-karim AR, Lombardi WB, Banerjee S, Brilakis ES. Contemporary outcomes of percuta-neous intervention in chronic total coronary occlusions due to in-stent restenosis. *Cardiovasc Revasc Med* 2011;**12**:170–6.

35. de la Torre Hernandez JM, Rumoroso JR, Subinas A, et al. Percutaneous intervention in chronic total coronary occlusions caused by in-stent restenosis. Procedural results and long term clini-cal outcomes in the TORO (Spanish registry of chronic Total occlusion secondary to an occlu-sive in stent Restenosis) multicenter registry. *EuroIntervention* 2017;**13**:e219–26.

36. Rinfret S, Ribeiro HB, Nguyen CM, Nombela-Franco L, Urena M, Rodes-Cabau J. Dissection and re-entry techniques and longer-term outcomes following successful percutaneous coronary intervention of chronic total occlusion. *Am J Cardiol* 2014;**114**:1354–60.

37. Ohya H, Kyo E, Katoh O. Successful bypass restenting across the struts of an occluded subintimal stent in chronic total occlusion using a retrograde approach. *Catheter Cardiovasc Interv* 2013;**82**:E678–83.

38. Quevedo HC, Irimpen A, Abi Rafeh N. Succesful antegrade subintimal bypass restenting of in-stent chronic total occlusion. *Catheter Cardiovasc Interv* 2015;**86**:E268–71.

39. Roy J, Lucking A, Strange J, Spratt JC. The difference between success and failure: subintimal stenting around an occluded stent for treatment of a chronic total occlusion due to in-stent restenosis. *J Invasive Cardiol* 2016;**28**:E136–8.

40. Tasic M, Sreckovic MJ, Jagic N, Miloradovic V, Nikolic D. Knuckle technique guided by intravascular ultrasound for in-stent restenosis occlusion treatment. *Postepy Kardiol Interwencyjnej* 2015;**11**:58–61.

41. Capretti G, Mitomo S, Giglio M, Carlino M, Colombo A, Azzalini L. Subintimal crush of an occluded stent to recanalize a chronic total occlusion due to in-stent restenosis: insights from a multimodality imaging approach. *JACC Cardiovasc Interv* 2017;**10**:e81–3.

42. Papayannis A, Banerjee S, Brilakis ES. Use of the Crossboss catheter in coronary chronic total occlusion due to in-stent restenosis. *Catheter Cardiovasc Interv* 2012;**80**:E30–6.

43. Wilson WM, Walsh S, Hanratty C, et al. A novel approach to the management of occlusive in-stent restenosis (ISR). *EuroIntervention* 2014;**9**:1285–93.

44. Ntatsios A, Smith WHT. Exit of CrossBoss between stent struts within chronic total occlusion to subintimal space: completion of case via retrograde approach with rendezvous in coronary. *J Cardiol Cases* 2014;**9**:183–6.

45. Maeremans J, Dens J, Spratt JC, et al. Antegrade dissection and reentry as part of the hybrid chronic total occlusion revascularization strategy: a subanalysis of the RECHARGE registry (Registry of CrossBoss and Hybrid Procedures in France, the Netherlands, Belgium and United Kingdom). *Circ Cardiovasc Interv* 2017;**10**.

46. Sapontis J, Grantham JA, Marso SP. Excimer laser atherectomy to overcome intraprocedural obstacles in chronic total occlusion percutaneous intervention: case examples. *Catheter Cardiovasc Interv* 2015;**85**:E83–9.

47. Levine GN, Bates ER, Blankenship JC, et al. 2011 ACCF/AHA/SCAI guideline for percutaneous coronary intervention. A report of the American College of Cardiology Foundation/American heart association Task force on Practice guidelines and the Society for Cardiovascular angiography and interventions. *J Am Coll Cardiol* 2011;**58**:e44–122.

48. Brilakis E, Banerjee S, Lombardi W. Retrograde recanalization of native coronary artery chronic occlusions via acutely occluded vein grafts. *Catheter Cardiovasc Interv* 2010;**75**:109–13.

49. Katoh H, Nozue T, Michishita I. A case of giant saphenous vein graft aneurysm successfully treated with catheter intervention. *Catheter Cardiovasc Interv* 2016;**87**:83–9.

50. Nguyen-Trong PK, Alaswad K, Karmpaliotis D, et al. Use of saphenous vein bypass grafts for retrograde recanalization of coronary chronic total occlusions: insights from a multicenter registry. *J Invasive Cardiol* 2016;**28**:218–24.

51. Dautov R, Manh Nguyen C, Altisent O, Gibrat C, Rinfret S. Recanalization of chronic total occlusions in patients with previous coronary bypass surgery and consideration of retrograde access via saphenous vein grafts. *Circ Cardiovasc Interv* 2016:9.

52. Sachdeva R, Uretsky BF. Retrograde recanalization of a chronic total occlusion of a saphenous vein graft. *Catheter Cardiovasc Interv* 2009;**74**:575–8.

53. Takano M, Yamamoto M, Mizuno K. A retrograde approach for the treatment of chronic total occlusion in a patient with acute coronary syndrome. *Int J Cardiol* 2007;**119**:e22–4.

54. Ho PC, Tsuchikane E. Improvement of regional ischemia after successful percutaneous intervention of bypassed native coronary chronic total occlusion: an application of the CART technique. *J Invasive Cardiol* 2008;**20**:305–8.

55. Brilakis ES, Grantham JA, Thompson CA, et al. The retrograde approach to coronary artery chronic total occlusions: a practical approach. *Catheter Cardiovasc Interv* 2012;**79**:3–19.
56. Garg N, Hakeem A, Gobal F, Uretsky BF. Outcomes of percutaneous coronary intervention of chronic total saphenous vein graft occlusions in the contemporary era. *Catheter Cardiovasc Interv* 2014;**83**:1025–32.
57. Debski A, Tyczynski P, Demkow M, Witkowski A, Werner GS, Agostoni P. How should I treat a chronic total occlusion of a saphenous vein graft? Successful retrograde revascularisation. *EuroIntervention* 2016;**11**:e1325–8.
58. Morino Y, Abe M, Morimoto T, et al. Predicting successful guidewire crossing through chronic total occlusion of native coronary lesions within 30 minutes: the J-CTO (Multicenter CTO Registry in Japan) score as a difficulty grading and time assessment tool. *JACC Cardiovasc Interv* 2011;**4**:213–21.
59. Karacsonyi J, Karatasakis A, Karmpaliotis D, et al. Effect of previous failure on subsequent procedural outcomes of chronic total occlusion percutaneous coronary intervention (from a contemporary multicenter registry). *Am J Cardiol* 2016;**117**:1267–71.

Chapter 10

Radiation Management During Chronic Total Occlusion Percutaneous Coronary Intervention

Radiation skin injury (Fig. 10.1) is a rare complication of any invasive cardiac procedure, but is more likely to occur in the setting of complex procedures, such as chronic total occlusion (CTO) percutaneous coronary intervention (PCI), where large doses of radiation are often used. Radiation skin injury can lead to severe consequences for the patient, such as painful, nonhealing ulcers that may require months or even years to heal and in some cases may even require surgical debridement and plastic reconstruction.

A plan for radiation dose management from the outset of all PCI cases, especially CTO PCI, is essential. Such a plan will not only lead to a lower radiation dose to the patient, but it will also reduce physician and staff dosing. Interventional cardiologists and staff are exposed to ionizing radiation on a daily basis over many years, which can increase their risk for developing cancer (such as, but not limited to, left-sided brain tumors[1]), cataracts, and other ailments, as well as orthopedic problems associated with protective garments.[2]

Despite the obvious benefits in of limiting radiation exposure, observations from multiple cardiac catheterization laboratories have shown that sound radiation management practices are infrequently implemented,[3] although progress has been documented by some programs.[4–6] The goal of this chapter is to provide simple and practical tips and tricks for reducing both patient and operator radiation exposure.[2,7–9]

10.1 WHY RADIATION MANAGEMENT IS IMPORTANT

1. To prevent radiation injury to the patient.
2. To prevent radiation injury to the operator and the cardiac laboratory staff.
3. To prevent medico–legal consequences, since significant radiation exposure (>15 Gy air kerma (AK) dose) is considered a sentinel event by the Joint Commission for Hospital Accreditation.

Manual of Chronic Total Occlusion Interventions. http://dx.doi.org/10.1016/B978-0-12-809929-2.00010-7

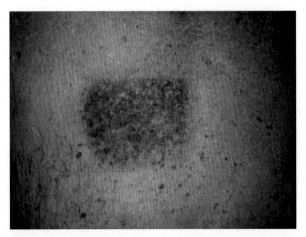

FIGURE 10.1 Example of radiation-induced skin injury after chronic total occlusion (CTO) percutaneous coronary intervention (PCI). Erythema and epilation developed on the patient's back 1 month after CTO PCI, during which he received 11.8 Gy air kerma dose. *Reproduced with permission from Chambers CE. Radiation dose in percutaneous coronary intervention OUCH did that hurt?* JACC Cardiovasc Interv *2011;4:344–6.*

4. Because there is increasing public and medical community concern about radiation exposure during medical procedures, regarding an individual procedure as well as the lifelong cumulative radiation exposure of patients.

10.2 ESSENTIALS OF RADIATION DOSE MANAGEMENT

It is recommended that operators wishing to develop a CTO (or any complex PCI) program consult with their institution's radiation officers to implement strict radiation management protocols and safe radiation management practices. It is then essential that these protocols, in conjunction with appropriate established thresholds, be incorporated into the cath lab quality assurance/quality improvement program.

> There are two ways to minimize radiation during CTO PCI procedures:
> 1. By acquiring the skill sets and expertise to perform safe and efficient CTO PCI procedures (as described throughout this text).
> 2. By implementing safe radiation management practices.

The rest of this chapter focuses on radiation management practices.

1. **Dose Assessment: Understand how radiation is measured in the cardiac cath lab and which radiation measure should be looked at.**
   Assessment of radiation dose in the cardiac cath lab is much more than fluoroscopy time (FT, measured in minutes). FT has several limitations, the

FIGURE 10.2 Screenshot from an X-ray machine screen, highlighting the fluoroscopy frames per second (*yellow box*: 15 fps in (A) vs. 7.5 fps in (B)). Also, the air kerma (494 mGy or 0.494 Gy) and the dose area product radiation dose are *highlighted*.

most obvious of which is its failure to include cine imaging; hence, FT alone is not adequate to assess patient radiation dose. The actual administered radiation dose depends on many other factors, such as the weight of the patient, the use of collimation, the positioning of the table and image intensifier, and the imaging angles. For this reason, since 2006, all fluoroscopic equipment sold in the United States have additional parameters to measure patient dose that are recorded and displayed during the procedure.

There are two standard parameters reported on interventional fluoroscopic equipment: cumulative AK at the interventional reference point (measured in Gray [Gy]) and dose area product (DAP, measured in Gycm², also called AK area product) (Fig. 10.2). DAP is used to monitor the potential for genetic defects or cancer risk over time, called stochastic effects, and is not used for intraprocedural radiation dose monitoring in the United States.

The AK dose is the number that the CTO operator should constantly monitor to determine the risk that the patient will develop radiation skin injury and adjust the procedural plan accordingly. Total AK is the procedural cumulative X-ray energy delivered to air at the interventional reference point (i.e., 15 cm on the X-ray tube side of isocenter), the point at which the primary X-ray beam intersects with the rotational axis of the C-arm gantry.[7] Kerma stands for Kinetic Energy Released in Matter. Although the AK dose is an approximation of the actual radiation that the patient receives during a procedure, it is a far better and physiologically relevant index as compared with FT. Deterministic radiation effects, such as skin injury, correlate directly with the AK dose to a particular skin area (Fig. 10.1).

The following AK dose thresholds are important to remember[3]:

>5 Gy: below this threshold skin injury is unlikely to occur.

>10 Gy: skin injury is likely, requiring physicist assessment of the case.

>15 Gy: considered a sentinel event by the Joint Commission for Hospital Accreditation and requires reporting to the regulatory authorities in the United States.

---

2. Laboratory Environment

All cardiac cath labs should have a radiation safety program with active participation of physicians, staff, and physicists with reports regularly reviewed in the cath lab QA program. All interventional cardiologists should apply two basic principles of radiation protection to their practice: reduce radiation exposure to as low as reasonably achievable (ALARA), and ensure procedure justification, such that no patient receives radiation without potential benefit.

Although only certain states mandate fluoroscopy training, it is important that everyone receives radiation dose management and safety training commensurate to their responsibilities. The National Council on Radiation Protection recommends both didactic and hands-on training. The didactic program should include initial training with periodic updates covering the topics of radiation physics and safety. Hands-on training should be provided for newly hired operators and all operators on newly purchased equipment.

It is the individual's responsibility to wear a dosimeter. Although a single dosimeter worn outside the collar can be used, two properly worn dosimeters—one at the waist under, and one at the collar outside, the protective garment—provide a better reflection of effective dose. However, one dosimeter worn correctly at collar height externally is better than two worn incorrectly. Protective garments stop approximately 95% of the scatter radiation. Radiation glasses must fit properly, have 0.25-mm lead equivalent protection, and have additional side shielding. Ceiling-mounted and below-table shielding are also effective; both should be used routinely.

Current fluoroscopic X-ray systems offer features for dose management including frame rate adjustment, virtual collimation, last image hold, X-ray store, and real-time dose display.[10,11] Image quality is a function of multiple patient, procedural, and equipment variables. As a general rule, image quality and radiation dose are tightly woven with higher dose often improving image quality: achieving the acceptable image quality for a procedure at the lowest dose is key. Automatic dose rate controls increase dose for a specific patient size in a specific projection to achieve adequate image quality. Knowing the equipment and working with a qualified physicist are essential for dose optimization. Also several X-ray equipment vendors are willing to work with hospitals to optimize the settings of the installed systems to reduce radiation dose.

3. **Procedure-Based Radiation Dose Management**

Table 10.1 provides a procedure-based dose management outline. Preprocedure planning is an essential component of radiation dose management. It is important to detect factors that place patients at high risk for radiation-induced skin injury, such as obesity or recent fluoroscopic procedures within the previous 30–60 days. Informed consent for CTO PCI should include radiation safety information.

TABLE 10.1 Procedure-Based Case Management of Radiation Dose

I. Preprocedure
    A. Radiation safety program for catheterization lab
    1. Dosimeter use, shielding, training/education
    B. Imaging equipment and operator knowledge
    1. On-screen dose assessment (air kerma, dose area product)
    2. Dose saving: Store fluoroscopy, adjustable pulse and frame rate and last image hold
    C. Preprocedure dose planning
    1. Assess patient and procedure, including patient's size and lesion(s) complexity; examine patient for potential skin injury from prior high-dose cases
    D. Informed patient with appropriate consent
II. Procedure
    A. Limit fluoroscopy: Step on pedal only when looking at screen
    B. Limit cine: Store fluoroscopy when high image quality is not required
    C. Limit magnification, frame rate, steep angles
    D. Use collimation and filters to the fullest extent possible
    E. Vary tube angle when possible to change skin area exposed
    F. Position table and image receptor: X-ray tube too close to patient increases dose; high image receptor increases scatter
    G. Keep patient and operator body parts out of field of view
    H. Maximize shielding and distance from X-ray source for all personnel
    I. Manage and monitor dose in real time from beginning of case
III. Postprocedure
    A. Document radiation dose in records (fluoroscopy time, $K_{a,r}$, P_{KA})
    B. Notify patient and referring physician when high dose delivered
    1. $K_{a,r} > 5\,Gy$, chart document; inform patient; arrange follow up
    2. $K_{a,r} > 10\,Gy$, qualified physicist should calculate skin dose
    3. PSD $> 15\,Gy$, Joint Commission Sentinel Event
    C. Assess and refer adverse skin effects to appropriate consultant

$K_{a,r}$, total air kerma at reference point; P_{KA}, air kerma area product; *PSD*, peak skin dose.
Modified from Chambers CE. Radiation dose in percutaneous coronary intervention OUCH did that hurt? *JACC Cardiovasc Interv* 2011;4:344–6.

The following tips and tricks can assist with intraprocedural radiation dose management (many of which are summarized in Time, Intensity, Distance, and Shielding):

A. Carefully monitor radiation exposure throughout the procedure.

First, monitor the radiation dose rate: if the radiation dose rate is high (>10–15 mGy per min), the operator should consider changing the working projection, angulation, and positioning of image intensifier to reduce the radiation dose rate.

Second, monitor the cumulative AK: the operator should consider stopping the procedure if 7–8 Gy AK dose is administered without the procedure nearly completed. It is recommended that each cath lab have a protocol for alerting the operator on radiation (e.g., announcing the dose used every 1000 mGy and/or every 30 min).

B. Do *not* place your hands in the direct radiation beam!

Operator and staff must maximize their distance from the X-ray tube (i.e., the inverse square law), which is of particular importance for radial access cases. All appendages—operators' and patients'—should be out of the imaging field!

If there are challenges while obtaining access, it is best to remove the operator's hands from the directly imaged field while X-ray imaging is on (Fig. 10.3). A device is available for obtaining access while keeping the physician's hands away from the peak radiation zone during puncture (Quick-Access Needle Holder, Spectranetics; Fig. 10.4). It is also important to exclude the patient's arm from the radiation field, since it

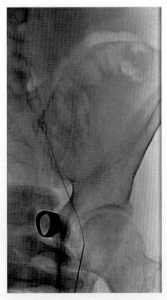

FIGURE 10.3 Example of what should *not* be done while obtaining access (i.e., placing the hand under the direct X-ray beam). *Reproduced from Brilakis ES, Patel VG. What you can't see can hurt you!* J Invasive Cardiol *2012;24:421 with permission from HMP Communications.*

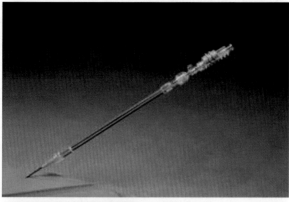

FIGURE 10.4   Image of Quick-Access Needle Holder. *Reproduced with permission from Spectranetics.*

increases the radiation delivered, as well as the risk of radiation injury to the patient's upper extremity.

C. **Minimize fluoroscopy frame rate**.

Most modern X-ray equipment (GE, Toshiba, Siemens, Shimadzu, and Philips) allow the operator to change the fluoroscopy frames per second (fps). Most machines default to 15 fps for cardiac procedures. Decreasing the fluoroscopy fps from 15 to 7.5 or 6 fps (Fig. 10.2) cuts fluoroscopy-related radiation in half or more. Although the images obtained at 7.5 fps are less pristine than those obtained at 15 fps, they are adequate for most if not all CTO PCI maneuvers. Many operators currently use 6 or 7.5 fps fluoroscopy for all cardiac procedures, not just CTO PCI. It is easiest for operators to get familiar with the 6 or 7.5 fps mode during conventional PCI procedures before adopting this practice in the more complex CTO PCI procedures.

D. **Do *not* step on the fluoroscopy pedal when not looking at the screen**.

Although this appears self-evident, it is amazing how often this simple principle is ignored! There is high prevalence of the "heavy-foot syndrome" (i.e., continuing to perform fluoroscopy when it is not needed).

E. Use techniques that limit fluoroscopy.

An example is the trapping technique for over-the-wire equipment exchanges, regardless of whether short or long (300 cm) wires are used (Section 3.7). Another example is using a torquer to mark the length of the wire that can be safely advanced through a microcatheter without exiting into the vessel, obviating the need for fluoroscopy during this maneuver.

F. Use techniques that limit cine-angiography.

Cine-angiography exposes the patient to 10× higher dose compared with fluoroscopy and is not reflected in the FT. The image store or fluoro save function is available in most modern X-ray equipment and should be used instead of cine to document balloon and stent inflations. Also, when performing dual injection (Section 3.2) the CTO collateral donor vessel can be injected before starting cine recording, as it takes time for the contrast to fill the CTO distal true lumen.

G. Use low magnification.

Lower magnification requires less radiation exposure. Similar to using 7.5 fps, using lower magnification requires a learning curve to adjust to the change in image size. Some X-ray equipment (Toshiba) have software for virtual magnification, which allows for images obtained in low magnification to be magnified and displayed on the screen as larger images. Although the angiographic definition is less sharp using this method, it is still adequate for most cases.

H. Use collimation.

Collimation reduces the size of the skin area exposed to radiation and reduces the overall DAP dose received by the patient, even though the total AK is not changed. Collimation allows a smaller skin exposure in any one projection, lessening potential skin injury from overlapping exposures when imaging angles are changed.

A caveat of collimation is that some equipment (for example the tips of guidewires or guide catheters) may not be included in the field of view, requiring intermittent monitoring to ensure that no significant changes have occurred (for example excessive distal migration of a guidewire that can lead to distal vessel perforation or deep engagement of the guide catheter that can lead to aortocoronary dissection).

I. Frequently rotate the C arm/X-ray tube to minimize the exposure to any one skin area.

Using multiple angles during fluoroscopy and cine-angiography is critical during long procedures to minimize radiation exposure to the same entry point. High radiation dose procedures may not be as deleterious if the radiation is applied to multiple areas of skin, because the dose to each particular area of skin is reduced.

J. Optimize the position of the table (as high as possible) and the image intensifier (as close to the patient as possible) (Fig. 10.5).

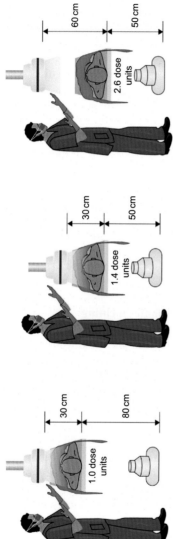

FIGURE 10.5   Example of optimal table positioning to minimize patient (and operator) radiation exposure. **Left:** the physician performs the procedure with the patient table elevated and the image intensifier close to the patient (total distance from the X-ray tube to the detector=110 cm). **Center:** the physician employs a lower table setting but maintains the image intensifier close to the patient's chest (total distance from the X-ray tube to the detector=80 cm). Because of the closer proximity to the X-ray tube, the dose rate to the patient at the beam entrance port will be about 40% higher. **Right:** the physician employs a low table height but has elevated the image intensifier (total distance from the X-ray tube to the detector=110 cm). The skin dose to the patient on the right is 160% higher than the patient on the left. (The image generated by the configuration on the right is 40%–50% larger owing to geometric magnification caused by the elevated image intensifier.) If the procedure on the left required a 3 Gy skin dose, the same procedure employing the center configuration would result in a 4.2 Gy dose, whereas the one performed employing the configuration on the right would result in 7.8 Gy. *Reproduced with permission from Hirshfeld Jr JW, Balter S, Brinker JA, et al. ACCF/AHA/HRS/SCAI clinical competence statement on physician knowledge to optimize patient safety and image quality in fluoroscopically guided invasive cardiovascular procedures. A report of the American College of Cardiology Foundation/American Heart Association/American College of Physicians Task Force on Clinical Competence and Training. J Am Coll Cardiol 2004;44:2259–82.*

K. Avoid steep angles of the image intensifier.

When performing CTO PCI the working angle should be minimized: steep angles, for example greater than 30 degrees from anteroposterior (AP), are associated with significantly higher radiation exposure due to the penetration through more tissue; less steep angles are preferred. This increased dose is often not recognized by the operator and reflects the automatic dose increase by the equipment to maintain image quality. The AP may not be the optimal projection, because the spine is included in the field; angulated views are preferred.

The right anterior oblique projection can result in less radiation exposure, but can be challenging for mid right coronary artery wiring, although it is excellent for wiring septal collaterals or working in the mid left anterior descending artery.

L. Use the X-ray stand position memory.

X-ray machines can store multiple stand positions in memory and can automatically move to a selected position on command. This enables the operator to avoid the use of fluoroscopy to achieve a desired stand position; however, this might increase the dose of the X-ray to two or more skin areas under the exact same angles stored in the memory.[12]

M. Use radiation monitoring devices that provide real-time feedback on operator radiation exposure.

The use of one such system in the Radicure study resulted in reduction in operator dose by approximately 30%.[13] An example of one such device is shown in Fig. 10.6.

N. Use all available above- and below-table shielding.

In addition to standard ceiling-attached and personal lead protection, adding a 1.0-mm lead flap below the lead glass and lead top along the undercouch lead shield (Fig. 10.7) can effectively reduce radiation exposure to the operator to <1% of typical levels by reducing scatter leakages. Also, the use of lead caps might reduce head radiation dose,[14] which is of particular concern given reports of higher frequency of left-sided brain tumors among interventionalists.[1]

Additional shielding can be placed in the sterile field to reduce scatter radiation, such as the RadPad shields (Worldwide Innovations & Technologies) (Section 2.11), which should be placed over the patient's abdomen.[15]

New radiation protection systems such as the Zero Gravity (CFI Medical Solutions, Fenton, MI) can further reduce or eliminate operator radiation exposure, while also obviating the adverse orthopedic consequences of wearing lead aprons. Robotic PCI (CorPath, Corindus) allows near elimination of operator radiation dose,[16] but its use for CTO PCI has been limited to date.

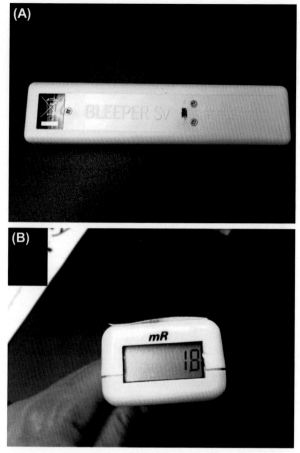

FIGURE 10.6 Example of a radiation monitoring device that provides real-time auditory feedback to the operator on the level of radiation exposure.

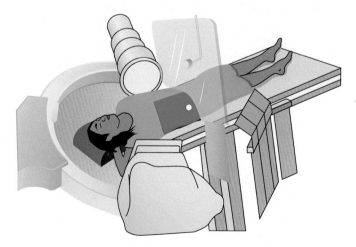

FIGURE 10.7 Position of undercouch top, flap below the glass sheet, and lead shield on the patient's abdomen. *Reproduced with permission from Kuon E, Schmitt M, Dahm JB. Significant reduction of radiation exposure to operator and staff during cardiac interventions by analysis of radiation leakage and improved lead shielding. Am J Cardiol 2002;89:44–9 (Fig. 2 of the paper).*

O. Always examine the patient's back prior to starting a CTO PCI procedure.

Patients with coronary CTOs often have had multiple prior diagnostic or interventional procedures. They tend to have extensive atherosclerotic disease and may have had several prior non-CTO PCIs. In large CTO referral centers it is not uncommon to treat patients with previously failed CTO PCI attempts that might have resulted in large amounts of radiation exposure. Performing a CTO PCI procedure in a patient with an unrecognized preexisting radiation burn can be catastrophic.

P. Use additional precautions on patients with prior radiation-induced skin injury that need repeat procedures.

Among patients with prior radiation-induced dermatitis, avoidance of repeat exposure is preferred although it may be unavoidable in some clinical circumstances. In such instances, methods to avoid and/or limit focal radiation exposure to the injured site are essential. A radiation shield may be constructed from a commercially available shielding product (RadPad, Worldwide Innovations & Technologies). Specifically, the shield is tailored to the area of skin injury by cutting the shield and placing it over the skin at the site of previous injury (Fig. 10.8A and B). The shield will attenuate approximately 90% of direct beam radiation and permit a visible shade to the operator under fluoroscopic guidance that obscures visualization and requires alteration of the beam angle.

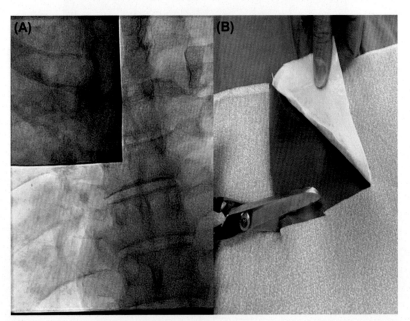

FIGURE 10.8 Example of a technique for performing repeat cardiac catheterization among patients with prior skin radiation injury. A piece of a disposable radiation shield is cut out (B) in the shape of the prior radiation injury skin area and placed over the injured skin area (*dark area on fluoroscopy as shown in* (A)) to minimize radiation exposure of the same area. *Courtesy of Dr. Karmpaliotis.*

4. Postprocedure: Follow-Up and Treatment

Cardiac catheterization reports should include all available radiation parameters: FT, AK dose, and DAP. Patient notification, chart documentation, and communication with the primary care provider should routinely occur following procedures with high radiation dose.

Patients who receive >5 Gy AK dose should be counseled about the possibility of radiation-induced skin injury on their back. Fig. 10.1

Patients who receive >5 Gy AK dose should be followed up within a month and their skin examination documented in the medical record. During physical examination the back of the patients should be inspected to detect any radiation injury; if such an injury is diagnosed, patients should be referred to specialists (dermatologist, plastic surgeons) for further evaluation and treatment. This is critical because if the patient is not aware of this risk, and develops erythema and local discomfort (Fig. 10.1), he/she may see a dermatologist who may biopsy the lesion, potentially leading to a nonhealing ulcer.

For >10 Gy AK dose, a qualified physicist should promptly calculate peak skin dose and the patient's skin should be examined at 2–4 weeks.

The Joint Commission identifies peak skin doses >15 Gy as a sentinel event; hospital risk management and regulatory agencies need to be contacted within 24 h.

10.3 CONCLUSIONS

Starting a CTO PCI program is a great opportunity to improve our radiation management habits, for all cardiac procedures. We strongly believe that radiation safety should be a top priority of a successful CTO PCI program. Once the appropriate routines are engrained, they become second nature and one performs them automatically.[3]

Radiation reduction is a win–win situation for both the patient and the operator. Investing the time and effort required to develop sound radiation reducing habits can provide high dividends for patients, physicians, and staff.[3] Every small step in radiation reduction helps, and in turn will allow the operator to spend more time opening the CTO to improve the patient's well-being, which is the main goal of the procedure.

The interventional cardiologist performing CTO PCI, as the person responsible for all aspects of patient care in the cath lab, must be actively involved in managing radiation dose to maximize patient and staff safety. Wilhelm Roentgen died from radiation-induced cancer because he did not know the devastating consequences of X-ray exposure. We do not have the same excuse: we should never forget that being unable to see an immediate effect of radiation on our body or the patients' bodies does not mean that we are safe.[3]

REFERENCES

1. Roguin A, Goldstein J, Bar O, Goldstein JA. Brain and neck tumors among physicians performing interventional procedures. *Am J Cardiol* 2013;**111**:1368–72.
2. Chambers CE, Fetterly KA, Holzer R, et al. Radiation safety program for the cardiac catheterization laboratory. *Catheter Cardiovasc Interv* 2011;**77**:546–56.
3. Brilakis ES, Patel VG. What you can't see can hurt you!. *J Invasive Cardiol* 2012;**24**:421.
4. Simard T, Hibbert B, Natarajan MK, et al. Impact of Center Experience on Patient Radiation Exposure During Transradial Coronary Angiography and Percutaneous Intervention: A Patient-Level, International, Collaborative, Multi-Center Analysis. *J Am Heart Assoc* 2016:5.
5. Werner GS, Glaser P, Coenen A, et al. Reduction of radiation exposure during complex interventions for chronic total coronary occlusions: implementing low dose radiation protocols without affecting procedural success rates. *Catheter Cardiovasc Interv* 89, 2017, 1005–12.
6. Olcay A, Guler E, Karaca IO, et al. Comparison of fluoro and cine coronary angiography: balancing acceptable outcomes with a reduction in radiation dose. *J Invasive Cardiol* 2015;**27**:199–202.
7. Chambers CE. Radiation dose in percutaneous coronary intervention OUCH did that hurt? *JACC Cardiovasc Interv* 2011;**4**:344–6.
8. Fetterly KA, Lennon RJ, Bell MR, Holmes Jr DR, Rihal CS. Clinical determinants of radiation dose in percutaneous coronary interventional procedures: influence of patient size, procedure complexity, and performing physician. *JACC Cardiovasc Interv* 2011;**4**:336–43.
9. Christopoulos G, Makke L, Christakopoulos G, et al. Optimizing radiation safety in the cardiac catheterization laboratory: a practical approach. *Catheter Cardiovasc Interv* 2016;**87**:291–301.
10. Ten Cate T, van Wely M, Gehlmann H, et al. Novel X-ray image noise reduction technology reduces patient radiation dose while maintaining image quality in coronary angiography. *Neth Heart J* 2015;**23**:525–30.
11. Christopoulos G, Christakopoulos GE, Rangan BV, et al. Comparison of radiation dose between different fluoroscopy systems in the modern catheterization laboratory: results from bench testing using an anthropomorphic phantom. *Catheter Cardiovasc Interv* 2015;**86**:927–32.
12. Hirshfeld Jr JW, Balter S, Brinker JA, et al. ACCF/AHA/HRS/SCAI clinical competence statement on physician knowledge to optimize patient safety and image quality in fluoroscopically guided invasive cardiovascular procedures. A report of the American College of Cardiology Foundation/American Heart Association/American College of Physicians Task Force on Clinical Competence and Training. *J Am Coll Cardiol* 2004;**44**:2259–82.
13. Christopoulos G, Papayannis AC, Alomar M, et al. Effect of a real-time radiation monitoring device on operator radiation exposure during cardiac catheterization: the radiation reduction during cardiac catheterization using real-time monitoring study. *Circ Cardiovasc Interv* 2014;**7**:744–50.
14. Reeves RR, Ang L, Bahadorani J, et al. Invasive cardiologists are exposed to greater left sided cranial radiation: the brain study (brain radiation exposure and attenuation during invasive Cardiology procedures). *JACC Cardiovasc Interv* 2015;**8**:1197–206.
15. Shorrock D, Christopoulos G, Wosik J, et al. Impact of a disposable sterile radiation shield on operator radiation exposure during percutaneous coronary intervention of chronic total occlusions. *J Invasive Cardiol* 2015;**27**:313–6.
16. Madder RD, VanOosterhout S, Mulder A, et al. Impact of robotics and a suspended lead suit on physician radiation exposure during percutaneous coronary intervention. *Cardiovasc Revasc Med* 18, 2017, 190–6.
17. Kuon E, Schmitt M, Dahm JB. Significant reduction of radiation exposure to operator and staff during cardiac interventions by analysis of radiation leakage and improved lead shielding. *Am J Cardiol* 2002;**89**:44–9.

Chapter 11

Stenting of Chronic Total Occlusion Lesions

11.1 STENT TYPE

Restenosis rates after chronic total occlusion (CTO) stenting can be relatively high. Bare metal stents (BMS) significantly reduce restenosis compared to balloon angioplasty alone,[1] yet the incidence of restenosis and reocclusion remains very high. In the Total Occlusion Study of Canada (TOSCA) 1 trial, the 6-month incidence of restenosis and reocclusion with BMS exceeded 50% and 10%, respectively.[2]

First-generation drug-eluting stents (DES) significantly reduce restenosis compared with BMS (Table 11.1). The first randomized-controlled trial comparing BMS and DES was the Primary Stenting of Totally Occluded Native Coronary Arteries (**PRISON** II) trial that compared BMS with the sirolimus-eluting stent (SES; Cypher, Cordis). The SES significantly reduced the 6-month incidence of binary angiographic restenosis (from 41% to 11%, $P < .001$), vessel reocclusion (from 13% to 4%, $P < .04$), and the need for new revascularization procedures (from 22% to 8%, $P < .001$) compared with BMS.[3] The benefit persisted at 5-year follow-up angiography, although some late catch-up in lumen diameter loss was observed in the SES group.[4] Two other studies, the Gruppo Italiano di Studio sullo Stent nelle Occlusioni Coronariche Societá Italiana di Cardiologia Invasiva (**GISSOC II-GISE**)[5] and the **CORACTO** trial[6] showed similar results. Four metaanalyses on DES versus BMS in CTOs were published in 2010–2011,[7–10] all reporting significant reduction in the risk for restenosis, reocclusion, and repeat revascularization with DES. DES appeared to be safe in CTOs, although the risk for stent thrombosis was higher with DES in one metaanalysis.[7]

Second-generation DES have been shown to provide incremental benefit compared with the first-generation paclitaxel-eluting stent in non-CTO lesions.[19,20] Three randomized clinical trials have compared the first-generation SES with the second-generation everolimus-eluting stent (EES; Xience V, Abbott Vascular)[13] and zotarolimus-eluting stent (ZES; Medtronic)[14] in CTOs. The Chronic Coronary Occlusion Treated by Everolimus-Eluting Stent (**CIBELES**) compared EES with SES and demonstrated similar restenosis and repeat revascularization rates, with a trend for lower stent thrombosis risk in the EES group (3% vs. 0%, $P = .075$). The Catholic Total Occlusion study (**CATOS**) trial showed similar angiographic and clinical outcomes with the SES and the Endeavor ZES (Medtronic).[14] In contrast,

Manual of Chronic Total Occlusion Interventions. http://dx.doi.org/10.1016/B978-0-12-809929-2.00011-9

TABLE 11.1 Comparison of Published Prospective Studies on the Clinical and Angiographic Outcomes With Drug-Eluting Stents in Coronary Chronic Total Occlusions

Study Acronym	Year	Stent	n	FU Angio Time (months)	Prior CABG (%)	Total Stent Length (mm)	In-Stent Restenosis (%)	In-Segment Restenosis (%)	TLR (%)	TVR (%)
PRISON II[3,12]	2006	SES	100	6	3	32±15	7	11	4	8
ACROSS-TOSCA 4[11]	2009	SES	200	6	8.5	45.9 (30.2, 62.1)	9.5	12.4	9.8	11.4
GISSOC II – GISE[5]	2010	SES	78	8	6.7	41±18	8.2	9.8	8.1	14.9
CORACTO[6]	2010	SES	48	6	NR	45.5±24.8	NR	17.4	NR	10.8
CIBELES[13]	2012	SES	101	9	4	47±24	NR	10.5	7.5	11.6
		EES	106	9	4.7	50±23	NR	9.1	6.0	7.9
CATOS[14]	2012	SES	80	9	NR	44.6±20.2	NR	13.7	NR	13.8
		Endeavor ZES	80	9	NR	43.4±21.5	NR	14.1	NR	7.5
PRISON III[15]	2012	SES	60	8	5.0	38.4±18.4	2.0	12.0	6.7	8.3
		Endeavor or Resolute ZES	62	8	8.1	41.0±19.2	5.5	10.9	4.8	4.8

Study Acronym	Year	Stent	n	FU Angio Time (months)	Prior CABG (%)	Total Stent Length (mm)	In-Stent Restenosis (%)	In-Segment Restenosis (%)	TLR (%)	TVR (%)
ACE-CTO[16]	2015	EES	100	8	27	85 ± 34	46	46	37	39
EXPERT-CTO[17]	2015	EES	222	12	9.9	52 ± 27	NR	NR	6.3	NR
PRISON IV[18]	2017	Orsiro SES	165	12	3.6	52 ± 28	NR	NR	10.5	10.5
		EES	165	12	6.7	52 ± 27	NR	NR	4.0	6

ACE-CTO, Angiographic Evaluation of the Everolimus-Eluting Stent in Chronic Total Occlusion; *CABG*, coronary artery bypass graft surgery; *CATOS*, Catholic Total Occlusion study; *CIBELES*, Chronic Coronary Occlusion study Treated by Everolimus-Eluting Stent; *CTO*, chronic total occlusion; *EES*, everolimus-eluting stent; *EXPERT-CTO*, Evaluation of the XIENCE Coronary Stent, Performance, and Technique in Chronic Total Occlusions; *FU*, follow-up; *GISSOC II – GISE*, Gruppo Italiano di Studio sullo Stent nelle Occlusioni Coronariche Società Italiana di Cardiologia Invasiva; *NR*, not reported; *PRISON*, Primary Stenting of Totally Occluded Native Coronary Arteries; *SES*, sirolimus-eluting stent; *TLR*, target lesion revascularization; *TOSCA*, Total Occlusion Study of Canada; *TVR*, target vessel revascularization; *ZES*, zotarolimus-eluting stent.

the **PRISON III** trial reported higher in-segment late lumen loss at 8-month follow-up angiography with the Endeavor ZES compared with the SES, although rates were similar with the Resolute ZES.[14]

In an Italian registry of 802 patients undergoing CTO percutaneous coronary intervention (PCI) over 8 years, use of EES was associated with a significantly lower reocclusion rate compared with first-generation DES (3.0% vs. 10.1%; $P < .001$).[21] However, in the Angiographic Evaluation of the Everolimus-Eluting Stent in Chronic Total Occlusion (**ACE-CTO**) study that included very long lesion and stent length and a high proportion of patients with prior coronary artery bypass graft surgery, target lesion revascularization rates were significantly higher (37% at 12 months).[16] In the Evaluation of the XIENCE Coronary Stent, Performance, and Technique in Chronic Total Occlusions (**EXPERT-CTO**) study of 250 patients (with CTO PCI success in 222) from 20 US centers target lesion revascularization at 1 year was 6.3%.[17]

The **PRISON IV** trial randomized 330 patients to either an ultrathin-strut sirolimus-eluting stent (Orsiro SES, Biotronik) with biodegradable polymer or the durable polymer Xience EES, and failed to show noninferiority for the primary endpoint of in-segment late lumen loss (0.13 ± 0.63 mm for Orsiro SES vs. 0.02 ± 0.47 mm for EES; $P = .08$ for noninferiority).[18] In-stent and in-segment binary restenosis was significantly higher with Orsiro SES as compared with EES (8.0% vs. 2.1%; $P = .028$), with trend for higher risk for target-lesion and target-vessel revascularization (9.2% vs. 4.0% [$P = .08$] and 9.2% vs. 6.0% [$P = .33$]).

Bioabsorbable scaffolds have been used in CTOs, where they held great promise given the frequently long length of stent implantation (full metal jacket) and the ability to restore physiologic vasomotion and allow vessel remodeling. Early studies showed promising results with low incidence of adverse events.[22–27] The Bioresorbable Scaffolds versus Drug-Eluting Stents in Chronic Total Occlusions (BONITO) international multicenter registry compared 153 subjects who were treated with Absorb, with 384 patients who received a second-generation DES.[28] At a median follow-up of 703 days, there were no differences in target-vessel failure (cardiac death, target-vessel myocardial infarction, ischemia-driven target-lesion revascularization) between bioresorbable scaffold and DES (4.6% vs. 7.7%; $P = .21$). Excellent lesion preparation, high-pressure postdilation, and use of intravascular imaging are recommended to obtain optimal final results.[28,29] In particular, intravascular imaging is strongly advised in any bioabsorbable scaffold procedure, and particularly in CTO PCI, due to the difficulty of adequately assessing the diameter of chronically underperfused vessels, as well as the need for stenting of long segments with multiple overlaps. Given concerns for increased rates of very late stent thrombosis with bioabsorbable scaffolds additional adequately powered studies with long-term follow-up and comparison with DES-treated subjects are needed.

In patients with prior coronary bypass graft surgery, treatment of a native coronary artery CTO is preferable to treatment of a saphenous vein graft

(SVG) CTO supplying the same territory because of very high (>50%) restenosis rates after SVG CTO PCI.[30] Occasionally, occluded SVGs can be used for retrograde access to the native coronary artery CTO. If native CTO PCI is not possible, PCI of the SVG CTO may be a reasonable treatment option.[30–36]

In summary, durable polymer second-generation DES (EES and Resolute ZES) provide better outcomes compared to first-generation DES and are currently the preferred options for CTO PCI. Novel technologies are needed to address the limitations of current metallic stents in CTOs (i.e., late acquired malapposition, due to significant enlargement of the CTO target vessel after successful recanalization).[37]

11.2 SUBINTIMAL VERSUS TRUE LUMEN STENTING

As described in detail in Chapter 5, extensive dissection/reentry crossing strategies (such as the subintimal tracking and reentry technique) are associated with high restenosis and reocclusion rates and should be used only as a last resort.[21,38] However, outcomes with limited antegrade[39–43] or retrograde[44,45] dissection/reentry techniques have been similar with those of true-to-true crossing with few exceptions,[46] supporting the use of these techniques in contemporary CTO PCI.

11.3 STENT OPTIMIZATION

Maximizing stent expansion is important for decreasing the risk for restenosis, which remains relatively high, even with second-generation DES (approximately 10%–15% at 1 year; Table 11.1) and possibly stent thrombosis. Optimal stent expansion can be accomplished using the following techniques.

1. **Appropriate stent sizing**
    The CTO vessel often increases in size in the months following successful CTO PCI. In one study, 69% of the recanalized vessels had a mean increase in lumen diameter of 0.4 mm over a period of 6 months.[47] Intravascular imaging may help optimize stent sizing at the index procedure: stent undersizing can lead to stent malapposition and higher long-term restenosis and stent thrombosis rates, whereas stent oversizing may lead to perforation. In cases where intravascular imaging is not feasible, the size of the loop of a knuckled guidewire could serve as a gross estimate of the vessel size.

    Optical coherence tomography may also be very useful for selecting the optimum stent length and detecting areas of dissection that require stenting (Fig. 11.1).[48]

    Overall, it is recommended to avoid stenting excessively long vessel segments,[49] as longer stent length may increase the risk for restenosis and possibly stent thrombosis. This is frequently referred to as "mission creep,"

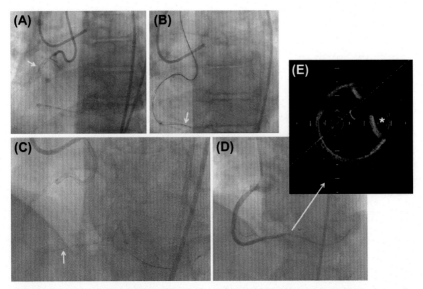

FIGURE 11.1 Example of distal dissection confirmed with optical coherence tomography.
A right coronary artery chronic total occlusion (*arrow*, A) was crossed subintimally using antegrade wire escalation (B). The Stingray system (*arrow*, C) was used, enabling successful reentry into the distal true lumen, however subsequent angiography demonstrated suboptimal distal result (D). Optical coherence tomography demonstrated a distal dissection with compression of the true lumen (*asterisk*, E), which was treated with implantation of an additional stent.

or the tendency to make the distal vessel look perfect, rather than just treating the CTO. If compressive hematomas form in the vessel wall distal to the stent, cutting balloons could be considered for treatment (for hematoma relief) instead of placing another stent, although in most cases additional stents are implanted.

2. **High-pressure balloon inflation**

   High-pressure balloon inflations are critical for adequate stent expansion, especially in calcified vessels. Routine postdilation of stents placed in CTOs should be performed using noncompliant balloons at high (20 atm or higher) pressures. However, high pressure postdilation of oversized stents should be avoided (especially when dissection/reentry techniques are used and in heavily calcified vessels), as it can cause perforation (see Online Case 17 and 40).

3. **Intravascular imaging**

   In addition to helping with crossing,[50] intravascular imaging with intravascular ultrasonography (IVUS) or optical coherence tomography (OCT) can help with stent optimization. Specifically, imaging can help detect areas of underexpansion (see Online Case 18), malapposition, dissection (Fig. 11.1), or incomplete stent coverage of ostial lesions (see

Online Case 18), treatment of which may improve outcomes of CTO PCI. OCT can be very useful in limiting the extent of stenting by confirming that the small caliber of the distal coronary vessel is due to chronic hypoperfusion and not due to dissection or atherosclerotic lesions (indicated by the presence of a trilaminar arterial structure, no evidence of intimal hyperplasia, and no atheroma).[37,48]

Two randomized-controlled trials demonstrated improved long-term outcomes with use of IVUS in CTO PCI. The Korean Chronic Total Occlusion Intervention with Drug-eluting Stents guided by IVUS (**CTO-IVUS**) trial randomized 402 patients after successful guidewire crossing of the CTO to IVUS versus angiographic guidance, and reported lower 12-month incidence of major adverse cardiac events in the IVUS group (2.6% vs. 7.1%; $P = .035$).[51] Similarly, the Comparison of Angiography—versus IVUS—guided Stent Implantation for Chronic Total Coronary Occlusion Recanalization (AIR CTO) trial randomized 230 patients after successful guidewire crossing to IVUS-guided versus angiography-guided stenting, and showed lower 12-month in-stent late lumen loss (0.28 ± 0.48 mm vs. 0.46 ± 0.68 mm; $P = .025$) and in true lumen restenosis (3.9% vs. 13.7%; $P = .021$) rates in the IVUS group.[52]

In a contemporary CTO PCI registry intravascular imaging (mainly IVUS) was performed in approximately one of three cases, both to facilitate crossing and for stent optimization.[53] Expanded use of intravascular imaging is highly encouraged as it is likely to significantly improve the long-term outcomes after CTO PCI.

11.4 ANTIPLATELET THERAPY AFTER CHRONIC TOTAL OCCLUSION STENTING

The optimal duration of dual antiplatelet therapy (DAPT) remains poorly defined: at least 6 months are currently recommended poststenting in patients with stable ischemic heart disease,[54] but many operators empirically recommend longer DAPT duration after CTO PCI in view of the often long stent lengths ("full metal jacket"), especially in the right coronary artery. Longer DAPT duration may also be needed if bioabsorbable scaffolds are used. Although the Resolute ZES has received Conformité Européenne mark for 1 month DAPT duration and the Xience EES and Synergy EES for 3 months DAPT duration, it remains unknown whether such short DAPT durations are safe in the context of CTO PCI.

11.5 CONCLUSIONS

In summary, implantation of a second-generation DES and ensuring good stent expansion and apposition are important for maximizing the likelihood of long-term patency after successful CTO recanalization.

REFERENCES

1. Agostoni P, Valgimigli M, Biondi-Zoccai GG, et al. Clinical effectiveness of bare-metal stenting compared with balloon angioplasty in total coronary occlusions: insights from a systematic overview of randomized trials in light of the drug-eluting stent era. *Am Heart J* 2006;**151**:682–9.

2. Buller CE, Dzavik V, Carere RG, et al. Primary stenting versus balloon angioplasty in occluded coronary arteries: the Total Occlusion Study of Canada (TOSCA). *Circulation* 1999;**100**:236–42.

3. Suttorp MJ, Laarman GJ, Rahel BM, et al. Primary Stenting of Totally Occluded Native Coronary Arteries II (PRISON II): a randomized comparison of bare metal stent implantation with sirolimus-eluting stent implantation for the treatment of total coronary occlusions. *Circulation* 2006;**114**:921–8.

4. Teeuwen K, Van den Branden BJ, Rahel BM, et al. Late catch-up in lumen diameter at five-year angiography in MACE-free patients treated with sirolimus-eluting stents in the Primary Stenting of Totally Occluded Native Coronary Arteries: a randomised comparison of bare metal stent implantation with sirolimus-eluting stent implantation for the treatment of total coronary occlusions (PRISON II). *EuroIntervention* 2013;**9**:212–9.

5. Rubartelli P, Petronio AS, Guiducci V, et al. Comparison of sirolimus-eluting and bare metal stent for treatment of patients with total coronary occlusions: results of the GISSOC II-GISE multicentre randomized trial. *Eur Heart J* 2010;**31**:2014–20.

6. Reifart N, Hauptmann KE, Rabe A, Enayat D, Giokoglu K. Short and long term comparison (24 months) of an alternative sirolimus-coated stent with bioabsorbable polymer and a bare metal stent of similar design in chronic coronary occlusions: the CORACTO trial. *EuroIntervention* 2010;**6**:356–60.

7. Colmenarez HJ, Escaned J, Fernandez C, et al. Efficacy and safety of drug-eluting stents in chronic total coronary occlusion recanalization: a systematic review and meta-analysis. *J Am Coll Cardiol* 2010;**55**:1854–66.

8. Saeed B, Kandzari DE, Agostoni P, et al. Use of drug-eluting stents for chronic total occlusions: a systematic review and meta-analysis. *Catheter Cardiovasc Interv* 2011;**77**:315–32.

9. Niccoli G, Leo A, Giubilato S, et al. A meta-analysis of first-generation drug-eluting vs bare-metal stents for coronary chronic total occlusion: effect of length of follow-up on clinical outcome. *Int J Cardiol* 2011;**150**:351–4.

10. Ma J, Yang W, Singh M, Peng T, Fang N, Wei M. Meta-analysis of long-term outcomes of drug-eluting stent implantations for chronic total coronary occlusions. *Heart Lung* 2011;**40**:e32–40.

11. Kandzari DE, Rao SV, Moses JW, et al. Clinical and angiographic outcomes with sirolimus-eluting stents in total coronary occlusions: the ACROSS/TOSCA-4 (approaches to chronic occlusions with sirolimus-eluting stents/total occlusion study of coronary Arteries-4) trial. *JACC Cardiovasc Interv* 2009;**2**:97–106.

12. Van den Branden BJ, Rahel BM, Laarman GJ, et al. Five-year clinical outcome after primary stenting of totally occluded native coronary arteries: a randomised comparison of bare metal stent implantation with sirolimus-eluting stent implantation for the treatment of total coronary occlusions (PRISON II study). *EuroIntervention* 2012;**7**:1189–96.

13. Moreno R, Garcia E, Teles R, et al. Randomized comparison of sirolimus-eluting and everolimus-eluting coronary stents in the treatment of total coronary occlusions: results from the chronic coronary occlusion treated by everolimus-eluting stent randomized trial. *Circ Cardiovasc Interv* 2013;**6**:21–8.

14. Park HJ, Kim HY, Lee JM, et al. Randomized comparison of the efficacy and safety of zotarolimus-eluting stents vs. sirolimus-eluting stents for percutaneous coronary intervention in chronic total occlusion–CAtholic Total Occlusion Study (CATOS) trial. *Circ J* 2012;**76**:868–75.

15. Van den Branden BJ, Teeuwen K, Koolen JJ, et al. Primary Stenting of Totally Occluded Native Coronary Arteries III (PRISON III): a randomised comparison of sirolimus-eluting stent implantation with zotarolimus-eluting stent implantation for the treatment of total coronary occlusions. *EuroIntervention* 2013;**9**:841–53.

16. Kotsia A, Navara R, Michael TT, et al. The AngiographiC Evaluation of the Everolimus-Eluting Stent in Chronic Total Occlusion (ACE-CTO) study. *J Invasive Cardiol* 2015;**27**:393–400.

17. Kandzari DE, Kini AS, Karmpaliotis D, et al. Safety and effectiveness of everolimus-eluting stents in chronic total coronary occlusion revascularization: results from the EXPERT CTO multicenter trial (evaluation of the XIENCE coronary stent, performance, and technique in chronic total occlusions). *JACC Cardiovasc Interv* 2015;**8**:761–9.

18. Teeuwen K, van der Schaaf RJ, Adriaenssens T, et al. Randomized multicenter trial investigating angiographic outcomes of hybrid sirolimus-eluting stents with biodegradable polymer compared with everolimus-eluting stents with durable polymer in chronic total occlusions: the PRISON IV trial. *JACC Cardiovasc Interv* 2017;**10**:133–43.

19. Kedhi E, Joesoef KS, McFadden E, et al. Second-generation everolimus-eluting and paclitaxel-eluting stents in real-life practice (COMPARE): a randomised trial. *Lancet* 2010;**375**:201–9.

20. Lanka V, Patel VG, Saeed B, et al. Outcomes with first – versus second-generation drug-eluting stents in coronary chronic total occlusions (CTOs): a systematic review and meta-analysis. *J Invasive Cardiol* 2014;**26**:304–10.

21. Valenti R, Vergara R, Migliorini A, et al. Predictors of reocclusion after successful drug-eluting stent-supported percutaneous coronary intervention of chronic total occlusion. *J Am Coll Cardiol* 2013;**61**:545–50.

22. Vaquerizo B, Barros A, Pujadas S, et al. Bioresorbable everolimus-eluting vascular scaffold for the treatment of chronic total occlusions: CTO-ABSORB pilot study. *EuroIntervention* 2015;**11**:555–63.

23. Lesiak M, Lanocha M, Araszkiewicz A, et al. Percutaneous coronary intervention for chronic total occlusion of the coronary artery with the implantation of bioresorbable everolimus-eluting scaffolds. Poznan CTO-Absorb Pilot Registry. *EuroIntervention* 2016;**12**:e144–51.

24. Goktekin O, Yamac AH, Latib A, et al. Evaluation of the safety of everolimus-eluting bioresorbable vascular scaffold (BVS) implantation in patients with chronic total coronary occlusions: acute procedural and short-term clinical results. *J Invasive Cardiol* 2015;**27**:461–6.

25. Vaquerizo B, Barros A, Pujadas S, et al. One-year results of bioresorbable vascular scaffolds for coronary chronic total occlusions. *Am J Cardiol* 2016;**117**:906–17.

26. Wiebe J, Hoppmann P, Kufner S, et al. Impact of stent size on angiographic and clinical outcomes after implantation of everolimus-eluting bioresorbable scaffolds in daily practice: insights from the ISAR-ABSORB registry. *EuroIntervention* 2016;**12**:e137–43.

27. Fam JM, Ojeda S, Garbo R, et al. Everolimus eluting bioresorbable vascular scaffold for treatment of complex chronic total occlusions. *EuroIntervention* 2017.

28. Azzalini L, Giustino G, Ojeda S, et al. Procedural and long-term outcomes of bioresorbable scaffolds versus drug-eluting stents in chronic total occlusions: the BONITO registry (bioresorbable scaffolds versus drug-eluting stents in chronic total occlusions). *Circ Cardiovasc Interv* 2016:9.

29. Mitomo S, Naganuma T, Fujino Y, et al. Bioresorbable vascular scaffolds for the treatment of chronic total occlusions: an international multicenter registry. *Circ Cardiovasc Interv* 2017:10.

30. Brilakis ES, Banerjee S, Lombardi WL. Retrograde recanalization of native coronary artery chronic occlusions via acutely occluded vein grafts. *Catheter Cardiovasc Interv* 2010;**75**:109–13.

31. Sachdeva R, Uretsky BF. Retrograde recanalization of a chronic total occlusion of a saphenous vein graft. *Catheter Cardiovasc Interv* 2009;**74**:575–8.

32. Takano M, Yamamoto M, Mizuno K. A retrograde approach for the treatment of chronic total occlusion in a patient with acute coronary syndrome. *Int J Cardiol* 2007;**119**:e22–4.

33. Ho PC, Tsuchikane E. Improvement of regional ischemia after successful percutaneous intervention of bypassed native coronary chronic total occlusion: an application of the CART technique. *J Invasive Cardiol* 2008;**20**:305–8.

34. Brilakis ES, Grantham JA, Thompson CA, et al. The retrograde approach to coronary artery chronic total occlusions: a practical approach. *Catheter Cardiovasc Interv* 2012;**79**:3–19.

35. Nguyen-Trong PK, Alaswad K, Karmpaliotis D, et al. Use of saphenous vein bypass grafts for retrograde recanalization of coronary chronic total occlusions: insights from a multicenter registry. *J Invasive Cardiol* 2016;**28**:218–24.

36. Debski A, Tyczynski P, Demkow M, Witkowski A, Werner GS, Agostoni P. How should I treat a chronic total occlusion of a saphenous vein graft? Successful retrograde revascularisation. *EuroIntervention* 2016;**11**:e1325–8.

37. Galassi AR, Tomasello SD, Crea F, et al. Transient impairment of vasomotion function after successful chronic total occlusion recanalization. *J Am Coll Cardiol* 2012;**59**:711–8.

38. Kandzari DE, Grantham JA, Lombardi W, Thompson C. Not all subintimal chronic total occlusion revascularization is alike. *J Am Coll Cardiol* 2013;**61**:2570.

39. Mogabgab O, Patel VG, Michael TT, et al. Long-term outcomes with use of the Bridgepoint Medical System for the recanalization of coronary chronic total occlusions. *J Invasive Cardiol* 2013;**25**, 579–85.

40. Rinfret S, Ribeiro HB, Nguyen CM, Nombela-Franco L, Urena M, Rodes-Cabau J. Dissection and re-entry techniques and longer-term outcomes following successful percutaneous coronary intervention of chronic total occlusion. *Am J Cardiol* 2014;**114**:1354–60.

41. Amsavelu S, Christakopoulos GE, Karatasakis A, et al. Impact of crossing strategy on intermediate-term outcomes after chronic total occlusion percutaneous coronary intervention. *Can J Cardiol* 2016;**32**:1239e1–7.

42. Carlino M, Figini F, Ruparelia N, et al. Predictors of restenosis following contemporary subintimal tracking and reentry technique: the importance of final TIMI flow grade. *Catheter Cardiovasc Interv* 2016;**87**:884–92.

43. Azzalini L, Dautov R, Brilakis ES, et al. Procedural and longer-term outcomes of wire- versus device-based antegrade dissection and re-entry techniques for the percutaneous revascularization of coronary chronic total occlusions. *Int J Cardiol* 2017;**231**:78–83.

44. Muramatsu T, Tsuchikane E, Oikawa Y, et al. Incidence and impact on midterm outcome of controlled subintimal tracking in patients with successful recanalisation of chronic total occlusions: J-PROCTOR registry. *EuroIntervention* 2014;**10**:681–8.

45. Saito S, Maehara A, Yakushiji T, et al. Serial intravascular ultrasound findings after treatment of chronic total occlusions using drug-eluting stents. *Am J Cardiol* 2016;**117**:727–34.

46. Hasegawa K, Tsuchikane E, Okamura A, et al. Incidence and impact on midterm outcome of intimal versus subintimal tracking with both antegrade and retrograde approaches in patients with successful recanalisation of chronic total occlusions: J-PROCTOR 2 study. *EuroIntervention* 2017;**12**:e1868–73.

47. Park JJ, Chae IH, Cho YS, et al. The recanalization of chronic total occlusion leads to lumen area increase in distal reference segments in selected patients: an intravascular ultrasound study. *JACC Cardiovasc Interv* 2012;**5**:827–36.

48. Jaguszewski M, Guagliumi G, Landmesser U. Optical frequency domain imaging for guidance of optimal stenting in the setting of recanalization of chronic total occlusion. *J Invasive Cardiol* 2013;**25**:367–8.

49. Gasparini GL, Oreglia JA, Milone F, Presbitero P. Avoid overtreatment in the setting of chronic total occlusions: the role of blood flow restoration in positive vascular remodeling. *Int J Cardiol* 2015;**184**:414–5.

50. Galassi AR, Sumitsuji S, Boukhris M, et al. Utility of intravascular ultrasound in percutaneous revascularization of chronic total occlusions: an overview. *JACC Cardiovasc Interv* 2016;**9**:1979–91.

51. Kim BK, Shin DH, Hong MK, et al. Clinical impact of intravascular ultrasound-guided chronic total occlusion intervention with zotarolimus-eluting versus biolimus-eluting stent implantation: randomized study. *Circ Cardiovasc Interv* 2015;**8**:e002592.

52. Tian NL, Gami SK, Ye F, et al. Angiographic and clinical comparisons of intravascular ultrasound- versus angiography-guided drug-eluting stent implantation for patients with chronic total occlusion lesions: two-year results from a randomised AIR-CTO study. *EuroIntervention* 2015;**10**:1409–17.

53. Karacsonyi J, Alaswad K, Jaffer FA, et al. Use of intravascular imaging during chronic total occlusion percutaneous coronary intervention: insights from a contemporary multicenter registry. *J Am Heart Assoc* 2016;**5**.

54. Levine GN, Bates ER, Bittl JA, et al. 2016 ACC/AHA guideline focused update on duration of dual antiplatelet therapy in patients with coronary artery disease: a report of the American College of Cardiology/American Heart Association Task Force on clinical practice guidelines: an Update of the 2011 ACCF/AHA/SCAI Guideline for percutaneous coronary intervention, 2011 ACCF/AHA Guideline for coronary artery bypass graft surgery, 2012 ACC/AHA/ACP/AATS/PCNA/SCAI/STS Guideline for the Diagnosis and Management of patients with stable ischemic heart disease, 2013 ACCF/AHA Guideline for the Management of ST-Elevation myocardial infarction, 2014 AHA/ACC Guideline for the Management of patients with non-ST-Elevation acute coronary syndromes, and 2014 ACC/AHA Guideline on Perioperative Cardiovascular Evaluation and Management of patients undergoing noncardiac surgery. *Circulation* 2016;**134**:e123–55.

Chapter 12

Complications

From all that has been discussed in the previous chapters of this book the reader will have already realized that chronic total occlusion (CTO) interventions are among the most complex percutaneous coronary interventions (PCIs). In this chapter we perform a thorough review of coronary and noncoronary complications that may occur in the course of CTO PCI. Awareness of the potential complications constitutes the cornerstone of their prevention. Furthermore, the various alternative techniques for complication management will be discussed.

Complications of CTO PCI can be classified according to timing (as acute and long-term) and according to location (cardiac coronary, cardiac noncoronary, and noncardiac). The acute complications of CTO PCI are summarized in Fig. 12.1.[1]

The frequency of acute outcomes of CTO PCI from a 2013 metaanalysis of 65 studies with 18,061 patients is shown in Table 12.1.[2]

12.1 ACUTE COMPLICATIONS

12.1.1 Acute Coronary Complications

12.1.1.1 Acute Vessel Closure

12.1.1.1.1 Donor Vessel Injury During Retrograde Chronic Total Occlusion Percutaneous Coronary Intervention

See Online Cases 22, 50, 66, 69, and 97.

Although, by definition, CTO signifies that the target vessel is occluded for greater than 3 months, CTO PCI can be complicated by occlusion of a collateral-donor vessel instrumented for contralateral angiography or for the retrograde approach (Fig. 12.2). This is one of the most serious complications of CTO PCI and requires prompt identification and management, since it is frequently followed by extensive ischemia and hemodynamic decompensation. Unless this complication is rapidly and diligently treated with PCI or coronary artery bypass graft surgery (CABG), it may result in death, particularly when the donor vessel is the last remaining vessel (a common situation in patients with prior CABG).[3]

Similar to planning of any PCI, a pretreatment plan should be in place in case donor vessel occlusion occurs. For example, when performing retrograde recanalization of the right coronary artery in a patient with left main disease, pre-PCI intravascular ultrasound (IVUS) of the left main coronary artery is

Manual of Chronic Total Occlusion Interventions. https://doi.org/10.1016/B978-0-12-809929-2.00012-0

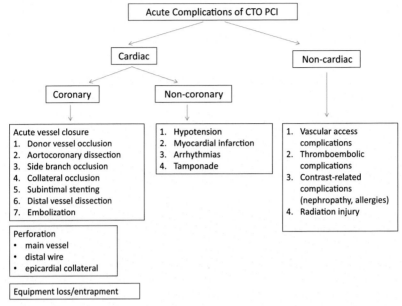

FIGURE 12.1 Classification of acute complications of chronic total occlusion percutaneous coronary intervention.

important. Similarly, prophylactic hemodynamic support may increase procedural safety in patients with multivessel disease and low ejection fraction or those undergoing the retrograde approach through the last remaining vessel (Section 12.1.2.1) (see Online Case 51). In contrast to conventional PCI, management of donor vessel occlusion may be complicated by the presence of hardware in the acutely occluded segment (microcatheters, externalized wires, etc.) that can hinder stenting as an emergency bailout solution.

Causes

1. Catheter-induced vessel injury. This may occur with either diagnostic or guide catheters, especially during equipment withdrawal, which may cause the guide to deeply engage the vessel, or with forceful pulling of the snared retrograde guidewire.
2. Donor-vessel thrombosis can occur during long procedures requiring coronary artery intubation by microcatheters or guidewires, especially when the level of anticoagulation is suboptimal (Section 3.5).

Prevention

1. Pay close attention to the position of diagnostic and guiding catheters (especially during equipment manipulations) and to the pressure waveform (dampening of the pressure waveform should be avoided and promptly corrected if it occurs). When dual injection is performed, the donor vessel catheter should be disengaged or removed as soon as it is not needed.

TABLE 12.1 Frequency of Angiographic Success and Complications in Chronic Total Occlusion Percutaneous Coronary Intervention

Outcome	Pooled Estimate Rate, %	95% CI	Reported Rate Min, Max %
Angiographic success	77.0	74.3–79.6	41.2–100.0
MACE	3.1	2.4–3.7	0–19.4
Death	0.2	0.1–0.3	0.0–3.6
Emergent CABG	0.1	0–0.2	0–2.3
Stroke	<0.01	0–0.1	0–0.7
Myocardial infarction	2.5	1.9–3.0	0–19.4
Q-wave myocardial infarction	0.2	0.1–0.3	0–2.6
Coronary perforation (per lesion)	2.9	2.2–3.6	0–11.9
Tamponade	0.3	0.2–0.5	0–4.7
Acute stent thrombosis	0.3	0.1–0.5	0–2.0
Vascular complication	0.6	0.3–0.9	0–2.8
Major bleed	0.4	0–0.7	0–3.7
Contrast nephropathy	3.8	2.4–5.3	2.4–18.1
Radiation skin injury	<0.01	0–0.1	0–11.1

CABG, coronary artery bypass graft surgery; MACE, major adverse cardiac events (composite of death, emergency CABG, stroke, and myocardial infarction).
Modified with permission from Patel VG, Brayton KM, Tamayo A, et al. Angiographic success and procedural complications in patients undergoing percutaneous coronary chronic total occlusion interventions: a weighted meta-analysis of 18,061 patients from 65 studies. *JACC Cardiovasc Interv* 2013;**6**:128–36.

2. Never use a catheter with side holes to engage the CTO donor vessel, as it may mask suboptimal catheter position and flow compromise.
3. Maintain high activated clotting time (ACT) (the authors suggest >350s during retrograde CTO PCI and >300s during antegrade CTO PCI) to prevent donor vessel thrombosis (Section 3.5). The ACT should be checked every 20–30 min, which is more easily accomplished by inserting a small venous sheath and delegating this task to a nurse, with a clear message that any drop in ACT below the prespecified safety level should be communicated and corrected with additional boluses of unfractionated heparin.
4. Avoid the retrograde approach through diffusely diseased donor vessels. Consider intracoronary imaging to investigate the anatomy of the donor

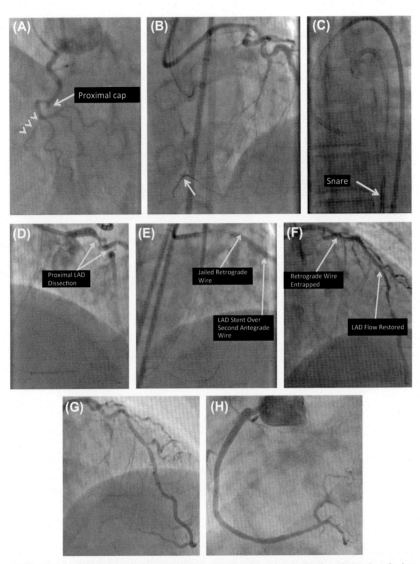

FIGURE 12.2 Example of donor vessel dissection during retrograde chronic total occlusion (CTO) percutaneous coronary intervention (PCI). PCI of a right coronary artery (RCA) CTO (A). After a failed antegrade crossing attempt, retrograde crossing was performed (B) and the retrograde guidewire was externalized (C). During RCA stenting over the externalized guidewire (D), the patient developed severe chest pain and hypotension due to proximal left anterior descending artery (LAD) dissection (D). The LAD was immediately stented (E) with restoration of antegrade flow and stabilization of the patient (F, G). After removal of the entrapped retrograde guidewire and stenting of the right coronary artery an excellent final angiographic result was achieved (H) (see Online Case 22).

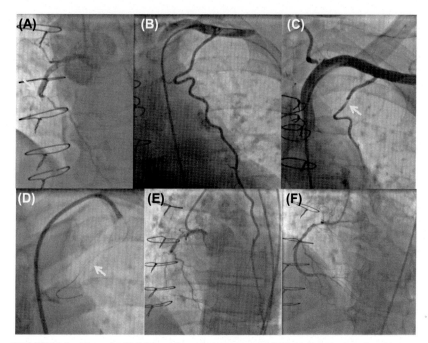

FIGURE 12.3   **Example of left internal mammary artery (LIMA) graft dissection.** Percutaneous coronary intervention of a right coronary artery chronic total occlusion (A) was attempted with dual injection via a catheter inserted in the LIMA graft (B). Engagement of the LIMA graft was challenging due to an acute take-off angle. A GuideLiner catheter was used to engage the LIMA, causing ostial LIMA dissection with decreased antegrade flow (C). LIMA flow was restored after stenting (D), enabling continuation of the procedure (E), with a final successful outcome (F) (see Online Case 50).

vessel in advance. If the donor artery requires PCI, it should be performed before CTO PCI.

5. Keep the retrograde guidewire encased by a microcatheter or over-the-wire balloon during all manipulations.[4]

6. Consider inserting a workhorse "safety" guidewire into the donor vessel, both to stabilize the guide catheter, but also to allow rapid treatment of the donor vessel in case of acute donor vessel occlusion.

7. Guidewire externalization leads to the creation of a wire loop with outstanding support for PCI devices, at the expense of communicating to the contralateral guiding catheter any traction during PCI, causing deep guide engagement and possibly vessel damage. During externalization meticulous attention should be paid to the contralateral catheter at all times.

8. Flush regularly all guiding and diagnostic catheters to prevent in-catheter thrombosis. Back-bleed the guide catheter after performing balloon trapping to minimize the risk of air embolization.

9. Avoid use of the left internal mammary artery (LIMA) graft for retrograde CTO PCI because LIMA dissection can occur (Fig. 12.3; see Online Case 50)

and/or LIMA wiring may cause acute closure or severe ischemia if the LIMA tortuosity is straightened by the guidewire.[5]

10. Convert the retrograde system to a fully antegrade system either by using the kissing microcatheter technique,[6] or by using the tip-in technique; that is, advancing an antegrade microcatheter over the retrograde wire into the distal vessel and then exchanging the retrograde wire for an antegrade workhorse guidewire (Step 8 in Chapter 6).

Treatment

1. Successful treatment of abrupt occlusion of the collateral donor artery becomes the highest priority of PCI. Treatment of the CTO, in most cases, should be aborted.

2. In anticipation of hemodynamic collapse, notify medical staff to prepare a hemodynamic support device, such as intraaortic balloon pump or an Impella, ensure femoral artery access (in radial procedures), and prepare drugs, while you concentrate in solving the abrupt occlusion.

3. Stenting is usually required to treat donor vessel dissections, as it is the fastest way to prevent its extension. If this complication occurs in the context of retrograde CTO PCI, the operator faces the problem of hardware in the donor vessel. Treatment options include:

   A. Withdrawing the retrograde wire, leaving the retrograde microcatheter beyond the acute occlusion, and exchanging for a regular PCI wire for stenting, avoiding the advancement of a new wire through a dissected segment.

   B. Using the externalized retrograde wire as a platform for stenting the acutely occluded donor vessel (provided it is anatomically feasible, for example in a short, limited dissection).

   C. Performing PCI with a second wire, potentially jailing the externalized retrograde wire with the stent (Fig. 12.2).

   Options B and C are more likely to be followed if CTO PCI is virtually completed (pending only CTO stenting, for example), while option A is probably the best choice if the CTO has not yet been crossed. Option C requires great caution during withdrawal of the jailed retrograde wire through the collateral channels (it may require protection with an antegrade microcatheter). An overall assessment of the patient's safety is mandatory at the time of deciding the best possible choice (e.g., contralateral femoral artery access may be required for intraaortic balloon pump insertion, making option A the preferred option).

4. Aspiration of thrombus and administration of glycoprotein IIb/IIIa inhibitors may be needed for donor vessel thrombosis. The ACT should be checked, as thrombosis may occur in other locations of the instrumented collateral donor vessel or the CTO target vessel.

12.1.1.1.2 Aortocoronary Dissection

See Online Case 10.

Aortocoronary dissection is a rare complication that can occur with any PCI, but is more common with CTO PCI (especially retrograde procedures) (frequency was 0.8%–1.8% in two contemporary series[7,8]), and most commonly occurs in the right coronary artery[9] (Figs. 12.4 and 12.5). Dissection may be limited to the coronary sinus, but may extend to the proximal ascending aorta or even beyond the ascending aorta.[10]

Prevention

1. Consider using anchor techniques (Section 3.6.5) as an alternative to aggressive guiding catheter intubation to enhance guide catheter support.
2. Use of guide catheters with side holes (Chapter 2, Section 2.3.3) in occluded right coronary arteries significantly decreases the risk of barotrauma. However, side-hole guide catheters may provide a false sense of security, as the pressure waveform may appear normal but aortocoronary dissection can still occur.

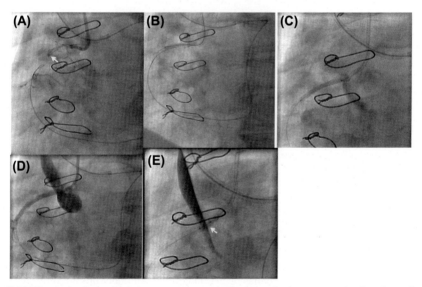

FIGURE 12.4   **Illustration of an aortocoronary dissection during retrograde chronic total occlusion (CTO) intervention.** Retrograde CTO intervention was performed to recanalize a proximal right coronary artery CTO (*arrow*, A), using the reverse controlled antegrade and retrograde tracking and dissection technique (B). Staining of the aortocoronary junction was observed with test injections during stent placement (C), which expanded when cine angiography was performed (D). Stenting of the right coronary artery ostium was performed (*arrow*, E) without further antegrade contrast injections. The patient had an uneventful recovery. This case illustrates the importance of stopping antegrade contrast injections and stenting the vessel ostium if aortocoronary dissection occurs, to seal the dissection flap at the entry point of the dissection. (*Courtesy of Dr. Parag Doshi.*)

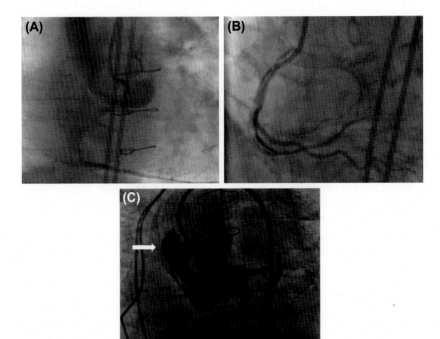

FIGURE 12.5 **Examples of aortocoronary dissection.** (A) Anteroposterior cranial view showing class 2 aortocoronary dissection caused by a retrograde approach of proximal chronic total occlusion (CTO) lesion of the right coronary artery (RCA). Guide catheter: left Amplatz 1 (Cordis, Miami, FL). Presumed mechanism: contrast injection with a wedged catheter. (B) Left anterior oblique view showing class 1 (limited to right sinus of Valsalva) aortocoronary dissection caused by an antegrade approach of the ostial CTO lesion of the RCA. Guide catheter: Judkins right 4 (Cordis). Presumed mechanism: catheter trauma. (C) Anteroposterior view showing class 2 aortocoronary dissection with a parietal hematoma (*arrow*) caused by a retrograde approach of the proximal CTO lesion of the RCA. Guide catheter: right Amplatz 1. Presumed mechanism: catheter trauma. *(Reproduced with permission from Boukhris M, Tomasello SD, Marza F, Azzarelli S, Galassi AR. Iatrogenic aortic dissection complicating percutaneous coronary intervention for chronic total occlusion. Can J Cardiol 2015;31:320–7, Elsevier.)*

3. Power injectors should also be avoided or used with caution after the proximal segment of the CTO vessel has been dilated; manual injections are preferred.

Causes (Fig. 12.5)

1. Deep coronary engagement and utilization of aggressive guide catheters, such as 8 Fr Amplatz catheters.
2. Guide pressure dampening.
3. Forceful contrast injection, especially through wedged guide catheters with dampened pressure waveform.
4. Predilation of the coronary ostium.

5. Balloon rupture.
6. Retrograde wire advancement into the subintimal and subaortic space during retrograde crossing attempts.

Treatment

1. **Stop injecting contrast into the coronary artery** (as injections can expand the dissection plane).
2. Stent the ostium of the dissected coronary artery with a stent that can expand to a diameter that will seal the dissection. The stent should be protruding 1 mm into the aorta) to cover the ostium of the dissected vessel.
3. Use intravascular ultrasonography to guide stent placement and ensure complete ostial coverage.[11]
4. If contrast injection is considered absolutely essential to check the status of the distal vessel, it is best performed through an aspiration catheter advanced to the distal part of the vessel (Fig. 12.6).[12]
5. If the aortocoronary dissection is large, perform serial noninvasive imaging (with computed tomography or transesophageal echocardiography) to ensure that the dissection has stabilized and resolved (Fig. 12.7). This is of particular importance if the dissection involves the ascending aorta.
6. Emergency surgery is rarely needed except in patients who develop aortic regurgitation, tamponade due to rupture into the pericardium, or extension of the dissection (Fig. 12.8).[7–9,13]

12.1.1.1.3 Side-Branch Occlusion

Side-branch occlusion can occur during CTO PCI, especially when subintimal dissection/reentry strategies are used, and is associated with higher frequency of post-PCI myocardial infarction (Fig. 12.9; Online Cases 100 and 102).[9,14]

Causes

1. Use of dissection/reentry strategies in vessels with side branches at the proximal or distal CTO cap.

Prevention

1. When treating CTOs that involve a bifurcation (e.g., those involving the right coronary artery crux) (Chapter 9, Section 9.5 and 9.6), a careful analysis of collateral support should be performed before the procedure to determine whether both branches have independent collateral support.
2. Avoid use of (antegrade or retrograde) dissection/reentry strategies when a bifurcation is present at the proximal or distal CTO cap.
3. Whenever possible, perform side-branch protection with a second guidewire.

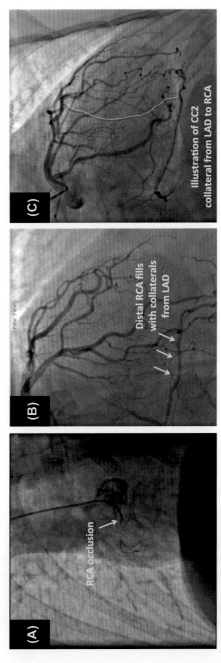

FIGURE 12.6 Use of an aspiration catheter for imaging a vessel distal to an aortocoronary dissection. Coronary angiography demonstrating a proximal right coronary artery (RCA) chronic total occlusion (A), with the distal vessel filling via septal collaterals (B, C). Retrograde guidewire crossing was successful via a septal collateral (D) followed by use of the reverse controlled antegrade and retrograde tracking and dissection technique (E) and wire externalization (F). After stenting, aortocoronary dissection became evident (G). A 7 Fr thrombectomy catheter was inserted into the distal RCA and contrast injection confirmed adequate distal angiographic result without propagating the aortocoronary dissection (H). *(Reproduced with permission from Al Salti Al Krad H, Kaminsky B, Brilakis ES. Use of a thrombectomy catheter for contrast injection: a novel technique for preventing extension of an aortocoronary dissection during the retrograde approach to a chronic total occlusion. J Invasive Cardiol 2014;26:E54–5.)*

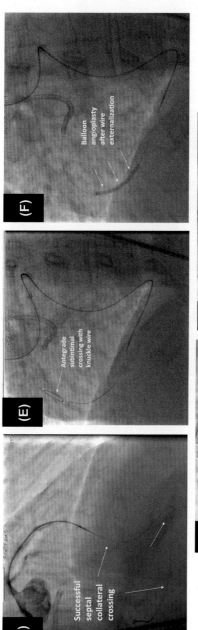

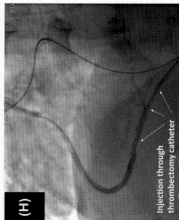

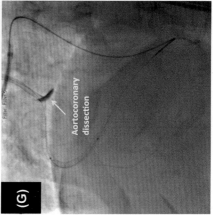

FIGURE 12.6 cont'd.

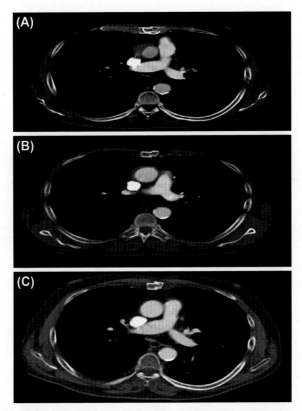

FIGURE 12.7 **Computed tomography (CT) follow-up of the aortocoronary dissection shown in (A) of Fig. 12.5.** (A) Twenty-four hours after the procedure, CT angiogram examination showing dislocation of intimal calcification with an eccentric double lumen. (B) One-month CT control examination demonstrating almost total resolution of the dissected thrombosed lumen. (C) Six-month CT control examination demonstrating total resolution of the dissection. *(Reproduced with permission from Boukhris M, Tomasello SD, Marza F, Azzarelli S, Galassi AR. Iatrogenic aortic dissection complicating percutaneous coronary intervention for chronic total occlusion.* Can J Cardiol *2015;31:320–7, Elsevier.)*

Treatment

1. Antegrade wiring of the occluded branch (which may be challenging if dissection/reentry strategies were used for crossing).
2. Retrograde recanalization of the occluded branch (if collaterals to that branch exist).
3. Use of IVUS may facilitate the identification of the cause of side branch occlusion (for example, the presence of a subintimal track at the level of the side branch ostium) and rewiring (Fig. 12.10).

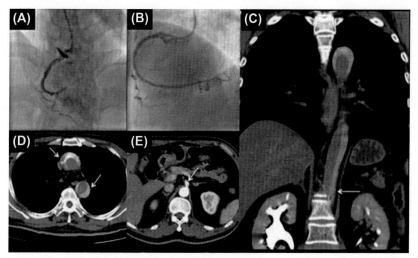

FIGURE 12.8 **Example of aortocoronary dissection extending into the descending aorta after chronic total occlusion intervention.** Angiography demonstrating proximal long segment dissection of the right coronary artery, extending to the sinus of Valsalva (A, B). After stenting with a 3.5 × 24 mm bare metal stent, the final angiogram revealed limited dissection to the sinus of Valsalva (B). Computed tomography imaging demonstrated a type A aortic dissection extending from the ascending aorta to the suprarenal abdominal level (C, D) with involvement of the aortic arch and celiac trunk (E). *(Reproduced with permission from Liao MT, Liu SC, Lee JK, Chiang FT, Wu CK. Aortocoronary dissection with extension to the suprarenal abdominal aorta: a rare complication after percutaneous coronary intervention. JACC Cardiovasc Interv 2012;5:1292–3.)*

If a dissection/reentry strategy is used it is important to minimize the extent of the subintimal dissection by reentering into the true lumen at the most proximal location possible, usually using the Stingray system, as described in Chapter 5.[15] Moreover, using the CrossBoss catheter may minimize the extent of subintimal dissection and facilitate reentry attempts. The presence of a coronary bifurcation at the distal CTO cap may favor use of a primary retrograde approach to minimize the risk for side-branch occlusion during antegrade crossing attempts.

12.1.1.1.4 Collateral Occlusion

Occlusion of a single large collateral (usually epicardial) vessel may cause severe ischemia and hemodynamic instability; therefore, there should be a high threshold for performing retrograde CTO PCI through such collaterals. Moreover, successful CTO recanalization causes rapid derecruitment of the collateral circulation, that may cause severe ischemia, if the vessel reoccludes.[16]

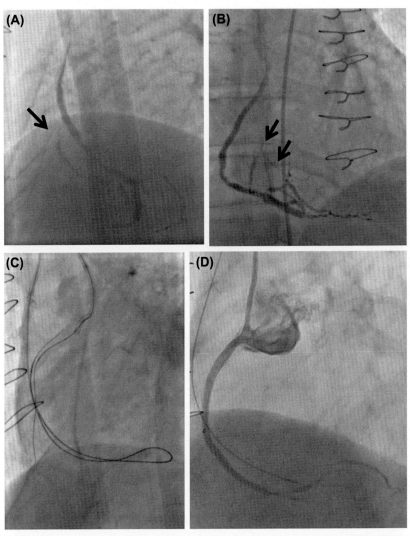

FIGURE 12.9   Acute side branch occlusion during a retrograde chronic total occlusion (CTO) intervention. CTO of the mid-right coronary artery that filled by a diffusely diseased saphenous vein graft (A). A large acute marginal branch originated at the distal CTO cap (*arrow*, A; *arrows*, B). Successful retrograde recanalization was achieved using a retrograde true lumen puncture technique (C). Stent placement restored antegrade flow to the distal right coronary artery but the acute marginal branch became occluded leading to inferolateral ST-segment elevation and postprocedural acute myocardial infarction. *(Reproduced with permission from Michael TT, Papayannis AC, Banerjee S, Brilakis ES. Subintimal dissection/reentry strategies in coronary chronic total occlusion interventions.* Circ Cardiovasc Interv *2012;5:729–38.)*

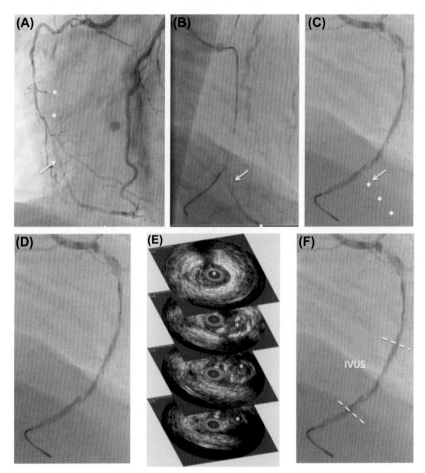

FIGURE 12.10 Use of intravascular ultrasonography to facilitate side branch occlusion assessment and treatment. Antegrade percutaneous coronary intervention was planned in a first attempt to recanalize a long chronic total occlusion (CTO) located in the mid-segment of the right coronary artery (RCA) (*asterisks*, A). Adequate progress was made using a parallel wire technique with a polymer-jacketed guidewire and a blunt-tip coil wire down to the posterolateral branch (B). However, antegrade injections after predilation with a 1.5 mm balloon revealed occlusion of the posterior descending artery (*asterisks*, C). Intravascular ultrasound (IVUS) of the mid-RCA and crux (E, F) revealed subadventitial course of the wire located in the posterolateral artery with compression of the vessel structures at the level of the RCA crux (*stars* in IVUS shown in E).

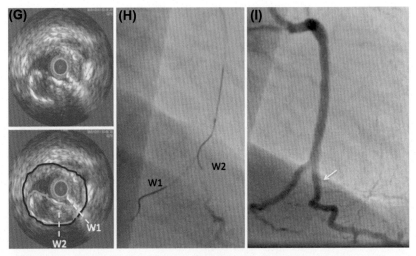

FIGURE 12.10 cont'd—IVUS-guided reentry to the posterior descending artery (PDA) with a new wire (W2) was performed (IVUS shadows of both guidewires are shown in G). This allowed successful advancement of W2 into the PDA (H) with a good result using a provisional stenting technique. *(Courtesy of Dr. Javier Escaned.)*

12.1.1.1.5 Subintimal Stenting

See Online Case 73.

Occasionally, subintimal distal position of the guidewire may not be appreciated and stents may be inadvertently deployed within the subintimal space, obstructing the outflow of the vessel (Fig. 12.11).[17] After this occurs, the patient may remain asymptomatic[18,19] or may develop ST-segment elevation due to side-branch loss.[17]

Causes

1. Misjudgment of the guidewire position before stent implantation (i.e., impression that the wire is located in the distal true lumen) when in reality it is located in the subintimal space.

Prevention

The first step in preventing this complication is to have a high threshold of suspicion. For example, apparent intraluminal position distal to the CTO when using a wire knuckle or very aggressive guidewires (≥12 gr tapered-tip wires, for example) should always raise the concern that the wire might actually be in the subintimal space close to the lumen, and be followed by an adequate check before proceeding to balloon dilation and stenting.

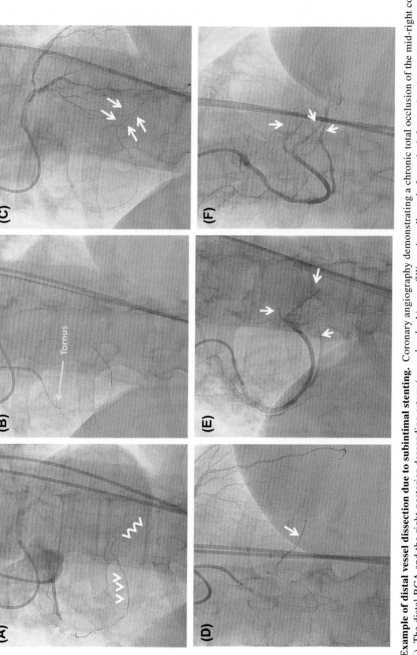

FIGURE 12.11 **Example of distal vessel dissection due to subintimal stenting.** Coronary angiography demonstrating a chronic total occlusion of the mid-right coronary artery (RCA) (*arrows*, A). The distal RCA and the right posterior descending artery (*arrowheads*, A) were filling via collaterals from the left anterior descending artery. The RCA was crossed antegradely with a Pilot 200 wire, however no other catheter, such as the Tornus catheter (*arrow*, B) could cross the occlusion. The RCA occlusion was crossed with a second Pilot 200 wire (*arrows*, C). Although contralateral injection suggested intraluminal distal wire position (*arrow*, D), after stenting antegrade flow in the acute marginal branch, right posterior descending artery and right posterolateral branch and right posterolateral branch (*arrows*, E) ceased. After rewiring and balloon angioplasty of the acute marginal branch, right posterior descending artery and right posterolateral branch (*arrows*, F), antegrade coronary flow was restored in all three vessels (see Online Case 73). (*Reproduced from Patel VG, Banerjee S, Brilakis ES. Treatment of inadvertent subintimal stenting during intervention of a coronary chronic total occlusion. Interv Cardiol 2013;5(2):165–9 with permission of Future Medicine Ltd.*)

To confirm that the wire has entered the distal true lumen *before* balloon dilation and stenting the following methods can be used (see also Chapter 4, Step 6):

1. Contralateral injection. This is the most commonly used method and is crucial for nearly all CTO procedures, even when most collaterals are ipsilateral, because ipsilateral collaterals may become compromised during crossing attempts.[20,21] After CTO crossing it is recommended to check intraluminal position of the guidewire in two orthogonal angiographic projections.

2. Contrast injection through a microcatheter. Routine use of this maneuver is discouraged, since antegrade contrast injection through a microcatheter always entails the risk of subintimal space staining and dissection propagation if the wire is not in the distal true lumen, which can then hinder subsequent reentry attempts. In selected cases, controlled microcatheter tip injections can be performed with care to verify intraluminal location, always checking for the back-bleeding sign (blood coming out of the microcatheter after waiting for at least 30 s from withdrawal of the guidewire).

3. Intravascular imaging. IVUS, particularly with a short-tip IVUS probe (Section 2.9) that is less likely to extend the suspected dissection, can be of great help in detecting subintimal guidewire position (Fig. 12.12).[22] Optical coherence tomography (OCT) has been reported as an alternative,[23] but is hampered by the limited penetration of OCT imaging, the distance of the OCT lens from the catheter tip, and the need to perform contrast or dextran[24] injections that may propagate a subintimal dissection.

4. Observing the wire movement into distal branches. This is suggestive of true lumen position,[25] but may also be misleading as the wire can also advance subintimally into side branches. Exchanging for a workhorse guidewire can facilitate the process, as the latter is less likely to advance into side branches.

Treatment

If subintimal guidewire position is confirmed, reentering the distal true lumen can be achieved using several strategies[14]:

1. Stingray system.[26]
2. Retrograde crossing of the target vessel.[21]
3. Wire-based techniques, such as wire redirection, the subintimal tracking and reentry (STAR) technique[27] or the limited antegrade subintimal tracking (LAST) and mini-STAR[28] techniques, in which the area of subintimal dissection is limited by reentering the true lumen as close as possible to the distal cap without propagating the dissection into the distal part of the vessel, as described in Chapter 5. However, wire-based reentry is discouraged due to unpredictability and potential for enlarging the dissection: use of the Stingray system or the retrograde approach are preferred instead.
4. IVUS guided reentry, following a similar technique as shown in Fig. 12.10.

These same techniques can also be employed to reenter the true lumen in cases of acute vessel closure due to dissection during non-CTO PCI (Figs. 12.13 and 12.14; Online Case 92).[29]

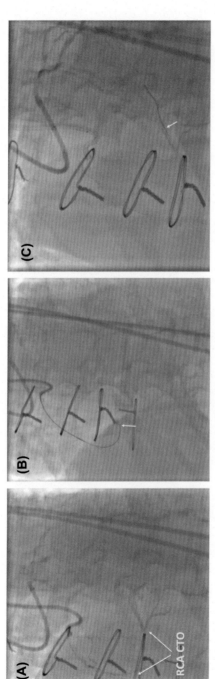

FIGURE 12.12   **Use of intravascular ultrasound (IVUS) to achieve and confirm successful reentry into the distal true lumen.** (A) Right coronary artery chronic total occlusion. (B) An antegrade knuckled guidewire is advanced past the distal cap. (C) Using the Stingray balloon a guidewire is advanced into the distal right coronary artery (*arrow*), however it is not 100% clear that it is located within the distal true lumen. (D, E) IVUS demonstrates that the guidewire is into the false lumen, not the distal true lumen. (F) Reentry is attempted again using the Stingray system. (G) A second guidewire is advanced distally and appears to be in a different location than the first guidewire. (H) IVUS confirms that the second guidewire is within the distal true lumen, whereas the first guidewire remains within the false lumen. (I) Final result after stenting (see Online Case 35).

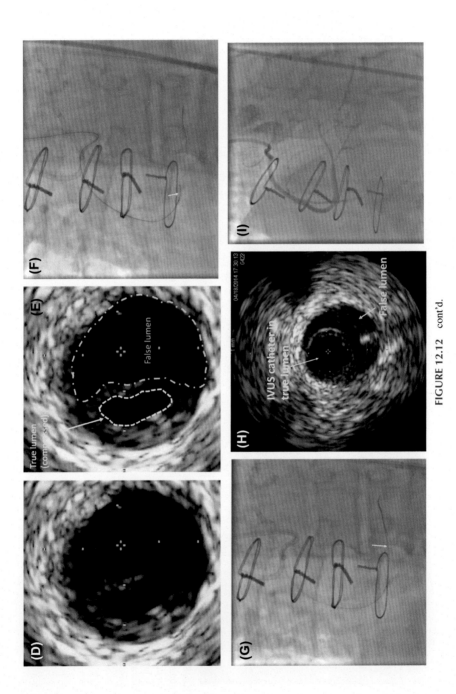

FIGURE 12.12 cont'd.

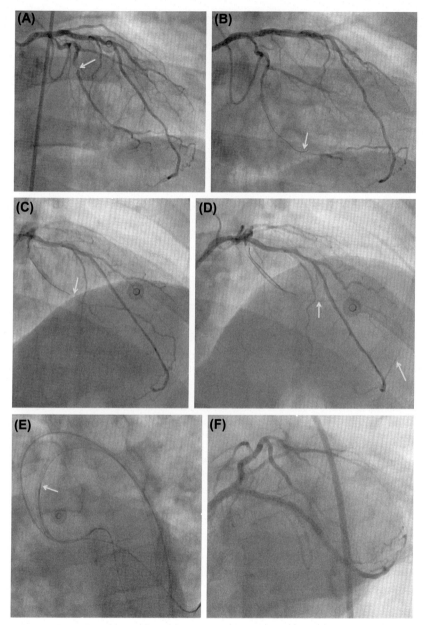

FIGURE 12.13   Use of the retrograde approach to treat acute vessel closure. (A) Severe lesion of the second obtuse marginal branch (*arrow*). (B) Subintimal guidewire crossing (*arrow*). (C) Acute vessel closure (*arrow*). (D) Successful retrograde crossing into the second obtuse marginal branch via an epicardial collateral from the distal left anterior descending artery (*arrows*). (E) After reverse controlled antegrade and retrograde tracking and dissection was performed the retrograde guidewire and the Corsair catheter was inserted into a second guide catheter (*arrow*), using the ping-pong guide catheter technique. (F) Successful recanalization of the second obtuse marginal branch after stenting (see Online Case 38).

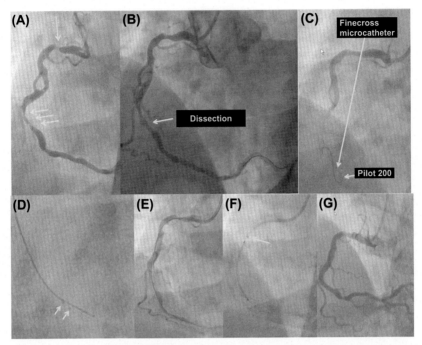

FIGURE 12.14   Use of antegrade dissection/reentry to treat acute vessel closure. Coronary angiography demonstrating a tortuous right coronary artery with a proximal (*arrow*, A) and a mid-lesion (*multiple arrows*, A). Mid-right coronary artery dissection after balloon predilation (*arrow*, B). Guidewire position and antegrade flow were lost after an unsuccessful attempt for stent delivery. After failure to advance a guidewire through the dissected segment, a knuckle was formed with a Pilot 200 guidewire (Abbott Vascular) (*arrow*, C) and advanced around the dissected segment. Using a Stingray balloon (*arrows*, D) and guidewire distal true lumen reentry was achieved (D). Using a GuideLiner catheter (*arrow*, F) two stents were successfully delivered with an excellent final angiographic result (G). *(Reproduced with permission from Martinez-Rumayor AA, Banerjee S, Brilakis ES. Knuckle wire and stingray balloon for recrossing a coronary dissection after loss of guidewire position. JACC Cardiovasc Interv 2012;5:e31–2.)*

12.1.1.1.6 Distal Vessel Dissection

Similar to inadvertent subintimal stenting described earlier, occasionally distal vessel dissection may occur, hindering further attempts to reenter into the distal true lumen.

Causes

1. Subintimal wire crossing with failed reentry attempts, often causing subintimal hematoma that compresses the distal true lumen.
2. Use of stiff and/or tapered-tip guidewires that may dissect the distal vessel.

Prevention

1. Avoid use of large loops during subintimal dissection techniques. It is best to perform the distal part of dissection with the CrossBoss catheter to minimize the extent of subintimal dissection.
2. Avoid antegrade contrast injections if the wire enters the subintimal space.
3. Prevent excessive movement of stiff and/or tapered-tip guidewires, for example by using the trapping technique for equipment exchanges.

Treatment

1. Use the subintimal transcatheter withdrawal (STRAW) technique (as described in Chapter 5, Figs. 5.38 and 5.39) to aspirate the subintimal hematoma and reexpand the distal true lumen.[30]
2. Use retrograde crossing into the distal true lumen.
3. Perform balloon angioplasty in the subintimal space and stop the procedure (investment procedure).[31] Coronary angiography can be repeated after 2–3 months to allow for healing of the dissection. Occasionally, flow into the distal true lumen is restored at follow-up angiography, especially when good antegrade flow is achieved during the index procedure.[32]

12.1.1.1.7 Embolization

Embolization, if severe, can cause profound hemodynamic compromise and requires prompt treatment.

Causes

1. Air. Air embolization is more likely to occur when the trapping technique is used without back-bleeding the guide catheter afterwards.
2. Aortic or iliac plaque, when the guide catheter is not thoroughly cleared after advancing through the aorta.
3. Thrombus, especially during prolonged cases if the ACT becomes low.
4. Coronary plaque, although this is more common when treating lesions causing acute coronary syndromes rather than CTOs.

Prevention

1. Always back-bleed the guide catheter after performing the trapping technique.
2. Always back-bleed the guide catheter after insertion, before injecting the coronary arteries.
3. Maintain high enough ACT throughout the case (>300s for antegrade cases and >350s for retrograde cases—Section 3.5).

Treatment

1. **Air embolization.** Administer 100% oxygen (helps with resorption of the air), aspirate (if large amount has been given); may require intracoronary epinephrine if the patient develops cardiac arrest.[33] Hemodynamic deterioration can be very rapid, requiring rapid intervention.
2. **Thrombus or plaque embolization.** Aspiration either through an aspiration catheter (such as the Export; see Online Case 19), through deep guide catheter engagement, or through a guide catheter extension. If unable to aspirate, laser or rheolytic thrombectomy may be useful.

12.1.1.2 Perforation

Coronary perforation is one of the most feared complications of CTO PCI, as it can lead to pericardial effusion and tamponade, sometimes necessitating emergency pericardiocentesis (and rarely cardiac surgery) to be controlled. Sometimes perforation may not lead to tamponade, but create a loculated effusion (especially in prior CABG surgery patients)[34–38] or intramyocardial hematoma.[39]

Although coronary perforations are common in CTO PCI (27.6% in one series[40]), most perforations do not have serious consequences, and the risk of tamponade is low, approximately 0.3%.[2] However, the risk is higher with retrograde CTO PCI (1.3%).[41,42] In contrast to PCI of non-CTO vessels, occlusion of a perforated target vessel in CTO PCI usually does not cause myocardial ischemia, allowing time for for testing sequential strategies, preparing hardware, performing an echocardiogram, etc.

12.1.1.2.1 Perforation Classification

Coronary perforations are best classified according to location, as location has important implications regarding management. There are three main perforation locations: (1) main vessel perforation, (2) distal artery perforation, and (3) collateral vessel perforation, in either a septal or an epicardial collateral (Figs. 12.15 and 12.16).[1]

The severity of coronary perforations has traditionally been graded using the Ellis classification[43]:

- Class 1: A crater extending outside the lumen only in the absence of linear staining angiographically suggestive of dissection.
- Class 2: Pericardial or myocardial blush without a ≥1 mm exit hole.
- Class 3: Frank streaming of contrast through a ≥1 mm exit hole.
- Class 3–cavity spilling: Perforation into an anatomic cavity chamber, such as the coronary sinus, or the right ventricle.

This classification has to be adapted to various scenarios discussed next, which were not contemplated at the time the Ellis classification was developed (i.e., perforation of epicardial and septal collateral channels).

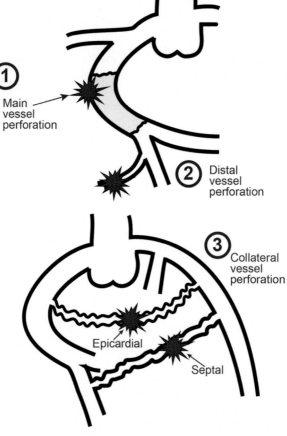

FIGURE 12.15   Types of coronary perforation.

Types of coronary perforation

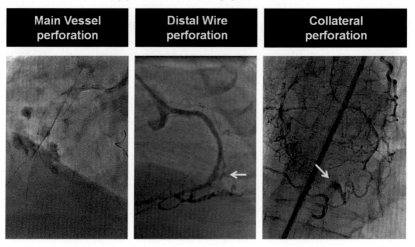

Main Vessel perforation	Distal Wire perforation	Collateral perforation

FIGURE 12.16   Examples of the different types of coronary perforation.

Prevention

1. Balloon advancement or dilation and microcatheter advancement should not be performed when the guidewire is not confirmed to be within the vessel architecture.
2. In some CTOs negative vessel remodeling may occur, potentially leading to vessel rupture during dilation if the balloon is sized according to proximal vessel dimensions (i.e., oversized for the CTO segment). IVUS imaging can be of great help in clarifying the vessel size distal to the CTO.
3. Whenever a large balloon is required in CART and reverse CART procedures to facilitate guidewire passage, use of IVUS can facilitate safe balloon size selection (75% of media to media vessel diameter).
4. During over-the-wire device exchanges, uncontrolled advancement of a hydrophilic or polymer-jacketed wire to a distal small branch may cause distal vessel perforation. Use of balloon trapping (first choice) or wire extensions (second choice) are preferred whereas Nanto's maneuver (saline injection through the microcatheter—hydraulic exchange; Section 3.7.2) should be avoided.
5. Outlining the anatomy of a collateral channel before and during its crossing, either with bilateral angiography or selective tip injections with the microcatheter, can help prevent guidewire exit and channel perforation.
6. Double-coil tip wires (Section 2.5.4 and 2.5.5) have excellent torque control and a blunt tip that, compared with polymer-jacketed guidewires, is less likely to cause collateral channel perforation. Hence double-coil tip wires should be considered the first choice for collateral crossing.
7. In general, perforation of an epicardial collateral channel is more difficult to control than a septal one, and this fact should be taken into consideration when choosing the most adequate interventional collateral channel.
8. Unfractionated heparin is preferred for anticoagulation, as it can be reversed in case of perforation, in contrast to bivalirudin.
9. A glycoprotein IIb/IIIa inhibitor should not be administered during CTO PCI, even after successful crossing and stenting, as it may cause an unrecognized perforation to bleed.

12.1.1.2.2 General Treatment of Perforations (Fig. 12.17)

Treatments specific to the perforation are described in the following section. General measures that can decrease the risk of continued bleeding into the pericardium include the following:

1. **Balloon inflation** proximal to the perforation to stop the bleeding. This should be performed *immediately* to prevent accelerated accumulation

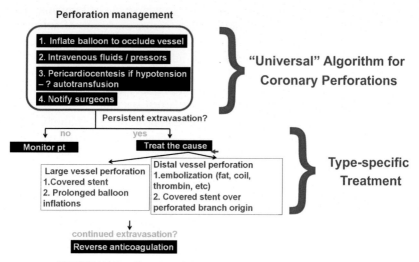

FIGURE 12.17 Overview of the management of coronary perforations.

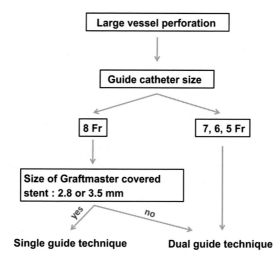

FIGURE 12.18 Determining the need for single or dual guide catheter technique to treat large vessel perforations.

of blood in the pericardial space and tamponade. Since a second arterial access is usually available in CTO PCI for contralateral injections, in many cases this can be used to introduce a second guide catheter with specific hardware to treat the perforation if needed although a single large guide catheter may suffice in many cases as shown in Fig. 12.18). The balloon should be the same size as the vessel and must be semicompliant and inflated to no more than 8–10 atm to ensure occlusion of antegrade flow, while avoiding stretching of the vessel. Hemostasis at the site of perforation can be maintained with an inflated balloon through the first guide catheter.

2. Administration of **intravenous fluids and pressors**, and possibly atropine, if the patient develops bradycardia due to a vagal reaction.

3. Appropriate timing for performing **pericardiocentesis**. Hemodynamic instability requires immediate pericardiocentesis, yet smaller size pericardial effusions may be best managed conservatively, as the elevated pericardial pressure due to the entrance of blood into the pericardial space may help tamponade the perforation site and decrease the risk for further bleeding. **Pericardiocentesis** can frequently be performed using X-ray guidance due to contrast exit into the pericardial space. Echocardiography remains important for assessing the size of pericardial effusion and the result of pericardiocentesis, and for determining whether pericardial bleeding continues. Use of an echocardiographic contrast agent can be useful for detecting ongoing bleeding into the pericardial space.[44]

4. **Cardiac surgery notification**. Notifying cardiac surgery early may facilitate subsequent treatment, if pericardial bleeding continues in spite of percutaneous management attempts.[44]

Reversal of anticoagulation in most cases should *not* be performed until after removal of interventional equipment, because reversing the heparin carries the risk of guide and/or target vessel thrombosis (see Online Case 39). Protamine dose for heparin reversal is 1 mg per 100 units of heparin, (max dose 50 mg) administered at a rate not to exceed 5 mg per minute. Protamine administration may cause anaphylactic reactions in patients treated with NPH insulin in the past or with a history of fish allergy.[45]

Platelet administration can help reverse the effect of abciximab, but glycoprotein IIb/IIIa inhibitors should not be administered during CTO PCI.

12.1.1.2.3 Large Vessel Perforation
See Online Cases 1, 3, 17, 39, 40, 63, 89, and 90).

Causes
1. Implantation of oversized stents or high-pressure balloon inflations (especially in heavily calcified vessels - Online Case 17).
2. Balloon rupture (when balloon rupture occurs, an angiogram should be obtained immediately after removal of the ruptured balloon to determine whether vessel perforation has occurred) (see Online Cases 3, 39, 40, and 90).

3. Wire exit from the vessel during CTO crossing attempts, followed by inadvertent advancement of equipment (such as balloons or microcatheters) into the pericardial space. Whereas wire perforation alone seldom causes blood extravasation and pericardial effusion (because it creates a very small, self-sealing hole), catheter/balloon advancement over the wire enlarges the hole, increasing the risk for blood extravasation. Occasionally the contrast extravasation may not occur until after a stent is placed over the perforated area.

Prevention

1. Avoid use of oversized stents and balloons.
2. Avoid very high pressure balloon inflations.
3. Always confirm guidewire position within the vessel architecture (true lumen or subintimal space) before advancing other equipment.

Treatment (Figs. 12.18–12.21)

1. Inflate a balloon proximal to the perforation to stop the bleeding.
2. If extravasation persists in spite of anticoagulation reversal and prolonged balloon inflations, place a covered stent (Graftmaster Rx in the United States; Section 2.10.1).[46,47]
3. Depending on the size of the guide catheter being used and the size of the covered stent (Fig. 12.18), delivery of the covered stent could be achieved using (a) a single guide catheter (also called block and deliver technique[48]) (Fig. 12.19) or (b) two guide catheters (dual guide catheter, also called ping- pong guide catheter, or dueling guide catheter technique) (Figs. 12.20 and 12.21). The goal of both techniques is to minimize bleeding into the pericardium while preparing for covered stent delivery and deployment. If the balloon used for hemostasis and the covered stent can fit through a single (usually 8 Fr) guide catheter then the single guide catheter technique is used; otherwise two guide catheters are required.
4. Single guide catheter technique for delivering a covered stent (Fig. 12.19).
5. Dual guide catheter technique (Fig. 12.20).[49]
6. Prolonged balloon inflations.

If a covered stent cannot be delivered to the perforation site, prolonged balloon inflation could often lead to hemostasis. Usually heparin is not reversed until after hemostasis is achieved and equipment is removed from the coronary artery. If pericardial bleeding continues despite prolonged balloon inflations, emergency cardiac surgery may be required (see Online Cases 17 and 40).

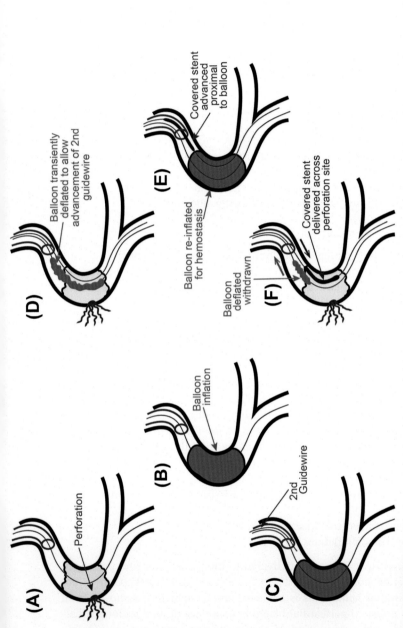

FIGURE 12.19 **Single guide catheter technique for delivering a covered stent.** (A) Large vessel perforation has occurred. (B) A balloon is inflated at the perforation site, stopping pericardial bleeding. (C) A second guidewire is advanced to the tip of the guide catheter. (D) The balloon achieving hemostasis is transiently deflated to allow advancement of the second guidewire. (E) The balloon that is over the first guidewire is reinflated and a covered stent is advanced to the tip of the guide catheter. The covered stent is advanced toward the inflated balloon that acts as a distal anchor (Section 3.6.5), facilitating delivery of the covered stent to the perforation site. (F) The balloon that is over the first guidewire is deflated and the covered stent is delivered across the perforation site. If the covered stent cannot be delivered to the perforation site, the balloon is reinflated to prevent pericardial bleeding until other techniques to facilitate delivery of the covered stent are employed. (G) The first guidewire and the hemostasis balloon are removed. (H) The covered stent is deployed and postdilated (the Graftmaster requires high-pressure postdilation to achieve hemostasis). (I) The perforation is sealed. Sometimes implantation of another stent proximal or distal may be necessary to seal any residual dissection that can serve as reentry point for bleeding (see Online Case 3).

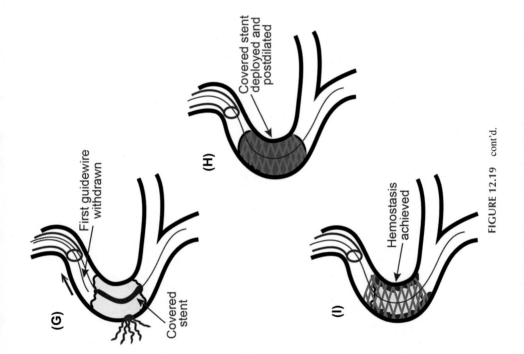

(G)

First guidewire withdrawn

Covered stent

(H)

Covered stent deployed and postdilated

(I)

Hemostasis achieved

FIGURE 12.19 cont'd.

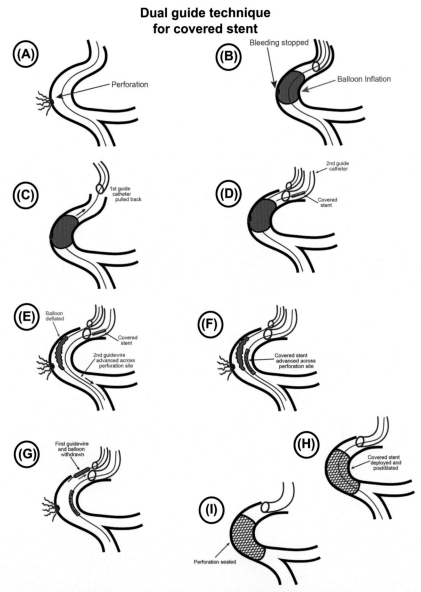

FIGURE 12.20 Illustration of the dual guide catheter technique for delivering a covered stent. (A) Large vessel perforation has occurred. (B) A balloon is inflated at the perforation site, stopping pericardial bleeding. (C) The guide catheter is pulled back into the aorta. (D) A second guide catheter (ideally 8 Fr) is advanced to the perforated vessel ostium. A second guidewire is inserted into the second guide catheter along with a covered stent. (E) The balloon that is over the first guidewire is deflated to allow advancement of the second guidewire across the perforation site. (F) The covered stent is delivered to the perforation site. If the covered stent cannot be delivered to the perforation site, the hemostasis balloon is reinflated to prevent pericardial bleeding until other techniques to facilitate delivery are employed. (G) The first guidewire and the balloon achieving hemostasis are removed. (H) The covered stent is deployed and postdilated (the Graftmaster requires high-pressure postdilation to achieve hemostasis). (I) The perforation is sealed.

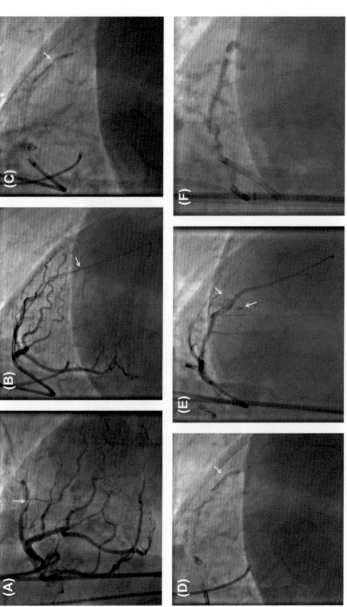

FIGURE 12.21 Illustrative case of large vessel perforation during chronic total occlusion (CTO) percutaneous coronary intervention. (A) CTO of a calcified mid-left anterior descending artery. (B) Successful crossing of the CTO with a Fielder XT guidewire. (C) Balloon undilatable lesion. (D) After orbital atherectomy the balloon undilatable lesion is expanding. (E) Perforation of the mid-left anterior descending artery after balloon rupture. (F) Balloon inflated proximal to the perforation, preventing pericardial bleeding. (G) In addition to the first guide catheter (*arrowhead*) through which the balloon covering the perforation site is inserted, a second guide catheter (*arrow*) is inserted, through which a second wire is advanced distal to the perforation site. (H) A 2.8 × 19 mm covered stent (*arrow*) is advanced over an 8 Fr GuideLiner (*arrowhead*) to the perforation site. (I) After deployment of the covered stent the perforation is sealed (*arrow*). (J) Echocardiogram showing a small pericardial effusion. (K) Thrombus formation (*arrow*) within stents placed in the left anterior descending artery. (L) Perforation of the proximal left anterior descending artery extending into the left main (*arrow*). (M) Excellent final angiographic result after thrombus aspiration and stenting of the left main. (N) Echocardiogram at the end of the procedure demonstrating a small pericardial effusion (see Online Case 39).

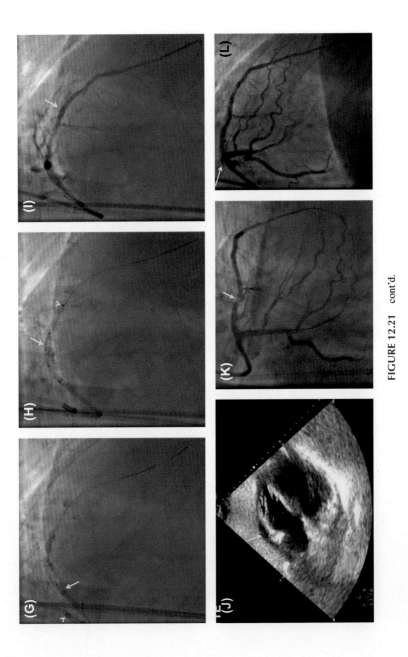

FIGURE 12.21 cont'd.

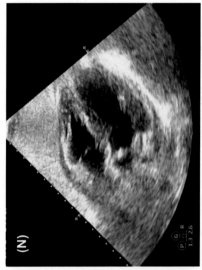

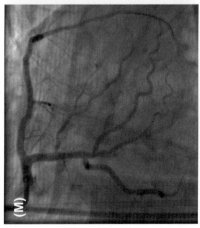

FIGURE 12.21 cont'd.

12.1.1.2.4 Distal Vessel Perforation

See Online Cases 26, 41, 42, and 43.

Distal vessel perforations (Figs. 12.22 and 12.23) can occasionally be difficult to diagnose, especially when collimation is used to minimize radiation exposure. Because blood flow into the pericardium may be slow, tamponade may not occur until several hours after the procedure has ended.[50] Patients with distal vessel perforation should be monitored closely and should not receive a glycoprotein IIb/IIIa inhibitor, as tamponade may not develop until hours after the end of PCI.

Causes

1. Inadvertent advancement of a guidewire and/or microcatheter into a distal small branch. Stiff, tapered, and polymer-jacketed guidewires are more likely to cause such perforations. Knuckled guidewires are not safe: they can still exit the vessel and cause a perforation.

Prevention

Distal wire perforation can be prevented by:

1. Paying meticulous attention to distal guidewire position during attempts to deliver equipment, especially when stiff and polymer-jacketed wires are used, as those are more likely to perforate than workhorse guidewires.
2. Using the trapping technique to minimize wire movement during equipment exchanges.
3. Exchanging a stiff or polymer-jacketed guidewire for a workhorse guidewire immediately after confirmation of successful crossing.

Treatment (Fig. 12.22)

Step 1 Inflate a Balloon Proximal to the Perforation

As in all coronary perforations the first management step is to inflate a balloon proximal to the perforation site to stop bleeding into the pericardium (Section 12.1.1.2.2 and Fig. 12.22A and B). Pericardiocentesis may be required if the patient develops hypotension. Notifying cardiac surgery could expedite management in case percutaneous treatment fails.

Step 2 Assess for Continued Pericardial Bleeding

If balloon inflation seals the perforation, observation and heparin reversal (after removal of equipment from the coronary artery) may be all that is needed. Sometimes, suction applied through a microcatheter may collapse the vessel and achieve hemostasis.[51] However, in most cases definitive treatment with embolization or a covered stent is preferred to minimize the risk for late reopening and late tamponade.

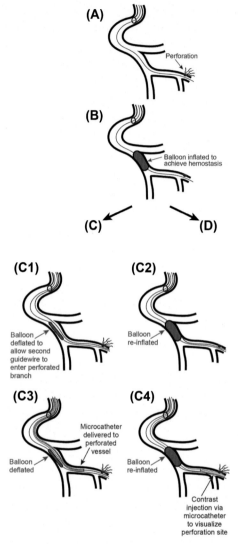

FIGURE 12.22 **Treatment of distal vessel perforations.** (A) Distal vessel perforation with active bleeding into the pericardium. (B) A balloon (blocking balloon) is inflated proximal to the perforation site to stop pericardial bleeding. **(C) Embolization:** (C1) The blocking balloon is temporarily deflated to allow advancement of a second guidewire into the perforated branch. (C2) The blocking balloon is reinflated to stop pericardial bleeding. (C3) The blocking balloon is transiently deflated to allow delivery of a microcatheter into the perforated vessel. (C4) The blocking balloon is reinflated. Injection of contrast is performed through the microcatheter to clarify the location of the perforation.

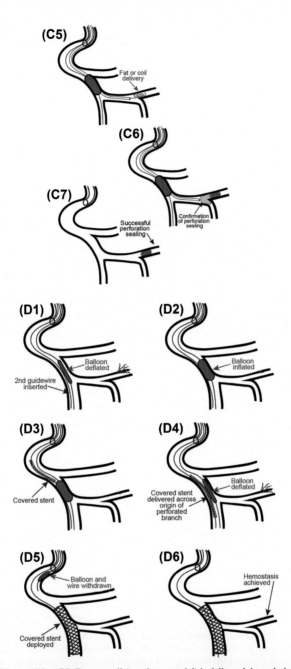

FIGURE 12.22 cont'd—(C5) Fat or a coil (or other material) is delivered through the microcatheter proximal to the perforation site. If the coil position is considered satisfactory the coil is released. (C6) Contrast is injected through the microcatheter to determine whether the perforation has been sealed (sometimes sealing is delayed for a few minutes after coil delivery). (C7) Successful sealing of the perforation. **(D) Covered Stent:** (D1) The blocking balloon is temporarily deflated to allow advancement of a second guidewire into the main vessel. (D2) The blocking balloon is reinflated to stop pericardial bleeding. (D3) A covered stent is advanced over the second guidewire proximal to the blocking balloon (which acts as distal anchor). (D4) The blocking balloon is deflated and the covered stent is advanced across the ostium of the perforated vessel. (D5) After removal of the blocking balloon and the first guidewire the covered stent is deployed. High-pressure postdilation should be performed in nearly all cases. (D6) Successful sealing of the perforated branch. This needs to be confirmed with contralateral injection to rule out retrograde filling of the perforated branch.

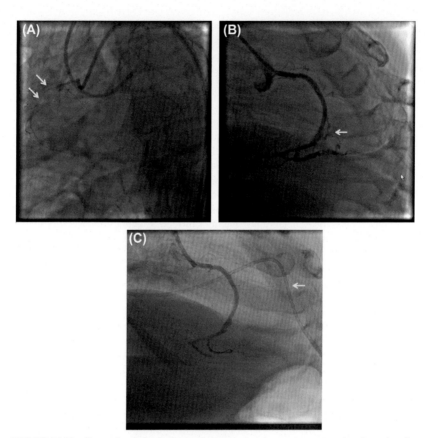

FIGURE 12.23 **Example of distal wire perforation.** A right coronary artery chronic total occlusion (*arrows*, A) was successfully crossed antegradely using a 12 g tip guidewire. Poststenting angiography demonstrated a side branch perforation with active bleeding into the pericardium (*arrow*, B). Emergency pericardiocentesis was performed (*arrow*, C). Delivery of a covered stent failed, but after prolonged balloon inflation and reversal of anticoagulation the side branch bleeding stopped. *(Reproduced with permission from Brilakis ES, Karmpaliotis D, Patel V, Banerjee S. Complications of chronic total occlusion angioplasty.* Interv Cardiol Clin *2012;1:373–89.)*

Step 3 Decide About Embolization or Covered Stent Implantation

Embolization is the most common treatment for distal vessel perforation and can usually be achieved using fat or coils (Fig. 12.22C1–C7). In rare cases thrombin, thrombus,[52] microparticles, portion of a guidewire,[53] or other materials are used for embolization. Embolization can in most cases be achieved through a single guide catheter using the block-and-deliver technique.[48,54]

Embolization may not be feasible in some cases, for example when the perforated branch is too small or too angulated to allow wiring and delivery of a microcatheter. In such cases an alternative treatment strategy is implantation of a **covered stent** across the ostium of the perforated branch.

In rare cases in which neither embolization nor covered stent delivery are feasible, **prolonged balloon inflations** may lead to hemostasis, otherwise cardiac surgery may be required. Cardiac surgery was required in only 3% of 1762 coronary perforations reported by the British Cardiovascular Society between 2006 and 2013.[55]

Embolization of Distal Vessel Perforation
(Fig. 12.22C1–C7)

Step 1 (C1) The blocking balloon is temporarily deflated to allow advancement of a second guidewire into the perforated branch.

Step 2 (C2) The blocking balloon is reinflated to stop pericardial bleeding.

Step 3 (C3) The blocking balloon is transiently deflated to allow delivery of a microcatheter into the perforated vessel.

Fat can be delivered through any microcatheter, but if the plan is to deliver a coil, the choice of microcatheter is critical. Most commercially available coils, such as the Azur (Terumo), Interlock (Boston Scientific), and Micronester (Cook) are compatible with 0.018 inch microcatheters, such as the Progreat (Terumo) or the Renegade (Boston Scientific), and *cannot* be delivered through the standard 0.014 inch microcatheters that are used for CTO crossing, such as the Corsair, Turnpike, and FineCross (although in bench testing we are able to deliver a Cook Micronester coil through a FineCross microcatheter). However, there are neurovascular coils (such as the Axium coil, Medtronic, Section 2.10.2) compatible with 0.014 inch microcatheters (Table 12.2).

Step 4 (C4) The blocking balloon is reinflated. Injection of contrast is performed through the microcatheter to clarify the location of the perforation.

If the microcatheter is too proximal, it can be repositioned so that the extent of the proximal vessel occluded with the coil is minimized.

Step 5 Embolization.

Embolization is most commonly done using fat or coils. Fat is preferred in most cases (except for very large perforations) because of universal availability, low cost, and biologic compatibility; however, delivery is not as controlled as when a coil is used (Table 12.2).

TABLE 12.2 Advantages and Disadvantages of Fat Versus Coil Embolization for Treating Distal Coronary Perforations

	Fat	Coil
Visibility	No (unless incubated with contrast)	Yes
Controlled delivery	No	Yes (if detachable coils are used)
Catheter needed for delivery	Any microcatheter	May need bigger microcatheter (0.018 inch) although any microcatheter can be used for neurovascular coils
Availability	Universal	Often limited
Cost	0	High

Fat Embolization

a. Fat can be harvested by advancing a hemostat in the femoral arteriotomy site (Fig. 12.24). Larger pieces can be cut into smaller ones using a scalpel.

b. Fat is then dipped into contrast for about a minute to absorb contrast and become visible under X-ray.

c. Loading the fat into the microcatheter can be challenging, because fat has low density and floats on water. Turning the microcatheter hub upside down can facilitate this step (Fig. 12.25).

d. The fat is injected through the microcatheter by flushing the microcatheter with saline.

e. Several fat pieces may need to be delivered to seal the distal vessel perforation.

Coil Embolization

a. Since coiling is very infrequent in the cardiac catheterization laboratory, obtaining familiarity with how to deliver and deploy a coil before a complication occurs can significantly facilitate management. Alternatively, obtaining help from an interventional radiologists (radiologists have significant experience with coiling and embolization) can be very helpful.

b. Having only one or two types of coils is usually sufficient.

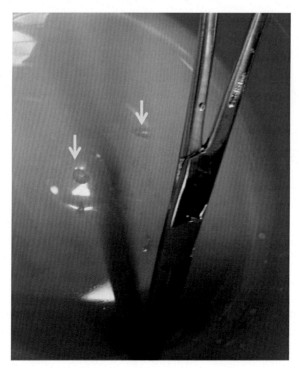

FIGURE 12.24   Harvested fat pieces (*arrows*) from the femoral arteriotomy site using a forceps.

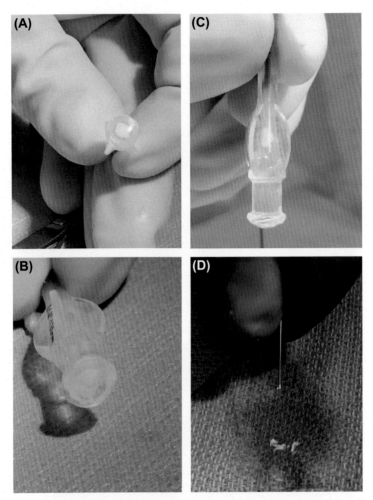

FIGURE 12.25 How to deliver fat for sealing a distal vessel perforation. Upon insertion into the hub of the microcatheter, fat floats (A). Upon turning the microcatheter hub upside down (B, C) the fat fragments advance into the microcatheter lumen (C). After injection with a syringe the fat particles are delivered through the tip of the microcatheter (D). *(Reproduced with permission from Shemisa K, Karatasakis A, Brilakis ES. Management of guidewire-induced distal coronary perforation using autologous fat particles versus coil embolization. Catheter Cardiovasc Interv 2016;89:253–8.)*

    c. Coil essentials. As described in Section 2.10.2, the most important character-
        istics of coils are (1) microcatheter compatibility (0.018 vs. 0.014 inch) and (2)
        mechanism of release (pushable vs. detachable). It is ideal to have 0.014 inch
        compatible coils (which can be deployed through the standard CTO microcath-
        eter) that are also detachable (which allows for accurate delivery to the desired
        location). Table 12.3 describes various 0.014 microcatheter-compatible coils

TABLE 12.3 Commercially Available Neurovascular Coils in the United States (Compatible With 0.014 inch Microcatheters)

Coil Name	Manufacturer	Description	Detachment System
Axium	Medtronic	Bare platinum coil with or without PGLA or nylon microfilaments enlaced through the coil	Axium I.D. (mechanical)
Hydrocoil (HES) MicroPlex (MCS)	Microvention	HES: Bare platinum coil combined with an expanding hydrogel polymer MCS: Bare platinum coil with various shapes and softness profiles	V-Grip (thermomechanical)
Orbit Cerecyte	Codman	Orbit: Bare platinum coil with various shapes and softness profiles Cerecyte: Bare platinum coil with PGA member within coil core	EnPower (thermomechanical)
Target	Stryker	Bare platinum coil with various shapes and softness profiles	InZone (electrolytic)

that are currently available in the United States. In our laboratory we currently use the Axium coil (Fig. 2.75). The smaller coil sizes are usually used for coronary perforations.

d. The coil is inserted into the delivery microcatheter and is advanced into the target vessel.

e. Pushable coils cannot be retrieved after delivery, whereas detachable coils can. If the location and configuration of the detachable coil is satisfactory, the coil is connected to the deployment system and released.

See Online video "How to deliver and deploy and Axium coil."

Step 6 (C6) Contrast is injected through the microcatheter to determine whether the perforation has been sealed (sometimes sealing is delayed for a few minutes after coil delivery).

If bleeding through the perforation site continues, additional fat pieces or coils are delivered.

Step 7 (C7) Angiography through the guide is performed to confirm complete sealing of the perforation.

An example of coil embolization is shown in Fig. 12.26.

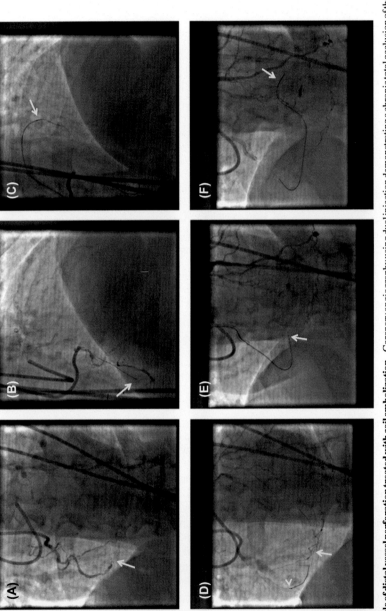

FIGURE 12.26 **Example of distal vessel perforation treated with coil embolization.** Coronary angiography using dual injection demonstrating a chronic total occlusion of the right coronary artery (*arrow*, A). Antegrade crossing attempts with a guidewire failed (*arrow*, B). Retrograde crossing attempts through a septal collateral (*arrow*, C) also failed. Repeat antegrade crossing with a CrossBoss catheter (*arrowhead*, D) and a knuckled Fielder XT guidewire (*arrow*, E) failed, but distal true lumen entry was achieved advancing the knuckled guidewire (*arrow*, F) (subintimal tracking and reentry—STAR technique). Coronary angiography after balloon predilation demonstrated distal vessel perforation (*arrow*, G). A balloon (*arrowhead*, H) was inflated, stopping pericardial bleeding, and a Progreat microcatheter (*arrow*, H) was delivered to the perforation site. Bleeding (*arrow*, I) continued after deployment of two 2 × 5 mm coils (*arrowheads*, I). Bleeding (*arrow*, J) slowed after deployment of a 2 mm × 2 cm detachable helical hydrocoil (*arrowhead*, J), and stopped 15 min later (*arrow*, K). Final angiography (L) revealed complete occlusion of the perforation site. Transthoracic echocardiography demonstrating a small pericardial effusion (*arrows*, M). No bleeding into the pericardium could be seen with administration of echocardiography contrast (N). (*Reproduced with permission from Tarar MN, Christakopoulos GE, Brilakis ES. Successful management of a distal vessel perforation through a single 8-French guide catheter: combining balloon inflation for bleeding control with coil embolization. Catheter Cardiovasc Interv 2015;86:412–6 (see Online Case 41).*)

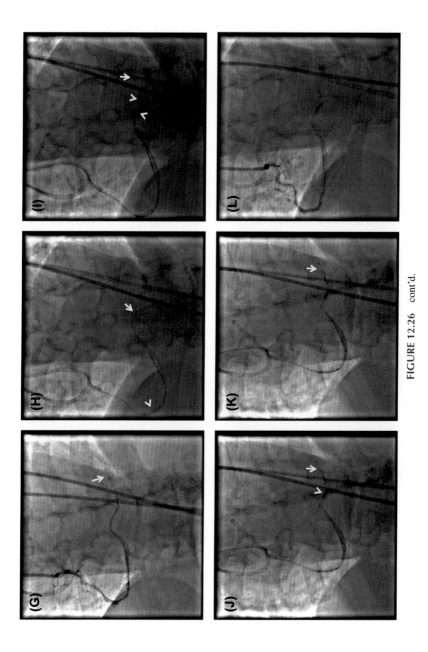

FIGURE 12.26   cont'd.

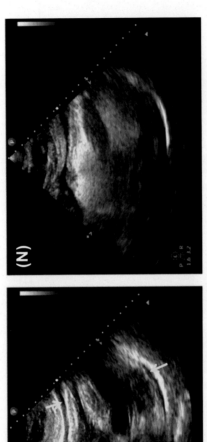

FIGURE 12.26 cont'd.

Covered Stent for Treating Distal Vessel Perforation (Online Case 86)
(Fig. 12.22D1–D6)

When the perforated vessel is too small or too tortuous, advancing a guidewire into it may not be feasible. Such cases could be treated with coiling of a more proximal larger branch, but if the perforated vessel is originating from a large vessel, occlusion of that vessel can be undesirable. An alternative solution is implantation of a covered stent over the origin of the perforated branch.[56]

Step 1 (D1) The blocking balloon is temporarily deflated to allow advancement of a second guidewire into the main vessel.

Step 2 (D2) The blocking balloon is reinflated to stop pericardial bleeding.

The goal of the blocking balloon is to stop bleeding from the perforation site into the pericardium, reducing the risk for tamponade. If a blocking balloon is used promptly, only a small pericardial effusion may develop that does not require pericardiocentesis.

Step 3 (D3) A covered stent is advanced over the second guidewire proximal to the blocking balloon (which acts as distal anchor).

The second guidewire allows for delivery of a covered stent, while maintaining hemostasis by the inflated (blocking) balloon. The blocking balloon provides extra support to the second guidewire (distal anchor), facilitating delivery of the bulky covered stent.

Step 4 (D4) The blocking balloon is deflated and the covered stent is advanced across the ostium of the perforated vessel.

The blocking balloon and its guidewire are usually not removed until after the covered stent has reached the perforation site.

Step 5 (D5) The covered stent is deployed. High pressure postdilation should be performed in nearly all cases.

High pressure dilation is important for covered stents (especially the Graftmaster) because they are hard to expand (the Graftmaster consists of two bare metal stents with a polytetrafluoroethylene (PTFE) sandwiched in-between).

Step 6 (D6) Successful sealing of the perforated branch. This needs to be confirmed with contralateral injection to rule out retrograde filling of the perforated branch.

Contralateral injection is required to confirm there is no continued bleeding into the perforation site via collaterals from the contralateral coronary artery.

Fig. 12.27 demonstrates a case of distal vessel perforation treated with covered stent implantation.

12.1.1.2.5 Collateral Vessel Perforation

Perforation of an epicardial collateral branch is a serious complication of retrograde CTO PCI, as it can rapidly lead to tamponade and may be particularly difficult to control.[57,58] In contrast, perforation of septal collaterals is unlikely to have adverse consequences,[3,59] although septal hematomas (Fig. 12.28)[60,61] and even tamponade[52] have been reported following septal wire perforation.

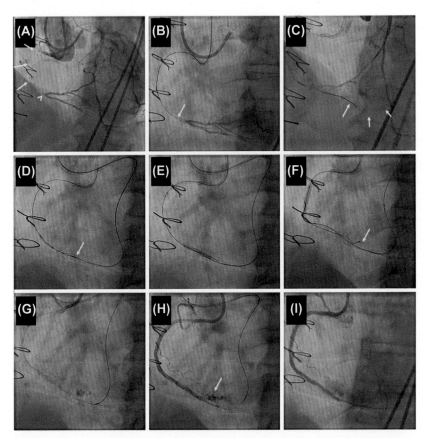

FIGURE 12.27 **Example of distal vessel perforation treated with covered stent implantation (see Online Case 42).** Coronary angiography demonstrating patent left anterior descending and circumflex arteries with a chronic total occlusion of the proximal right coronary artery
(*arrows*, A) with a calcified distal cap at the bifurcation of the right posterior descending and
posterolateral arteries (*arrowhead*, A). Antegrade wire escalation with multiple guidewires (Pilot
200, Gaia 2nd and 3rd and Confianza Pro 12, Abbott Vascular and Asahi Intecc) failed to penetrate
the distal cap (*arrow*, B). After multiple unsuccessful attempts retrograde crossing was performed
with a Sion guidewire (Asahi Intecc, *arrows*, C) through a septal collateral. After delivery of
the Corsair catheter to the distal cap (*arrow*, D) a retrograde knuckle was advanced through the
distal cap (E). GuideLiner-assisted reverse controlled antegrade and retrograde tracking and dissection was performed in the mid-right coronary artery (F) leading to externalization of an RG3
guidewire (Asahi Intecc) (G). After stenting, perforation of a small branch of the right posterior
descending artery (*arrow*, H) was seen, which was successfully sealed with implantation of a
2.8 × 19 mm Graftmaster Rx covered stent (Abbott Vascular) (I). *(Reproduced with permission
from Karatasakis A, Akhtar YN, Brilakis ES. Distal coronary perforation in patients with prior
coronary artery bypass graft surgery: the importance of early treatment. Cardiovasc Revasc Med
2016;17:412–7, Elsevier.)*

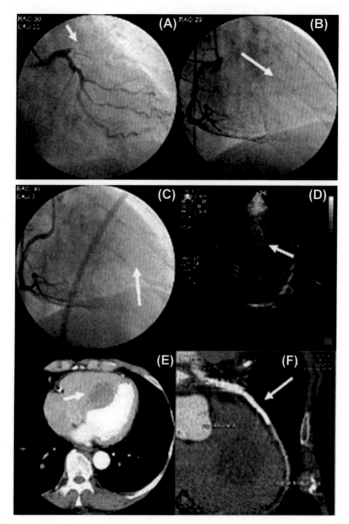

FIGURE 12.28   Example of septal hematoma due to septal collateral perforation. Left coronary angiography demonstrating a left anterior descending artery (LAD) chronic total occlusion (CTO) (*arrow*, A) and septal collaterals (*arrow*, B) from the right coronary artery to the LAD. Retrograde wiring was attempted (*arrow*, C) but was unsuccessful and the CTO was recanalized using the antegrade approach. Patient developed chest pain after the procedure and had increased cardiac biomarkers. Echocardiography demonstrated a mass (*arrow*, D) within the interventricular septum and computed tomography confirmed a septal hematoma (*arrow*, E) and patent stents in the LAD (*arrow*, F). *(Reproduced with permission from Lin TH, Wu DK, Su HM, et al. Septum hematoma: a complication of retrograde wiring in chronic total occlusion. Int J Cardiol 2006;113:e64–6.)*

Perforation in Patients With Prior Coronary Bypass Graft Surgery Carries Very High Risk

See Online Cases 42 and 43.

Although in the past prior CABG surgery was considered protective from tamponade in patients in whom perforation occurs, we currently know that loculated effusions can develop in these patients that can compress various cardiac structures[34] (such as the left atrium[35–37] or the right ventricle[38]). Such loculated effusions can be lethal, as they can be challenging to reach and drain percutaneously.

Therefore, perforations in prior CABG patients should be immediately treated (e.g., with covered stents, fat, or coils) to minimize the risk for loculated effusion development.[34]

Collateral vessel perforation may occur before, during, and after collateral vessel instrumentation with guidewires or devices. A meticulous technique is recommended to ensure its prevention, detection, and management.

Septal Collateral Perforation

Septal rupture/hematoma has been reported to occur in up to 6.9% of cases in a single series of patients treated with a retrograde approach.[62] In case reports, septal hematomas have caused asymptomatic bigeminy and severe chest pain, appear as an echo-free space in the interventricular septum on transthoracic echocardiography (Fig. 12.28), and resolve spontaneously.[63] Careful attention should be paid to the collateral branch course, as a collateral that appears to be septal, may in reality be epicardial. Moreover, perforation into the coronary sinus has been reported during attempts to cross a septal collateral.[64] Perforation into a cardiac chamber usually does not cause complications, however balloon dilation or advancement of additional equipment should be avoided.

Causes

1. Aggressive septal crossing guidewire maneuvers, especially advancing the microcatheter over the guidewire after it advances to an extraluminal location.
2. Selection of a very thin or tortuous septal channel.
3. Dilation of the septal channel.

Prevention

1. Selection of the most adequate interventional septal channel.
2. Exercise caution with tip injections of contrast in collateral channels if a wedged position of the microcatheter is suspected. The "back-bleeding sign" observed at the hub of the microcatheter may help in preventing injections that may cause barotrauma and rupture. In the absence of this sign it might be worth withdrawing the microcatheter tip to a slightly more proximal location.

3. Avoiding advancement of the Corsair microcatheter until the guidewire position (within the collateral vessel or the distal true lumen) has been ascertained.
4. Withdrawal of guidewires and microcatheters from collaterals after completion of CTO recanalization should be performed after collateral perforation has been ruled out. Bilateral injection while maintaining the retrograde wire position through the collateral vessel is useful.

Treatment

1. Usually no specific treatment is required, as septal perforation is self-limiting.[59]
2. Advancing the microcatheter is frequently enough to control bleeding.
3. Negative pressure applied from the wedged microcatheter contributes to collateral channel collapse and rupture sealing.
4. If tamponade occurs, the perforated collateral may need to be coiled.

Epicardial Collateral Perforation

Epicardial collateral perforation is riskier than septal collateral perforation, as it can rapidly lead to tamponade. Hence, wiring epicardial collaterals should be performed only by operators experienced in the retrograde approach. Epicardial collateral wiring is *not* safer in patients with prior CABG surgery or other surgery requiring opening of the pericardial sac, as bleeding may cause a loculated hematoma that compresses various cardiac chambers and causes hypotension, shock, or death[34,36,38]; such hematomas may require drainage under computed tomography guidance[36,38] or surgical evacuation.[65]

Causes

1. Aggressive guidewire advancement, especially through tortuous epicardial collaterals.
2. Agressive advancement of microcatheters or other equipment through small and tortuous collaterals.

Prevention

1. Epicardial collaterals should be wired using a contrast-guided technique with selective angiography using the microcatheter (tip injections). Surfing should *never* be performed in epicardial collaterals.
2. In contrast to septal collaterals, epicardial collaterals should *never* be dilated. However, microcatheters can (and should) be used in epicardial collaterals, paying careful attention to avoid catheter advancement in front of the guidewire.
3. Retrograde via ipsilateral collaterals can cause significant strain and injury of the collateral.[66] In such cases, the tip-in technique can be used to advance an antegrade guidewire across the CTO (hence avoiding wire externalization) and minimize the risk for perforation.

4. Before removal of the guidewire from a collateral, angiography should be performed to rule out perforation, as perforation is much easier to treat if guidewire access to the collateral is maintained.

Treatment

1. General perforation treatment measures should be employed, as described in Section 12.1.1.2.2.
2. In general, epicardial channel perforation is treacherous, with high risk of causing tamponade; therefore this complication should be taken seriously, with rapid action to correct it.
3. Prolonged inflation of a small balloon (Fig. 12.29) as well as negative pressure applied from the wedged microcatheter can cause collateral channel collapse and rupture sealing.

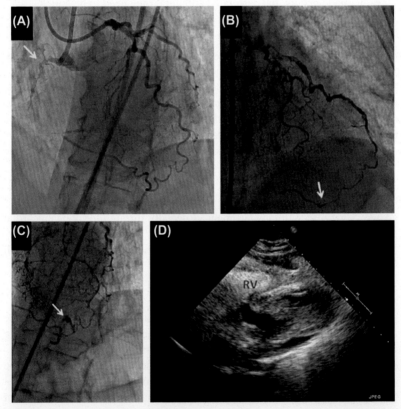

FIGURE 12.29 **Example of epicardial collateral perforation.** A retrograde crossing attempt was performed to treat a patient with a proximal right coronary artery chronic total occlusion (*arrow*, A). However, during attempts to cross the epicardial collateral from the diagonal to the right posterior descending artery, perforation occurred (*arrow*, B). After prolonged balloon inflation and reversal of anticoagulation bleeding through the perforation stopped (*arrow*, C), as confirmed by injection of echo-cardiographic contrast (D). (*Reproduced with permission from Brilakis ES, Karmpaliotis D, Patel V, Banerjee S. Complications of chronic total occlusion angioplasty.* Interv Cardiol Clin *2012;1:373–89.*)

4. Advancing the microcatheter is frequently enough to control bleeding through ruptured epicardial channels (see Online Case 13).
5. If bleeding continues, the perforation may need to be embolized/coiled (Fig. 12.30). Embolization or coiling should ideally be performed on both sides of the perforation, as blood flow can continue retrogradely in spite of occluding the antegrade limb of the collateral.[57]

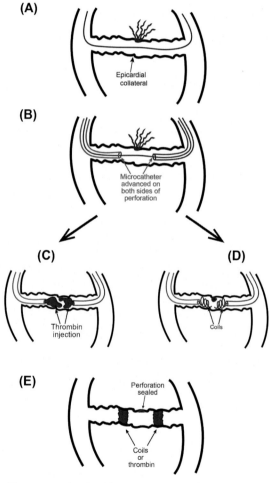

FIGURE 12.30 Treatment of epicardial collateral perforation. (A) Epicardial collateral perforation. Guidewire access within the collateral is preserved. (B) A microcatheter is advanced on each side of the perforated collateral. (C) Thrombin is injected from the two microcatheters or (D) coils are delivered from the two microcatheters. (E) The epicardial collateral perforation is sealed.

Step 1 Collateral Wiring (Fig. 12.30A)

The first step when epicardial collateral perforation is diagnosed is to maintain guidewire access (if guidewire is still within the collateral), otherwise the collateral is wired with a guidewire from both sides of the perforation.

What Can Go Wrong?
The collateral may not be able to be wired in case of extreme tortuosity or angulation. In such cases a covered stent could possibly be advanced, sealing the origin on both ends of the perforated collateral.

Step 2 Microcatheter Delivery (Fig. 12.30B)

A microcatheter is advanced from both ends of the collateral proximal to the perforation.

Step 3 Embolization (Fig. 12.30C and D)

Embolization can be achieved with thrombin or with coils.

Thrombin embolization[67]: Thrombin is mixed with a small amount of contrast to allow visualization of delivery within the perforated segment and injected slowly to avoid spilling over into the main vessel. A small volume (0.2–0.3 mL) of thrombin is injected slowly through each microcatheter (Fig. 12.31).

What Can Go Wrong?
Extreme care should be taken when injecting thrombin to seal collateral vessel perforation, as thrombin could cause a large myocardial infarction if it leaks into the main coronary vessel. Moreover, if the perforation occurs before recanalization of the CTO, occlusion of the collateral with thrombin might lead to ischemia or infarction of the myocardial territory supplied by the collateral, unless additional collaterals exist.

Coil embolization: Coils are delivered and deployed from both sides of the perforation, as described in detail in Section 12.1.1.2.4.

Step 4 Confirmation of Perforation Sealing (Fig. 12.30E)

A. Embolization from both sides of the perforation may not be feasible in cases in which the CTO cannot and has not been recanalized. Although in most of such cases selective embolization of the collateral channel proximal to the perforation is sufficient to cause hemostasis (after proximal coil implantation, retrograde intrachannel pressure from an occluded vessel is low (Fig. 12.32), bilateral angiography is always mandatory to rule out retrograde bleeding, which might eventually constitute an indication for cardiac surgery.

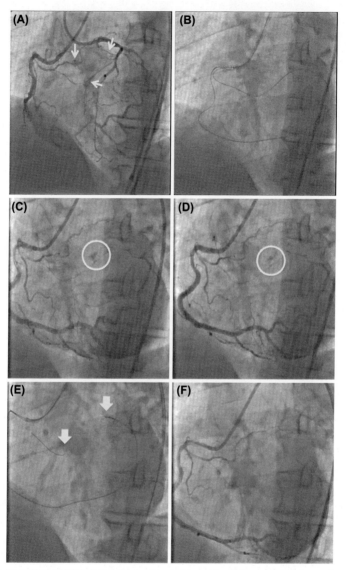

FIGURE 12.31 Epicardial collateral perforation treated with thrombin injection. Coronary angiography with dual injection demonstrating a right coronary artery chronic total occlusion (A). The distal right coronary artery filled by an ipsilateral atrial collateral (*arrows*, A). Antegrade crossing was unsuccessful. Following retrograde guidewire advancement through an atrial epicardial collateral, crossing was successful (B), followed by stent implantation over an externalized guidewire (C). Upon withdrawal of the Corsair catheter (Asahi Intecc), a collateral perforation was observed (C, D). After advancing a Corsair catheter in the antegrade direction and another Corsair catheter in the retrograde direction through the atrial collateral, thrombin was injected (E), resulting in sealing of the perforation (F). *(Reproduced with permission from Kotsia AP, Brilakis ES, Karmpaliotis D. Thrombin injection for sealing epicardial collateral perforation during chronic total occlusion percutaneous coronary interventions.* J Invasive Cardiol *2014;26:E124–6.)*

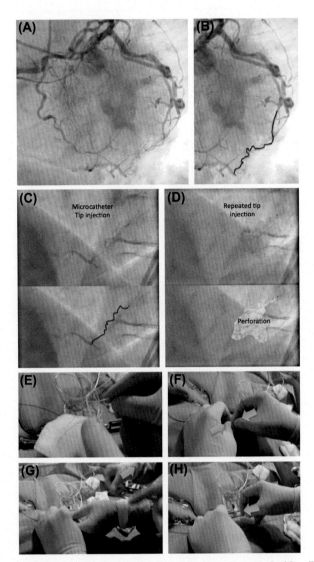

FIGURE 12.32 Example of management of rupture epicardial collateral channel with coil embolization. (A, B) Angiographic images obtained in a patient undergoing a second percutaneous coronary intervention (PCI) attempt of a right coronary artery (RCA) chronic total occlusion, several weeks after an antegrade PCI approach had failed to open the vessel. An interventional epicardial collateral with CC2 quality, connecting the circumflex artery and the posterolateral RCA branch (*blue line*, B), was chosen to perform retrograde PCI. Instrumentation of the collateral with a double coil, tapered, blunt-tip wire (Sion Blue) was performed under guidance with tip injections performed through a FineCross microcatheter (C). Despite this, a perforation with fast contrast spilling was noted after attempts to cross the tortuous channel (D). Negative pressure was applied to the microcatheter, without adequate hemostasis. The microcatheter was exchanged by a 0.021 inch Progreat microcatheter. An 0.018 inch Interlock coil (E) was inserted through the Progreat microcatheter while keeping tight the coil sleeve and the microcatheter hub (*arrows*, F). The images obtained during this PCI show how, once the coil was advanced inside the microcatheter, the sleeve was withdrawn (G), and the interlocked coil and its core wire was advanced under fluoroscopic control (F). The coil was positioned as close as possible to the perforation site (*red circle*), ensuring as much folding as possible of the coil (I) and eventually was released (J) (*blue arrow* shows how the coil is no longer locked once its junction with the core wire leaves the microcatheter tip). Complete interruption of bleeding was ensured by making simultaneous retrograde and antegrade injections (*blue lines*, K). Anticoagulation was not reversed during the treatment of collateral channel rupture. Antegrade PCI was then successfully performed during the same procedure (L) with implantation of a bioresorbable scaffold (Absorb) (*blue arrows*). The patient evolved favorably, with follow-up angiography performed 8 months later showing an excellent long-term result in the RCA. (*Courtesy of Dr. Javier Escaned.*)

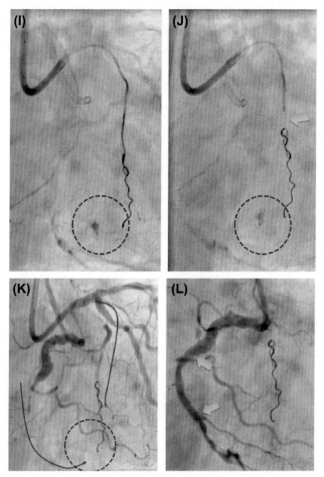

FIGURE 12.32   cont'd.

12.1.1.3 Equipment Loss or Entrapment

See Online Cases 74, 84, 87, and 104.

Equipment delivery may be challenging in CTO vessels that are frequently tortuous and calcified. Stent loss or wire entrapment may ensue.[68]

Causes

1. Attempting to deliver equipment via tortuous and calcified vessels.
2. Meeting of an antegrade balloon or stent with a retrograde Corsair catheter over the same guidewire.
3. Attempting to deliver equipment via a collateral during the retrograde approach to CTO interventions (that can predispose to both stent loss[69] and wire entrapment[70]).
4. Excessive torque applied to guidewires when the tip is entrapped in the CTO or when the wire fails to transmit its torque due to wire kinking.

5. Excessive rotation applied to a Tornus microcatheter, which may cause major distortion of its structure, with high risk of device entrapment.
6. Rotation applied to wire knuckles that might cause a wire knot, making its retrieval through microcatheters impossible.
7. Aggressive advancement of a rotational atherectomy burr.

Prevention

1. Careful preparation of the CTO target vessel before stenting (with balloon angioplasty, rotational atherectomy, etc.) before attempting to deliver stents.
2. Awareness of the force applied for advancing devices when using an externalized guidewire.
3. Regularly checking that the Tornus and other microcatheters can be withdrawn before further advancement.
4. Limiting the number of turns applied to the Tornus or other microcatheters to a maximum of 10 (either anticlockwise during Tornus advancement or clockwise during Tornus retrieval); then release the device to dissipate accumulated torque.
5. Checking visually the transmission of torque to the guidewire tip. In drilling maneuvers it is advisable to limit turns to 360 degrees, alternating clockwise and counterclockwise rotation.
6. Never allowing the tip of an antegrade microcatheter, balloon, or stent to meet with the tip of the retrograde microcatheter over the same guidewire.
7. Whenever possible, avoiding very tortuous epicardial collateral channels that may predispose to wire entrapment.
8. First delivering more distal and then more proximal stents, as attempting to deliver a stent through an already deployed stent may cause the stent to be dislodged or lost.[68]
9. If rotational atherectomy is used, slow advancement of the burr using a pecking motion and minimizing decelerations can help ablate more tissue before the lesion is actually crossed by the burr, hence minimizing the risk of entrapment.[71,72]

Treatment

1. First, determine whether it is best to attempt retrieval of the lost equipment or deploy/crush the lost equipment. Stent deployment in a coronary segment that is unlikely to be significantly affected by stenting may be the most time-efficient and low-risk strategy, as stent retrieval attempts can prolong the procedure, increase radiation exposure, and result in distal stent embolization or target vessel injury.[68,70,73]
2. If crushing or deployment of the lost stent (or encasing of a guidewire fragment with stents[70]) is selected, then intravascular imaging is important to ensure that there is adequate coverage of the lost equipment and that there is no wire unraveling extending proximally in the coronary artery or even into the aorta (see Online Case 74).[68]

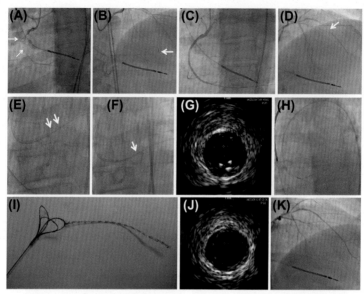

FIGURE 12.33 **Illustration of stent loss during retrograde chronic total occlusion (CTO) percutaneous coronary intervention.** Antegrade attempts for crossing a mid-right coronary artery CTO (*arrows*, A) failed due to subintimal wire passage. A Fielder FC guidewire was advanced retrogradely via a septal collateral over a Corsair catheter (*arrow*, B). The Corsair catheter was advanced distal to the CTO, followed by retrograde true lumen puncture, as confirmed by intravascular ultrasonography. Following retrograde balloon dilatation antegrade wiring was successful and after drug-eluting stent implantation in the right coronary artery thrombolysis in myocardial infarction 3 flow was restored (C). Imaging of the left anterior descending artery lesion revealed a lesion (*arrow*, D) at the site of the crossed collateral. During attempts to treat this lesion, a 2.5 × 28 mm stent was lost in the left main artery (*arrows*, E) and was snared by a Micro Snare Elite (*arrow*, F), but remained partially in the aorta and partially in the left main, as confirmed by intravascular ultrasound (*arrows*, G). After snaring with an En Snare (H) the stent was successfully retrieved (I), as confirmed by intravascular ultrasound (J). The left anterior descending artery patency was restored after stenting (K) (see Online Case 74). *(Reproduced from Iturbe JM, Abdel-Karim AR, Papayannis A, et al. Frequency, treatment, and consequences of device loss and entrapment in contemporary percutaneous coronary interventions.* J Invasive Cardiol *2012;**24**:215–21 with permission from HMP Communications.)*

3. If retrieval of lost devices (usually stents) is selected, it can be accomplished using various techniques, such as the small balloon technique, and snaring (see Online Case 74) (Fig. 12.33).
4. The small balloon technique can be used to retrieve a lost stent if wire position is maintained within the stent. A small balloon is advanced through the stent, inflated distal to the stent, and then withdrawn together with the lost stent into the guide catheter.
5. Various snares can be used for retrieving the lost equipment, most commonly the 3-loop snares, as described in Section 2.7.
6. Advancement of several wires in parallel to an entrapped one, followed by multiple turns to wrap it, can sometimes allow retrieval.

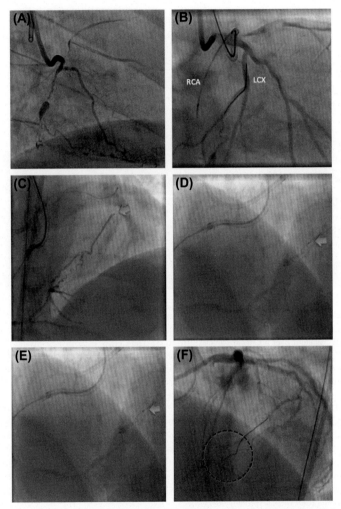

FIGURE 12.34 **Example of wire entrapment during retrograde chronic total occlusion (CTO) percutaneous coronary intervention.** A complex right coronary artery CTO was approached with a combined antegrade and retrograde technique (A). Being unable to make adequate progress antegradely, a retrograde approach through CC2 epicardial channels originating in the circumflex and following a left atrial course to the posterolateral right coronary artery branch was attempted (B). Outlining of the anatomy with tip injections performed through a Finecross microcatheter revealed a very tortuous anatomy (C–E). A double-coil wire (Sion Blue) was advanced with caution, but eventually became entrapped in the collateral channels (*red circle*, F). Eventually, the Finecross microcatheter was exchanged for a Corsair channel dilator and multiple rotations of the wire and the Corsair were performed with the tip as close as possible to the location of the wire entrapment, in an attempt to perform a controlled rupture of the guidewire (G). This was achieved, with two segments of the coil joined by a thin filament broken from the guidewire shaft (H). CTO revascularization was eventually not possible. Multidetector computed coronary angiography confirmed that the wire filament had not reached a major branch that would require additional actions (I, J). *(Courtesy of Dr. Javier Escaned.)*

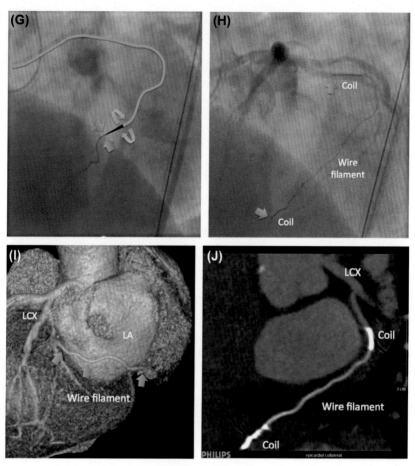

FIGURE 12.34 cont'd.

7. Controlled rupture of the wire might be considered in some cases, taking into account that frequently this leads to unfolding of the tip coil into a small filament (Fig. 12.34).

8. In general, strong traction should be avoided and, if required, potential associated risks should be taken into account and minimized. For example, since left main or proximal vessel dissection may occur as a result of trauma during device traction it is important to ensure that trapped wires are covered by a microcatheter.

9. Trapping the target device with a balloon close to the guiding catheter tip, followed by traction using the guiding catheter, should be considered (Online Case 84); this maneuver decreases the length of the device to which traction is applied and the risk of device rupture.

10. If a rotational atherectomy burr becomes entrapped, a second guidewire is passed across the area of the entrapped burr and balloon inflations are performed over the second wire to free the burr.[71] The wire and balloon can be advanced through the same guide catheter (if 8 Fr guide catheters are used or if the Rotablator catheter is cut and the drive shaft sheath is removed), through a second guide catheter,[74] or by exchanging the original guide catheter over the Rotablator catheter shaft for a larger guide catheter.[75]

12.1.2 Acute Cardiac Noncoronary Complications

Noncoronary cardiac complications include hypotension, periprocedural myocardial infarction, arrhythmias, and tamponade. Tamponade is the result of coronary perforation and has been discussed in detail in Section 12.1.1.2. Arrhythmias can complicate CTO PCI, but are infrequent and are usually caused by ischemia.

12.1.2.1 Hypotension During Chronic Total Occlusion Percutaneous Coronary Intervention

Continuous careful monitoring of the pressure and electrocardiographic tracing is critical for enhancing the safety of CTO (and non-CTO) PCI (Section 3.8). A differential diagnostic algorithm is critical to have in case hypotension occurs during CTO PCI. Potential causes of hypotension include:

1. **Pseudohypotension** (systemic pressure is normal but appears low on the arterial pressure tracing).
   a. Hemostatic valve (touhy) is open.
   b. Connection of catheter with pressure transducer is open (this should be readily apparent, since CTO PCI is almost always performed using two guide catheters and it is unlikely that both pressure transducers will be affected at the same time).
   c. Pressure dampening. This is common, especially in patients with ostial coronary lesions and with use of large guide catheters, such as 8 Fr. Dampening can be masked when side-hole guide catheters are used, hence the latter should never be used to engage the donor vessel, as they can mask ischemia and lead to patient hemodynamic collapse.
2. **True hypotension**
   a. Guide catheter-induced aortic regurgitation. This is common, especially when Amplatz guide catheters are used. The Amplatz guide may push the aortic cusp, causing severe aortic regurgitation. Simple guide catheter repositioning can correct the hypotension.
   b. Administration of vasodilators, such as nitroglycerin.
   c. Perforation and tamponade.

    d. Donor vessel ischemia.

    e. Other bleeding (for example, retroperitoneal bleeding). If retroperitoneal bleeding is suspected, fluoroscopy of the bladder may reveal displacement and strongly suggest this diagnosis.

    f. Systemic anaphylactic reaction, for example due to contrast allergy.

    g. Vasovagal reaction.

Prevention and Treatment

Prophylactic use of hemodynamic support devices (Section 2.13) during CTO PCIs at high risk for causing hemodynamic collapse could help reduce the associated procedural risk. Hemodynamic support devices can also provide support in patients who develop cardiogenic shock during the procedure.

    The following parameters could help with deciding whether to use hemodynamic support in a high risk CTO PCI.

    Anatomic parameters, such as:

1. Retrograde CTO PCI via last remaining vessel (see Online Case 11).
2. Retrograde CTO PCI via left internal mammary grafts (see Online Cases 29 and 46).
3. CTO PCI (especially retrograde) in patients with multivessel disease (see Online Cases 14, 18, 20, 31, and 51).

    Hemodynamic parameters:

1. Low ejection fraction.
2. High pulmonary capillary wedge pressure (risk heart catheterization before and during high-risk CTO PCI can help evaluate and monitor procedural risk).

    Comorbidities, such as:

1. Severe pulmonary disease.
2. Renal failure.

    Potential benefits with use of hemodynamic support devices should be balanced against potential risks, such as vascular access complications.

12.1.2.2 Periprocedural Myocardial Infarction

Postprocedural myocardial infarction is a potential complication of CTO PCI. However, its incidence is likely underdiagnosed due to lack of systematic screening with postprocedural serial cardiac troponin or CK-MB measurements.[76] Several mechanisms may lead to periprocedural myocardial injury or infarction during CTO PCI[1]:

1. Side branch occlusion (Online Case 102).[77]
2. Collateral vessel occlusion or injury, especially when collateral flow is provided by a single collateral (usually epicardial).

3. Injury of the target vessel distal to the CTO due to subintimal wire passage.
4. Donor vessel injury and/or thrombus and air embolization, as discussed in Section 12.1.1.
5. Septal hematoma formation.

Compared with antegrade only, retrograde CTO PCI is associated with higher rates of postprocedural cardiac biomarker elevation, but its impact on acute and long-term clinical outcomes remains controversial.[76,78,79] Meticulous attention to prevent vessel occlusion and perforation and equipment loss can help minimize the risk for post-PCI myocardial infarction.

12.1.2.3 Heart Block During Chronic Total Occlusion Percutaneous Coronary Intervention

Underlying conduction disease can pose a risk during CTO PCI. Ischemia or collateral manipulation can lead to increased vagal tone, which may convert a diseased but stable conduction system into an unstable one, potentially even progressing to cardiac arrest. In the setting of underlying trifascicular block, retrograde CTO PCI via septal collaterals can lead to block of the remaining fascicle and thus, complete heart block (Fig. 12.35).

Identification of underlying conduction disease should be added to the CTO operator's checklist when planning CTO PCI. Insertion of a temporary pacing wire at the start of the case could help prevent heart block and hemodynamic compromise during the intervention.

12.1.3 Other Acute General Complications

CTO interventions are subject to the same risks as non-CTO interventions.[1]

Vascular access complications can occur, especially given the frequent use of large sheaths in both femoral arteries (Section 3.4, see Online Case 104). Using unilateral or bilateral radial access[80,81] could potentially reduce the risk of vascular access complications, but may not provide adequate backup support and limits use of the trapping technique for equipment exchanges (Section 3.7). Moreover, femoral access can provide more support for retrograde collateral channel crossing and retrograde CTO PCI. Using fluoroscopy or ultrasonography to choose the femoral arterial puncture site may reduce the risk for vascular access complications.[82,83] Ultrasound guidance can also facilitate radial access.[84] Routine performance of iliofemoral angiography at the time of diagnostic angiography before referral for CTO PCI can also be very useful for optimal selection of the vascular access sites.

Systemic thromboembolic complications can complicate any cardiac catheterization, including CTO PCI. Careful attention to aspiration of the guide catheters after advancement through the aorta and use of 0.065 inch guidewires may help minimize scraping of the aorta and reduce the risk for peripheral

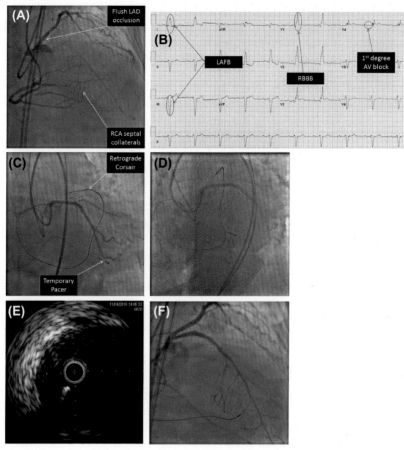

FIGURE 12.35   **Complete heart block during retrograde chronic total occlusion percuta-neous coronary intervention (PCI) in a patient with trifascicular block.** (A) A patient with ischemic cardiomyopathy was referred for PCI of an ostial occlusion of the left anterior descending artery. (B) Baseline electrocardiogram showed trifascicular block. (C) During retrograde crossing attempts via a septal collateral the patient developed complete heart block and cardiac arrest, requir-ing compressions and implantation of a temporary pacemaker. (D) The retrograde guidewire was advanced into the left main. (E) Intravascular ultrasound of the left main confirming intraluminal position of the retrograde guidewire. (F) Successful recanalization of the left anterior descending artery after stenting. *(Courtesy of Dr. Michael Luna.)*

embolization. Moreover, careful attention to anticoagulation (which is espe-cially important for retrograde CTO PCI given the risk for donor vessel throm-bosis) can minimize the risk of catheter thrombus formation and subsequent embolization.

Noncardiac bleeding may develop during long CTO procedures with profound anticoagulation (ACT >350s) if a known or unknown concomitant pathology (for example, a gastrointestinal tumor) is present. This possibility should be kept in mind if repeated vasovagal episodes occur during or after the procedure in the absence of other triggering causes.

Contrast allergic reactions may be prevented by using a premedication regimen. Such a regimen usually includes a steroid, an H1 antihistamine blocker (usually diphenhydramine), and an H2 antihistamine blocker such as cimetidine.

Contrast nephropathy could be prevented by adequate preprocedural hydration, limiting the volume of contrast administered (ideally the total contrast volume should be < 3.7× the creatinine clearance of the patient[85]) and administration of isoosmolar contrast media.[86]

With continuous improvement in X-ray systems that use lower radiation doses, contrast volume is becoming the more common limiting factor for needing to stop CTO PCI. The volume of contrast administered during CTO PCI can be reduced by:

1. Using a microcatheter to inject contrast instead of injecting through the guide catheter.
2. Using various markers on the vessel wall (such as calcification) of bypass graft clips and markers to determine equipment position.
3. Using IVUS and physiologic guidance: no-contrast or ultra-low contrast CTO PCI can be performed.[87]

Finally, **radiation injury** is of particular importance to CTO PCI and is discussed in detail in Chapter 10.

12.2 LONG-TERM COMPLICATIONS

The long-term durability and outcomes of CTO PCI require further study, as most studies have only reported acute procedural results. Similar to non-CTO interventions, patients who undergo CTO PCI may subsequently experience in-stent restenosis or stent thrombosis.

The incidence of **in-stent restenosis** and the need for repeat revascularization has decreased with drug-eluting stents (DES), especially second generation DES, as described in Chapter 11. It is important to avoid stent undersizing, as the target vessel is frequently negatively remodeled due to chronic hypoperfusion and increases in diameter over time.

Similarly, the risk of **stent thrombosis** post-CTO stenting has received limited study but appears to be lower with second- versus first-generation DES, as described in Chapter 11. Use of multiple and undersized stents in CTO PCI may predispose to stent thrombosis. To minimize this risk many operators routinely administer >12 months of dual antiplatelet therapy in patients who undergo successful CTO PCI, although the safety and efficacy of this regimen in this high-risk patient group remains unknown.

Late **coronary artery aneurysm** formation may also complicate CTO PCI (Figs. 12.36 and 12.37). Coronary artery aneurysms were seen in 7.3% of patients in whom retrograde intervention was performed versus 2.6% of patients

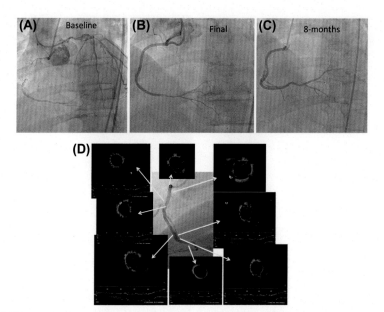

FIGURE 12.36   Late coronary artery aneurysm formation after chronic total occlusion interven-
tion. A right coronary artery chronic total occlusion was successfully crossed using a dissection/reentry
strategy and stented with three drug-eluting stents (B). Repeat angiography performed 8 months later
revealed a mid-right coronary artery aneurysm (C), as confirmed by optical coherence tomography (D).
The patient remained asymptomatic and repeat coronary angiography and long-term dual antiplatelet
therapy was recommended. *(Reproduced with permission from Brilakis ES, Karmpaliotis D, Patel V,*
Banerjee S. Complications of chronic total occlusion angioplasty. Interv Cardiol Clin 2012;1:373–89.)

in whom antegrade intervention was performed among 560 patients undergoing
CTO PCI in Japan.[88] Treatment of such aneurysms is controversial due to lack
of natural history data. At present a reasonable approach is to continue dual
antiplatelet therapy and perform serial angiographic and intravascular imaging
(with intravascular ultrasonography or ideally with optical coherence tomography, which has higher resolution), with aneurysm sealing limited to patients
with large aneurysms or aneurysms that enlarge during follow-up.[89] Some aneurysms may spontaneously resolve during follow-up (Fig. 12.37).

12.3 CONCLUSIONS

Every interventional procedure, including CTO PCI, is a balancing act between
potential risks and benefits (Chapter 1, Fig. 1.3). For patients in whom successful CTO PCI can provide significant benefit, the fear of complications should
not prevent us from performing the procedure. It should prompt us, however, to
meticulously and painstakingly prepare to prevent complications and to promptly
and efficiently treat them, should they occur. Our patients deserve no less.[90]

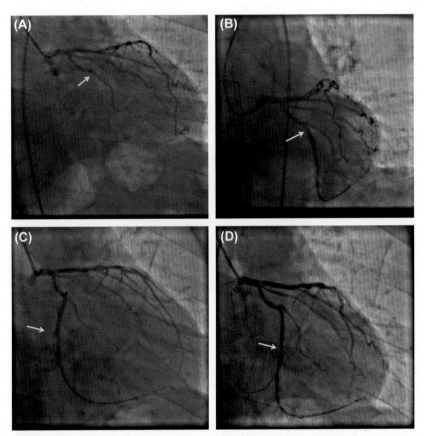

FIGURE 12.37 **Spontaneous aneurysm resolution after chronic total occlusion (CTO) intervention.** Percutaneous coronary intervention (PCI) of a mid-circumflex CTO (*arrow*, A) was successful (*arrow*, B) using an antegrade dissection/reentry technique. Follow-up angiography 3 months later demonstrated an aneurysm at the site of reentry (*arrow*, C) that resolved by 10 months post-PCI (D). (*Courtesy of Dr. Michael Luna.*)

REFERENCES

1. Brilakis ES, Karmpaliotis D, Patel V, Banerjee S. Complications of chronic total occlusion angioplasty. *Interv Cardiol Clin* 2012;**1**:373–89.
2. Patel VG, Brayton KM, Tamayo A, et al. Angiographic success and procedural complications in patients undergoing percutaneous coronary chronic total occlusion interventions: a weighted meta-analysis of 18,061 patients from 65 studies. *JACC Cardiovasc Interv* 2013;**6**:128–36.
3. Lee NH, Seo HS, Choi JH, Suh J, Cho YH. Recanalization strategy of retrograde angioplasty in patients with coronary chronic total occlusion – Analysis of 24 cases, focusing on technical aspects and complications. *Int J Cardiol* 2009.
4. Ge JB, Zhang F, Ge L, Qian JY, Wang H. Wire trapping technique combined with retrograde approach for recanalization of chronic total occlusion. *Chin Med J (Engl)* 2008;**121**:1753–6.

5. Lichtenwalter C, Banerjee S, Brilakis ES. Dual guide catheter technique for treating native coronary artery lesions through tortuous internal mammary grafts: separating equipment delivery from target lesion visualization. *J Invasive Cardiol* 2010;**22**:E78–81.

6. Papayannis A, Banerjee S, Brilakis ES. Use of the Crossboss catheter in coronary chronic total occlusion due to in-stent restenosis. *Catheter Cardiovasc Interv* 2012;**80**:E30–6.

7. Shorrock D, Michael TT, Patel V, et al. Frequency and outcomes of aortocoronary dissection during percutaneous coronary intervention of chronic total occlusions: a case series and systematic review of the literature. *Catheter Cardiovasc Interv* 2014;**84**:670–5.

8. Boukhris M, Tomasello SD, Marza F, Azzarelli S, Galassi AR. Iatrogenic aortic dissection complicating percutaneous coronary intervention for chronic total occlusion. *Can J Cardiol* 2015;**31**:320–7.

9. Carstensen S, Ward MR. Iatrogenic aortocoronary dissection: the case for immediate aortoostial stenting. *Heart Lung Circ* 2008;**17**:325–9.

10. Gomez-Moreno S, Sabate M, Jimenez-Quevedo P, et al. Iatrogenic dissection of the ascending aorta following heart catheterisation: incidence, management and outcome. *EuroIntervention* 2006;**2**:197–202.

11. Abdou SM, Wu CJ. Treatment of aortocoronary dissection complicating anomalous origin right coronary artery and chronic total intervention with intravascular ultrasound guided stenting. *Catheter Cardiovasc Interv* 2011;**78**:914–9.

12. Al Salti Al Krad H, Kaminsky B, Brilakis ES. Use of a thrombectomy catheter for contrast injection: a novel technique for preventing extension of an aortocoronary dissection during the retrograde approach to a chronic total occlusion. *J Invasive Cardiol* 2014;**26**:E54–5.

13. Liao MT, Liu SC, Lee JK, Chiang FT, Wu CK. Aortocoronary dissection with extension to the suprarenal abdominal aorta: a rare complication after percutaneous coronary intervention. *JACC Cardiovasc Interv* 2012;**5**:1292–3.

14. Michael TT, Papayannis AC, Banerjee S, Brilakis ES. Subintimal dissection/reentry strategies in coronary chronic total occlusion interventions. *Circ Cardiovasc Interv* 2012;**5**:729–38.

15. Lombardi WL. Retrograde PCI: what will they think of next? *J Invasive Cardiol* 2009;**21**:543.

16. Zimarino M, Ausiello A, Contegiacomo G, et al. Rapid decline of collateral circulation increases susceptibility to myocardial ischemia: the trade-off of successful percutaneous recanalization of chronic total occlusions. *J Am Coll Cardiol* 2006;**48**:59–65.

17. Patel VG, Banerjee S, Brilakis ES. Treatment of inadvertent subintimal stenting during intervention of a coronary chronic total occlusion. *Interv Cardiol* 2013;**5**(2):165–9.

18. Omurlu K, Ozeke O. Side-by-side false and true lumen stenting for recanalization of the chronically occluded right coronary artery. *Heart Vessels* 2008;**23**:282–5.

19. Krivonyak GS, Warren SG. Compression of a subintimal or false lumen stent by stenting in the true lumen. *J Invasive Cardiol* 2001;**13**:698–701.

20. Singh M, Bell MR, Berger PB, Holmes Jr DR. Utility of bilateral coronary injections during complex coronary angioplasty. *J Invasive Cardiol* 1999;**11**:70–4.

21. Brilakis ES, Grantham JA, Rinfret S, et al. A percutaneous treatment algorithm for crossing coronary chronic total occlusions. *JACC Cardiovasc Interv* 2012;**5**:367–79.

22. Banerjee S, Master R, Brilakis ES. Intravascular ultrasound-guided true lumen Re-entry for successful recanalization of chronic total occlusions. *J Invasive Cardiol* 2010;**22**:608–10.

23. Schultz C, van der Ent M, Serruys PW, Regar E. Optical coherence tomography to guide treatment of chronic occlusions? *J Am Coll Cardiol Intv* 2009;**2**:366–7.

24. Frick K, Michael TT, Alomar M, et al. Low molecular weight dextran provides similar optical coherence tomography coronary imaging compared to radiographic contrast media. *Catheter Cardiovasc Interv* 2014;**84**:727–31.

25. Hussain F. Distal side branch entry technique to accomplish recanalization of a complex and heavily calcified chronic total occlusion. *J Invasive Cardiol* 2007;**19**:E340–2.

26. Wosik J, Shorrock D, Christopoulos G, et al. Systematic review of the BridgePoint system for crossing coronary and peripheral chronic total occlusions. *J Invasive Cardiol* 2015;**27**: 269–76.

27. Colombo A, Mikhail GW, Michev I, et al. Treating chronic total occlusions using subintimal tracking and reentry: the STAR technique. *Catheter Cardiovasc Interv* 2005;**64**:407–11.

28. Galassi AR, Tomasello SD, Costanzo L, et al. Mini-STAR as bail-out strategy for percutaneous coronary intervention of chronic total occlusion. *Catheter Cardiovasc Interv* 2012;**79**: 30–40.

29. Martinez-Rumayor AA, Banerjee S, Brilakis ES. Knuckle wire and stingray balloon for recrossing a coronary dissection after loss of guidewire position. *JACC Cardiovasc Interv* 2012;**5**:e31–2.

30. Smith EJ, Di Mario C, Spratt JC, et al. Subintimal TRAnscatheter withdrawal (STRAW) of hematomas compressing the distal true lumen: a novel technique to facilitate distal reentry during recanalization of chronic total occlusion (CTO). *J Invasive Cardiol* 2015;**27**:E1–4.

31. Wilson WM, Walsh SJ, Yan AT, et al. Hybrid approach improves success of chronic total occlusion angioplasty. *Heart* 2016;**102**:1486–93.

32. Visconti G, Focaccio A, Donahue M, Briguori C. Elective versus deferred stenting following subintimal recanalization of coronary chronic total occlusions. *Catheter Cardiovasc Interv* 2015;**85**:382–90.

33. Prasad A, Banerjee S, Brilakis ES. Images in cardiovascular medicine. Hemodynamic consequences of massive coronary air embolism. *Circulation* 2007;**115**:e51–3.

34. Karatasakis A, Akhtar YN, Brilakis ES. Distal coronary perforation in patients with prior coronary artery bypass graft surgery: the importance of early treatment. *Cardiovasc Revasc Med* 2016;**17**:412–7.

35. Aggarwal C, Varghese J, Uretsky BF. Left atrial inflow and outflow obstruction as a complication of retrograde approach for chronic total occlusion: report of a case and literature review of left atrial hematoma after percutaneous coronary intervention. *Catheter Cardiovasc Interv* 2013;**82**:770–5.

36. Wilson WM, Spratt JC, Lombardi WL. Cardiovascular collapse post chronic total occlusion percutaneous coronary intervention due to a compressive left atrial hematoma managed with percutaneous drainage. *Catheter Cardiovasc Interv* 2015;**86**:407–11.

37. Franks RJ, de Souza A, Di Mario C. Left atrial intramural hematoma after percutaneous coronary intervention. *Catheter Cardiovasc Interv* 2015;**86**:E150–2.

38. Adusumalli S, Morris M, Pershad A. Pseudo-pericardial tamponade from right ventricular hematoma after chronic total occlusion percutaneous coronary intervention of the right coronary artery: successfully managed percutaneously with computerized tomographic guided drainage. *Catheter Cardiovasc Interv* 2016;**88**:86–8.

39. Kawana M, Lee AM, Liang DH, Yeung AC. Acute right ventricular failure after successful opening of chronic total occlusion in right coronary artery caused by a large intramural hematoma. *Circ Cardiovasc Interv* 2017:10. pii:e004674.

40. Rathore S, Matsuo H, Terashima M, et al. Procedural and in-hospital outcomes after percutaneous coronary intervention for chronic total occlusions of coronary arteries 2002 to 2008: impact of novel guidewire techniques. *JACC Cardiovasc Interv* 2009;**2**:489–97.

41. Karmpaliotis D, Karatasakis A, Alaswad K, et al. Outcomes with the use of the retrograde approach for coronary chronic total occlusion interventions in a contemporary multicenter us registry. *Circ Cardiovasc Interv* 2016:9. pii:e003434.

42. Danek BA, Karatasakis A, Karmpaliotis D, et al. Development and Validation of a Scoring System for Predicting Periprocedural Complications During Percutaneous Coronary Interventions of Chronic Total Occlusions: the Prospective Global Registry for the Study of Chronic Total Occlusion Intervention (PROGRESS CTO) Complications Score. *J Am Heart Assoc* 2016:5.
43. Ellis SG, Ajluni S, Arnold AZ, et al. Increased coronary perforation in the new device era. Incidence, classification, management, and outcome. *Circulation* 1994;**90**:2725–30.
44. Bagur R, Bernier M, Kandzari DE, Karmpaliotis D, Lembo NJ, Rinfret S. A novel application of contrast echocardiography to exclude active coronary perforation bleeding in patients with pericardial effusion. *Catheter Cardiovasc Interv* 2013;**82**:221–9.
45. Stewart WJ, McSweeney SM, Kellett MA, Faxon DP, Ryan TJ. Increased risk of severe protamine reactions in NPH insulin-dependent diabetics undergoing cardiac catheterization. *Circulation* 1984;**70**:788–92.
46. Briguori C, Nishida T, Anzuini A, Di Mario C, Grube E, Colombo A. Emergency polytetrafluoroethylene-covered stent implantation to treat coronary ruptures. *Circulation* 2000;**102**:3028–31.
47. Romaguera R, Waksman R. Covered stents for coronary perforations: is there enough evidence? *Catheter Cardiovasc Interv* 2011;**78**:246–53.
48. Tarar MN, Christakopoulos GE, Brilakis ES. Successful management of a distal vessel perforation through a single 8-French guide catheter: combining balloon inflation for bleeding control with coil embolization. *Catheter Cardiovasc Interv* 2015;**86**:412–6.
49. Ben-Gal Y, Weisz G, Collins MB, et al. Dual catheter technique for the treatment of severe coronary artery perforations. *Catheter Cardiovasc Interv* 2010;**75**:708–12.
50. Stathopoulos IA, Kossidas K, Garratt KN. Delayed perforation after percutaneous coronary intervention: rare and potentially lethal. *Catheter Cardiovasc Interv* 2014;**83**:E45–50.
51. Yasuoka Y, Sasaki T. Successful collapse vessel treatment with a syringe for thrombus-aspiration after the guidewire-induced coronary artery perforation. *Cardiovasc Revasc Med* 2010;**11**:e1–3.
52. Matsumi J, Adachi K, Saito S. A unique complication of the retrograde approach in angioplasty for chronic total occlusion of the coronary artery. *Catheter Cardiovasc Interv* 2008;**72**:371–8.
53. Hartono B, Widito S, Munawar M. Sealing of a dual feeding coronary artery perforation with homemade spring guidewire. *Cardiovasc Interv Ther* 2015;**30**:347–50.
54. Garbo R, Oreglia JA, Gasparini GL. The Balloon-Microcatheter technique for treatment of coronary artery perforations. *Catheter Cardiovasc Interv* 2017;**89**:E75–83.
55. Kinnaird T, Kwok CS, Kontopantelis E, et al. Incidence, determinants, and outcomes of coronary perforation during percutaneous coronary intervention in the United Kingdom between 2006 and 2013: an analysis of 527 121 cases from the British Cardiovascular Intervention Society Database. *Circ Cardiovasc Interv* 2016:9.
56. Sandoval Y, Lobo AS, Brilakis ES. Covered stent implantation through a single 8-French guide catheter for the management of a distal coronary perforation. *Catheter Cardiovasc Interv* 2017;**90**:584–8.
57. Boukhris M, Tomasello SD, Azzarelli S, Elhadj ZI, Marza F, Galassi AR. Coronary perforation with tamponade successfully managed by retrograde and antegrade coil embolization. *J Saudi Heart Assoc* 2015;**27**:216–21.
58. Ngo C, Christopoulos G, Brilakis ES. Conservative management of an epicardial collateral perforation during retrograde chronic total occlusion percutaneous coronary intervention. *J Invasive Cardiol* 2016;**28**:E11–2.

59. Araki M, Murai T, Kanaji Y, et al. Interventricular septal hematoma after retrograde intervention for a chronic total occlusion of a right coronary artery: echocardiographic and magnetic resonance imaging-diagnosis and follow-up. *Case Rep Med* 2016;**2016**:8514068.
60. Lin TH, Wu DK, Su HM, et al. Septum hematoma: a complication of retrograde wiring in chronic total occlusion. *Int J Cardiol* 2006;**113**:e64–6.
61. Abdel-Karim AR, Vo M, Main ML, Grantham JA. Interventricular Septal Hematoma, Coronary-ventricular Fistula: a complication of retrograde chronic total occlusion intervention. *Case Rep Cardiol* 2016:8750603.
62. Sianos G, Barlis P, Di Mario C, et al. European experience with the retrograde approach for the recanalisation of coronary artery chronic total occlusions. A report on behalf of the euroCTO club. *EuroIntervention* 2008;**4**:84–92.
63. Fairley SL, Donnelly PM, Hanratty CG, Walsh SJ. Images in cardiovascular medicine. Interventricular septal hematoma and ventricular septal defect after retrograde intervention for a chronic total occlusion of a left anterior descending coronary artery. *Circulation* 2010;**122**:e518–21.
64. Sachdeva R, Hughes B, Uretsky BF. Retrograde approach to a totally occluded right coronary artery via a septal perforator artery: the tale of a long and winding wire. *J Invasive Cardiol* 2010;**22**:E65–6.
65. Marmagkiolis K, Brilakis ES, Hakeem A, Cilingiroglu M, Bilodeau L. Saphenous vein graft perforation during percutaneous coronary intervention: a case series. *J Invasive Cardiol* 2013;**25**:157–61.
66. Mashayekhi K, Behnes M, Akin I, Kaiser T, Neuser H. Novel retrograde approach for percutaneous treatment of chronic total occlusions of the right coronary artery using ipsilateral collateral connections: a European centre experience. *EuroIntervention* 2016;**11**:e1231–6.
67. Kotsia AP, Brilakis ES, Karmpaliotis D. Thrombin injection for sealing epicardial collateral perforation during chronic total occlusion percutaneous coronary interventions. *J Invasive Cardiol* 2014;**26**:E124–6.
68. Iturbe JM, Abdel-Karim AR, Papayannis A, et al. Frequency, treatment, and consequences of device loss and entrapment in contemporary percutaneous coronary interventions. *J Invasive Cardiol* 2012;**24**:215–21.
69. Utsunomiya M, Kobayashi T, Nakamura S. Case of dislodged stent lost in septal channel during stent delivery in complex chronic total occlusion of right coronary artery. *J Invasive Cardiol* 2009;**21**:E229–33.
70. Sianos G, Papafaklis MI. Septal wire entrapment during recanalisation of a chronic total occlusion with the retrograde approach. *Hellenic J Cardiol* 2011;**52**:79–83.
71. Rangan BV, Brilakis ES. Getting out of jail: creative solutions in a moment of crisis. *Catheter Cardiovasc Interv* 2011;**78**:571–2.
72. Kaneda H, Saito S, Hosokawa G, Tanaka S, Hiroe Y. Trapped Rotablator: kokesi phenomenon. *Catheter Cardiovasc Interv* 2000;**49**:82–4.
73. Brilakis ES, Best PJ, Elesber AA, et al. Incidence, retrieval methods, and outcomes of stent loss during percutaneous coronary intervention: a large single-center experience. *Catheter Cardiovasc Interv* 2005;**66**:333–40.
74. Grise MA, Yeager MJ, Teirstein PS. A case of an entrapped rotational atherectomy burr. *Catheter Cardiovasc Interv* 2002;**57**:31–3.
75. Hyogo M, Inoue N, Nakamura R, et al. Usefulness of conquest guidewire for retrieval of an entrapped rotablator burr. *Catheter Cardiovasc Interv* 2004;**63**:469–72.
76. Lo N, Michael TT, Moin D, et al. Periprocedural myocardial injury in chronic total occlusion percutaneous interventions: a systematic cardiac biomarker evaluation study. *JACC Cardiovasc Interv* 2014;**7**:47–54.

77. Paizis I, Manginas A, Voudris V, Pavlides G, Spargias K, Cokkinos DV. Percutaneous coronary intervention for chronic total occlusions: the role of side-branch obstruction. *EuroIntervention* 2009;**4**:600–6.

78. Werner GS, Coenen A, Tischer KH. Periprocedural ischaemia during recanalisation of chronic total coronary occlusions: the influence of the transcollateral retrograde approach. *EuroIntervention* 2014;**10**:799–805.

79. Stetler J, Karatasakis A, Christakopoulos GE, et al. Impact of crossing technique on the incidence of periprocedural myocardial infarction during chronic total occlusion percutaneous coronary intervention. *Catheter Cardiovasc Interv* 2016;**88**:1–6.

80. Rinfret S, Joyal D, Nguyen CM, et al. Retrograde recanalization of chronic total occlusions from the transradial approach; early Canadian experience. *Catheter Cardiovasc Interv* 2011;**78**:366–74.

81. Alaswad K, Menon RV, Christopoulos G, et al. Transradial approach for coronary chronic total occlusion interventions: insights from a contemporary multicenter registry. *Catheter Cardiovasc Interv* 2015;**85**:1123–9.

82. Seto AH, Abu-Fadel MS, Sparling JM, et al. Real-time ultrasound guidance facilitates femoral arterial access and reduces vascular complications: FAUST (Femoral Arterial Access with Ultrasound Trial). *JACC Cardiovasc Interv* 2010;**3**:751–8.

83. Abu-Fadel MS, Sparling JM, Zacharias SJ, et al. Fluoroscopy vs. traditional guided femoral arterial access and the use of closure devices: a randomized controlled trial. *Catheter Cardiovasc Interv* 2009;**74**:533–9.

84. Seto AH, Roberts JS, Abu-Fadel MS, et al. Real-time ultrasound guidance facilitates transradial access: RAUST (Radial Artery Access with Ultrasound Trial). *JACC Cardiovasc Interv* 2015;**8**:283–91.

85. Laskey WK, Jenkins C, Selzer F, et al. Volume-to-creatinine clearance ratio: a pharmacokinetically based risk factor for prediction of early creatinine increase after percutaneous coronary intervention. *J Am Coll Cardiol* 2007;**50**:584–90.

86. Levine GN, Bates ER, Blankenship JC, et al. 2011 ACCF/AHA/SCAI Guideline for percutaneous coronary intervention: executive summary: a report of the American College of Cardiology Foundation/American Heart Association Task Force on Practice Guidelines and the Society for Cardiovascular Angiography and Interventions. *Catheter Cardiovasc Interv* 2012;**79**:453–95.

87. Ali ZA, Karimi Galougahi K, Nazif T, et al. Imaging- and physiology-guided percutaneous coronary intervention without contrast administration in advanced renal failure: a feasibility, safety, and outcome study. *Eur Heart J* 2016;**37**:3090–5.

88. Tanaka H, Kadota K, Hosogi S, Fuku Y, Goto T, Mitsudo K. Mid-term angiographic and clinical outcomes from antegrade versus retrograde recanalization for chronic total occlusions. *J Am Coll Cardiol* 2011;**57**:E1628.

89. Brilakis ES, Banerjee S. Advances in the treatment of coronary artery aneurysms. *Catheter Cardiovasc Interv* 2011;**77**:1042–4.

90. Brilakis ES. Should the fear of complications stop you from doing CTO interventions? *Cardiology Today's Interv* July August 2013.

Chapter 13

How to Build a Successful Coronary Chronic Total Occlusion Program

13.1 IS CHRONIC TOTAL OCCLUSION PERCUTANEOUS CORONARY INTERVENTION FOR YOU?

Start with why.

Simon Sinek

Should you embark on the trip of learning chronic total occlusion (CTO) percutaneous coronary intervention (PCI)? This is a challenging question with no easy answer. It requires significant introspection and thought. Here are some factors that may be useful in making this decision.

1. **Passion**

   Passion is key for going through the learning (and the maintenance) curve of CTO PCI. The CTO operator is passionate to help each patient by achieving excellent results, even among very challenging cases. Although passion can be developed, CTO PCI should be an exciting proposition from the beginning, to power you through the various developmental stages required.

2. **Procedural skills**

   Procedural skills can and will be developed and refined while learning CTO PCI, but operators should already be performing complex PCI and have robust technical skills. For example, operators should not be attempting retrograde crossing via epicardial collaterals without being experienced in doing pericardiocentesis. Similarly, operators should not attempt retrograde crossing through the last remaining vessel or through an internal mammary graft unless they are experienced and have access to left ventricular support devices.

3. **Career stage**

   Late career stages may be less conducive to starting a CTO PCI program, given many engrained habits that may be difficult to change. This is relative, however, as many late career operators have achieved tremendous success in CTO PCI. Junior operators starting a CTO PCI program may consider working with a more senior partner.

Manual of Chronic Total Occlusion Interventions. http://dx.doi.org/10.1016/B978-0-12-809929-2.00013-2

4. **PCI volume**
   CTO PCI is for high-volume, not low-volume operators, because procedural volume does correlate with skills.
5. **Approach to failure and complications**
   Even the best operators in the world have failures and complications. Failure can be highly frustrating and demoralizing, especially given the often significant effort that goes into planning and executing each case. Being able to accept failure and learn and apply the lessons that failure provides you is a critical step for the CTO operator. Often reattempt cases will be successful.
6. **Improve overall PCI skills**
   CTO operators develop several skills that translate in all aspects of non-CTO PCI. CTO PCI can significantly enhance the operator's armamentarium for treating complex lesions.
7. **Time availability**
   Time is needed to attend courses, read, and get proctored. Also early in the learning curve CTO PCI cases can be long, often lasting 2–4h each.

There are *wrong* reasons for wanting to do CTO PCI:

1. **Boosting the ego**
   Being a competent CTO operator can improve self-esteem, but helping the patient should be the main driving force, especially since failures and complications are certain to occur.
2. **Income generation**
   Given procedure complexity and time and effort required, income generation is not good reason for doing CTO PCI, since successful procedures can be lengthy and unsuccessful procedures are billed at the diagnostic catheterization level in the United States. However, acquiring a new skill set can be valuable in today's job market.

13.2 LEARNING CHRONIC TOTAL OCCLUSION PERCUTANEOUS CORONARY INTERVENTION: THE GOAL

As with any training, there are four distinct stages for learning CTO PCI (Fig. 13.1).

Learning CTO PCI starts with mastering **antegrade** techniques, first antegrade wire escalation (stage 1) and then antegrade dissection and reentry (stage 2).

Retrograde techniques are initially learned by using septal collaterals and bypass grafts, which are safer and easier to cross (stage 3), followed by use of the more challenging (and risky) epicardial (and ipsilateral) collaterals (stage 4).

Many operators may initially or permanently choose to remain antegrade-only operators,[1] given the rapid increase in complexity and risk associated with use of retrograde techniques. As long as they understand their strengths and limitations, whether an operator is antegrade-only or antegrade and retrograde is a matter of personal choice. With continued practice some operators who initially chose to do only antegrade techniques may elect to do retrograde procedures and vice versa.

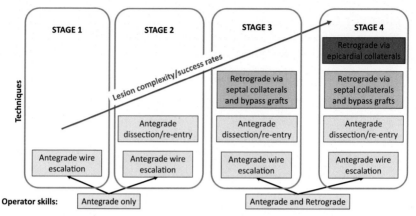

FIGURE 13.1 The four stages of learning chronic total occlusion percutaneous coronary intervention. *Reproduced with permission from CCI, Azzalini L, Brilakis ES. **Ipsilateral** vs. contralateral vs. no collateral (antegrade only) chronic total occlusion percutaneous coronary **interventions**: what is the right choice for your practice?* Catheter Cardiovasc Interv *2017;89:656–7.*

13.3 LEARNING CHRONIC TOTAL OCCLUSION PERCUTANEOUS CORONARY INTERVENTION: FELLOWSHIP AND ON-THE-JOB TRAINING

Learning CTO PCI can be achieved either through a formal fellowship program or through on-the-job training (Table 13.1). Most operators currently train for CTO PCI while practicing.[2]

The advantages of formal fellowship training include the concentrated experience and exposure to large case volume and highly complex cases, prolonged direct working relationship with advanced CTO operators, and opportunity to get heavily involved in CTO PCI research. Disadvantages include the still developing catheterization and angioplasty skills (most fellowships are done after conclusion of the formal interventional training), and limited availability of dedicated fellowships for CTO PCI and other complex and higher risk procedures.[5]

Both pathways can provide excellent training.

13.4 LEARNING CHRONIC TOTAL OCCLUSION PERCUTANEOUS CORONARY INTERVENTION: BOOKS, INTERNET, MEETINGS, PROCTORSHIPS

The following tools can assist an interventionalist to evolve into a successful CTO operator:

1. Reading CTO-related literature (all interventional journals; Catheterization and Cardiovascular Interventions, Journal of Invasive Cardiology, Eurointervention, Circulation: Cardiovascular Interventions, and JACC Cardiovascular Interventions provide detailed articles on the technical and clinical aspects of CTO PCI).

TABLE 13.1 Comparison of Chronic Total Occlusion Percutaneous Coronary Intervention Training Through a Formal Training Program[3,4] or Through On-the-Job Training

	Fellowship Program	On-the-Job Training
Availability	Limited	Wide
Flexibility	+	+++
Mastering of basic percutaneous coronary intervention skills	+	++
Concentrated experience	+++	+
Exposure to highly complex cases	+++	+
Development of mentoring relationships with advanced chronic total occlusion operators	+++	++
Research opportunities	+++	+

2. Participating in online CTO-related education: this book provides links to several recorded CTO PCI cases on YouTube (can be searched at: www.ctomanual.org). Also www.ctofundamentals.org, http://apcto.club/apcto-algorithm/, and www.incathlab.com are outstanding websites providing basic to advanced CTO PCI education; they also provide online physician communities that regularly share cases and expertise. In some cases success may hinge on a nuance of technique that an operator may never have done before and may be aware of it only through a course, the Internet, or the literature.

3. Observing CTO interventions at experienced CTO PCI centers.

4. Attending CTO PCI courses and meetings (such as the CTO Academy at CRT, the Cardiovascular Research Foundation CTO Summit, the SCAI Annual Meeting, the Japan CTO Club, the Cardiovascular Innovations meeting, TCT, and the EuroCTO Club).

5. Getting proctored by experienced CTO interventionalists: on-the-job training is invaluable for learning CTO PCI techniques.

6. Practicing: as with any procedure, the more CTO interventions you do, the better CTO operator you become!

7. Working with another interventionalist during CTO PCI, if feasible, allows for real-time feedback and adaptation of the procedural plan.

13.5 LEARNING CHRONIC TOTAL OCCLUSION PERCUTANEOUS CORONARY INTERVENTION: WHERE TO PUT PARTICULAR EMPHASIS

1. Meticulous procedural planning: understanding the CTO anatomy and the possible crossing strategies facilitates efficient and confident conversion within the hybrid algorithm (Chapter 7).

2. Carefully selecting patients who are likely to benefit from CTO PCI, as outlined in Chapter 1.
3. Focusing and practicing the basics of CTO PCI, as outlined in Chapter 2.
4. Persistence: committing time and energy is required for CTO PCI. Per Dr. Bill Lombardi, one of fathers of CTO interventions in North America, "you either do CTO PCI, or you don't–there is no such thing as trying." In other words some CTO interventions can be challenging and demanding, but the key to success is persistence. With increasing experience the procedures become faster and success rates increase.[6]
5. Being creative: every CTO is unique and may require a different, tailored, treatment approach (although an overall standardization of CTO PCI techniques, such as the hybrid approach, can facilitate planning as described in Chapter 7).[7]
6. Learning from failures: unlike non-CTO interventions, CTO PCI failure is not uncommon, especially early in the learning curve. Failed procedures should not be a source of discouragement, but should rather stimulate constructive evaluation and learning. Discussing failed cases with other operators can be fruitful, as can be reattempting these cases with a proctor or referring them to more experienced centers. Knowing when to fail is also important: it is better to fail without complication than try too hard and have a (sometimes) catastrophic complication.
7. Publishing challenging or unique CTO PCI cases, or the overall outcomes of the CTO PCI program.
8. Keeping track of procedural outcomes, for example by creating a local CTO PCI database or by joining the Prospective Global Registry for the Study of Chronic Total Occlusion Percutaneous Coronary Interventions (PROGRESS-CTO, clinicaltrials.gov Identifier: NCT02061436, www.progressscto.org).
9. Participating in new studies on CTO interventions.

13.6 CREATING A CHRONIC TOTAL OCCLUSION PERCUTANEOUS CORONARY INTERVENTION TEAM

The importance of building a CTO team, procuring the necessary equipment, and implementing appropriate policies cannot be overemphasized, and consists of:

1. Staff education, including:
   a. Lectures for cath lab staff on the indications and complexity of CTO PCI.
   b. Educating the non-cath team about the process and outcomes of CTO PCI.
   c. Identifying specific cath lab personnel champions who are:
      - Interested in developing further expertise in CTO PCI.
      - Interested in routinely being involved in CTO PCI cases (which helps in building experience and achieving excellent outcomes).
2. Obtaining the necessary infrastructure and equipment (Chapter 2, Table 2.1).
   a. At least two cath lab rooms (so that emergencies can go to the second room if CTO PCI is performed in the first room).

  b. Cardiac computed tomography and magnetic resonance imaging.
  c. On site cardiac surgery.
3. Establishing CTO-specific protocols for:
  a. Radiation (as described in detail in Chapter 10):
    - Utilizing 6 to 7.5 frame-per-second fluoroscopy.
    - Continuously monitoring radiation dose.
    - Stopping the procedure if crossing has not been achieved after approximately 7–10 Gy air kerma dose.
    - Following up patients who receive >5 Gy air kerma dose to detect any skin injury.
  b. Anticoagulation:
    - Repeating ACT every 30 min.
    - Goal ACT >300 s for antegrade cases.
    - Goal ACT >350 s for retrograde cases.
4. Establishing CTO days, which allows uninterrupted and concentrated focus on CTO PCI procedures: it is important for the operator to know that he/she has no other commitments for several hours, allowing prolonged treatment attempts if necessary. Moreover, dedicated CTO days can improve staff acceptance of starting a CTO program, facilitate visits by proctors or clinical specialists, and also allow referring cardiologists to visit.
5. Performing challenging cases as a team: having two interventionalists in the procedure improves the likelihood of success.

13.7 SECURING SUPPORT FROM THE ADMINISTRATION

The following steps can assist with securing support from the administration to build a CTO PCI program:

1. Highlighting the need for such a program by showing the number of patients who could benefit from CTO PCI.
2. Demonstrating to administration that CTO PCI is both feasible and not a money-losing proposition. Pilot economic analyses from the Piedmont Heart Institute demonstrate similar contribution margins for CTO and non-CTO PCI.[2]
3. Highlighting the institutional benefits of a CTO PCI program:
  a. Developing regional and/or national reputation for doing complex interventions.
  b. Increasing internal procedural volume.
  c. Increasing outside referrals.

13.8 INCREASING AWARENESS OF YOUR CHRONIC TOTAL OCCLUSION PROGRAM

Once the CTO program has been developed and good outcomes are achieved, increasing the awareness of the CTO PCI program can increase referrals of patients who may benefit from these procedures.

However, premature promotion may hurt the program. Projecting yourself as a CTO PCI expert and tackling very complex occlusions before reaching maturity can have the opposite of the desired effects, as it can be frustrating for referring MDs to see repeat failure of cases they referred. It is best to work quietly behind the scene, building the CTO PCI skills before aggressively promoting the program. At the same time it is important to share all successes with colleagues, effectively retraining them on what can be achieved with CTO PCI.

Increasing awareness of the CTO PCI program can be achieved by:

1. Educating referring physicians (both cardiologists and general practitioners). It is imperative for referring physicians to understand the rationale and potential clinical benefits of CTO PCI.
2. Presenting CTO PCI cases at case conferences, Grand Rounds, and Roundtables. These presentations can illustrate that many of the previously considered undoable procedures are actually feasible and can provide significant benefit to the patients.
3. Educating the patients. Many patients with severe angina are very motivated to find treatment options themselves. Online posting of patient brochures focusing on CTO intervention, as well as video testimonials (with appropriate patient consent) can be powerful educational tools.

13.9 WHY CHRONIC TOTAL OCCLUSION PERCUTANEOUS CORONARY INTERVENTION WILL MAKE YOU A BETTER INTERVENTIONALIST

The goal of CTO interventions (and any procedure or intervention) is to benefit the patient. In addition to benefiting the patient, performing CTO PCI can make you a better interventionalist. This section describes how.

This section was reproduced with permission from Cardiology Today's Intervention. http://www.healio.com/cardiac-vascular-intervention/chronic-total-occlusion/news/print/cardiology-today-intervention/%7B4f30f5ee-f5 50-4085-86d7-2ab9f06e0a36%7D/top-10-reasons-why-doing-cto-interventions-will-make-you-a-better-interventionalist.

13.9.1 The Importance of Growth

Doing CTO interventions offers a refreshing new perspective on PCI: there is a definite (and pretty substantial, especially in the beginning) chance of failure and there is a need to master multiple skills and techniques to succeed. CTO PCI trains interventionalists to treat patients with complex coronary anatomy safely and effectively. As a result CTO PCI becomes a powerful motivator for learning and applying new (and often challenging) techniques.

13.9.2 Better Angiogram Evaluation Skills

Angiographic evaluation can often be brief. Given the superb deliverability of current equipment we often focus mainly on determining the diameter and the length of the stent we are going to implant. When we are approaching a CTO, however, things may not be quite as simple. We look at the vessel proximal to the lesion to determine if there is tortuosity and calcification that may hinder advancement of equipment; whether there are lesions that need to be treated before attempting to treat the CTO; and whether there are proximal side branches that could be used to perform the side-branch anchor technique to facilitate equipment delivery. We use multiple projections to determine where the CTO actually starts (proximal cap) and whether the entry point to the CTO is tapered or blunt. We evaluate the length of the occlusion, as the longer it is the more likely that prolonged crossing attempts and use of advanced crossing methods (retrograde and antegrade dissection/ reentry) will be required. We examine the vessel distal to the occlusion to determine if reentry would be easy in case of subintimal guidewire crossing and whether distal side branches are at risk for occlusion. Finally, we examine the presence, size, and tortuosity of collaterals to determine if the retrograde approach is feasible and safe.

Given the frequent complexity of CTO PCI we are forced to look at the angiogram in multiple projections using dual injections to better understand the anatomy and devise several alternative approaches in case the initially chosen one fails. CTO PCI teaches us the value of careful and detailed planning and the value of understanding in depth each patient's unique coronary anatomy. It is not uncommon to spend 15–30 min looking at the angiogram. This is time well spent as it can make the difference between success and failure later in the case. Learning the CTO way of angiogram evaluation spills over when evaluating other coronary lesions and enhances the likelihood of successful, efficient, and safe treatment.

13.9.3 Familiarity With Complex Lesions and Techniques

The days after a CTO day are so much fun–because everything feels so easy! When you learn how to deal with the complexities of CTO PCI, treating non-occlusive lesions becomes pretty straightforward! The CTO interventionalist knows how to achieve excellent guide support by using 8 Fr guides (and is not afraid to use 8 Fr guides when needed), or by using guide catheter extensions and anchor techniques. He or she often uses 45 cm long sheaths to minimize the impact of iliac tortuosity on guide catheter support. He or she understands how to optimally use microcatheters (including using them for non-CTO complex cases) and which guidewire works best for each task. The subintimal space stops being a forbidden zone—it becomes an ally, especially for treating long calcified lesions.

13.9.4 Better Understanding of Equipment

Not all guidewires are created equal! This is especially true when it comes to CTO guidewires. Understanding the differences between stiff tip, highly penetrating guidewires (such as the Miracle and Confianza family of guidewires) and polymer-jacketed guidewires (such as the Pilot 200 and Fielder FC) is important for their optimal use. Such wires are often invaluable for crossing high-grade subtotal lesions, or even 100% occlusions in the setting of acute myocardial infarction. The basic principles of CTO interventions (never advance balloons and microcatheter before you confirm that the guidewire is in the distal true lumen or at least in the vessel architecture) apply to all interventions, including acute myocardial infarction interventions. In several primary PCI cases dual injections have been used to confirm the guidewire position during attempts to cross the acute occlusion.

13.9.5 Increased Percutaneous Coronary Intervention Volume

The PCI (and coronary artery bypass graft surgery) volumes have been steadily declining over the last few years. CTO PCI is probably the only area that has potential to grow. Given that the more cases you perform the better you get, performing CTO PCI is more important now than ever before.

13.9.6 Improved Workflow

By its very nature CTO PCI requires continuous and prompt adjustment of the procedural strategy. If one technique is not successful within a short period of time an adjustment is performed (rather than continuing with the same failing technique). The adjustment could be small (such as changing the shape of the guidewire tip) or major (such as changing from an antegrade to a retrograde approach or vice versa). Early change is at the heart of the hybrid algorithm and maximizes the likelihood of success by minimizing the time (and radiation) used in failing approaches. However, the basic principle of trying an alternative strategy if the initially selected one is not working applies to *any* interventional cardiology procedure. Constantly thinking about the steps ahead allows preparation of the necessary equipment and smooth and timely transitions. Workflow is improved and outcomes are improved.

13.9.7 Participation in a Community

The CTO PCI community is very strong, perhaps because everyone realizes that they need the help and advice of everyone else to be successful. As a result, it is not uncommon to call for advice from colleagues in the middle of a challenging case and there is constant interaction and learning, both at meetings and also through www.ctofundamentals.org. There is always more to learn and there is always room for improvement.

13.9.8 Complication Management

Planning for CTO PCI means also planning for treating complications, should they occur. Although perforation is of special concern with CTO PCI, it can occur with any PCI. Hence, the planning and emphasis on preventing and treating complications required for CTO PCI directly helps improve the safety of non-CTO PCI. It forces operators to familiarize themselves with use of covered stents and coils, thrombin, or microcatheters. There are cases of acute vessel closure during non-CTO PCI that were salvaged using CTO PCI techniques, such as antegrade dissection/reentry (Online Case 92, 98)[8] or the retrograde approach (Online Case 38).[9,10] There are also cases of equipment entrapment that resolved by using antegrade dissection/reentry techniques.[11] These patients otherwise would have required emergent cardiac surgery. They definitely would have been grateful had they realized that the reason emergent surgery was avoided was operator experience with CTO PCI techniques!

13.9.9 Radiation Management

Minimizing radiation exposure is critical for the success of CTO interventions. If too much radiation is used early in the case, crossing attempts will have to stop and the procedure will fail. Expert CTO operators are obsessive/compulsive about minimizing patient radiation exposure. They routinely use 6–7.5 frames per second fluoroscopy and continually monitor the amount of radiation administered to the patient. They also know that the air kerma is the number to watch, and know how to make sense of the numbers (after 5 Gy air kerma radiation dose, the risk of radiation skin injury increases; if crossing has not been achieved within 7–10 Gy, the case must stop). These skills become second nature and directly affect non-CTO interventions. Many operators currently use 7.5 frames per second fluoroscopy in every case, not just CTO PCI. They also minimize use of cine angiography and use the fluorostore function to document balloon and stent inflations.

13.9.10 Humility

Last, but not least, CTO PCI provides a constant lesson on the importance of humility and respect. Respect for the lesion and respect for the patient. CTO PCI teaches us that failure is always a distinct possibility and that we never know it all. Acknowledging our limitations and failures in front of the patients and their families helps us be humble and also grateful for what we can accomplish.

In the end, who benefits most from the improved procedural skills that CTO PCI operators develop? The patients themselves, who receive better, safer, and more efficient treatment. Perhaps one day the question, "Do you treat coronary CTOs?" will be increasingly asked by patients in need of a coronary intervention. That day may be closer than we think.

13.10 CONCLUSIONS

In summary, development of a successful CTO PCI program requires a concerted and coordinated effort that combines operator and staff development and education of administration and referring physicians. There remains great need for CTO PCI in the United States and worldwide and developing a high-level program can offer an excellent therapeutic option to an undertreated patient population.

REFERENCES

1. Rinfret S, Joyal D, Spratt JC, Buller CE. Chronic total occlusion percutaneous coronary intervention case selection and techniques for the antegrade-only operator. *Catheter Cardiovasc Interv* 2015; **85**: 408–15.
2. Karmpaliotis D, Lembo N, Kalynych A, et al. Development of a high-volume, multiple-operator program for percutaneous chronic total coronary occlusion revascularization: procedural, clinical, and cost-utilization outcomes. *Catheter Cardiovasc Interv* 2013;**82**:1–8.
3. Kalra A, Bhatt DL, Kleiman NS. A 24-month interventional cardiology fellowship: learning motor skills through blocked repetition. *JACC Cardiovasc Interv* 2017;**10**:210–1.
4. Kalra A, Bhatt DL, Pinto DS, et al. Accreditation and funding for a 24-month advanced interventional cardiology fellowship program: a call-to-action for optimal training of the next generation of interventionalists. *Catheter Cardiovasc Interv* 2016;**88**:1010–5.
5. Kirtane AJ, Doshi D, Leon MB, et al. Treatment of higher-risk patients with an indication for revascularization: evolution within the field of contemporary percutaneous coronary intervention. *Circulation* 2016;**134**:422–31.
6. Brilakis ES. The why and how of CTO interventions. *Cardiol Today's Interv* January/February 2012.
7. Brilakis ES, Grantham JA, Rinfret S, et al. A percutaneous treatment algorithm for crossing coronary chronic total occlusions. *JACC Cardiovasc Interv* 2012;**5**:367–79.
8. Martinez-Rumayor AA, Banerjee S, Brilakis ES. Knuckle wire and stingray balloon for recrossing a coronary dissection after loss of guidewire position. *JACC Cardiovasc Interv* 2012;**5**:e31–2.
9. Azemi T, Fram DB, Hirst JA. Bailout antegrade coronary reentry with the stingray balloon and guidewire in the setting of an acute myocardial infarction and cardiogenic shock. *Catheter Cardiovasc Interv* 2013; **82**: E211–4.
10. Patel VG, Zankar A, Brilakis E. Use of the retrograde approach for primary percutaneous coronary intervention of an inferior ST-segment elevation myocardial infarction. *J Invasive Cardiol* 2013;**25**:483–4.
11. Tanaka Y, Saito S. Successful retrieval of a firmly stuck rotablator burr by using a modified STAR technique. *Catheter Cardiovasc Interv* 2016;**87**:749–56.

Appendix 1

Equipment Commonly Utilized in CTO Interventions

Name	Type	Manufacturer	Pages
Amplatz gooseneck snare	Snare	Covidien	76
Angiosculpt	Scoring balloon	Spectranetics	281
Astato XS 20	Very stiff (20 gr tip load), tapered tip guidewire	Asahi Intecc	63
Atlantis SR Pro	Intravascular ultrasound catheter	Boston Scientific	85
Atrieve	Snare (3 loop)	Angiotech	76
Axium	0.014-inch coils	Medtronic	87
Azur	0.018-inch coils	Terumo	89
BHW	Support guidewire	Abbott Vascular	73
Blimp	Scoring balloon catheter	Interventional Medical Device Solutions	84
Caravel	Microcatheter (low profile)	Asahi Intecc	37
CenterCross	Support catheter (self-expanding scaffold, with large central lumen)	Roxwood Medical	40
Choice PT floppy	Guidewire (polymer-jacketed)	Boston Scientific	58
Confianza	Guidewire (stiff with tapered tip)	Asahi Intecc	63
Confianza Pro 12	Guidewire (stiff with tapered tip and hydrophilic coating)	Asahi Intecc	57
Co-pilot	Y-connector with hemostatic valve	Abbott Vascular	28
Corsair	Microcatheter—channel dilator	Asahi Intecc	40
Corsair Pro	Microcatheter—channel dilator (more tip flexibility than Corsair)	Asahi Intecc	40
CrossBoss	Blunt-tip microcatheter for antegrade dissection and reentry	Boston Scientific	164
Cross-it 100XT	Guidewire	Abbott Vascular	58
Crosswire NT	Guidewire	Terumo	58
Crusade	Dual lumen microcatheter	Kaneka Medix Corporation	50
Diamondback 360	Orbital artherectomy system	CSI	82
Eagle Eye	Intravascular ultrasound catheter (solid state)	Philips Volcano	85
Eagle Eye Short Tip	Intravascular ultrasound catheter with short tip	Philips Volcano	85
Eaucath	Sheathless guide catheters	Asahi Intecc	24
ELCA	Excimer laser coronary artherectomy	Spectranetics	81
Ensnare	Snare (3 loop)	Merit Medical	76
Fielder FC	Guidewire (polymer-jacketed with soft, nontapered tip)	Asahi Intecc	63
Fielder XT	Guidewire (polymer-jacketed with soft, tapered tip)	Asahi Intecc	63

Name	Description	Manufacturer	Page
Fielder XT-A	Guidewire (polymer-jacketed, composite core technology, for antegrade crossing)	Asahi Intecc	62, 63
Fielder XT-R	Guidewire (polymer-jacketed, composite core technology, for retrograde collateral crossing)	Asahi Intecc	62
Fighter	Guidewire (polymer-jacketed with tapered tip)	Boston Scientific	57
Finecross	Microcatheter	Terumo	41
FineDuo	Dual lumen microcatheter	Terumo	50
Gaia (1st, 2nd, 3rd)	Guidewire (stiff, tapered tip, composite core technology)	Asahi Intecc	63
Gladius	Guidewire (polymer-jacketed, composite core technology)	Asahi Intecc	57
Glidesheath Slender	Thin wall sheath for radial access	Terumo	21
Graftmaster Rx	Covered stent (rapid-exchange)	Abbott Vascular	86
Grand Slam	Support guidewire	Asahi Intecc	73
Guardian	Y-connector with hemostatic valve	Vascular Solutions	28
GuideLiner Navigation Catheter	Dedicated dilator to facilitate delivery of the GuideLiner guide catheter extension	Vascular Solutions	32
GuideLiner V3	Guide catheter extension	Vascular Solutions	28
Guidezilla II	Guide catheter extension	Boston Scientific	28
Guidion	Guide catheter extension	Interventional Medical Device Solutions	28
Heartmate PHP	Left ventricular assist device (*PHP*, percutaneous heart pump)	Abbott Vascular	93
Hornet 10, 14	Stiff, tapered-tip guidewire	Boston Scientific	58
Impella (2.5, CP, 5.0)	Left ventricular assist device	Abiomed Inc.	93
Interlock	0.018-inch coils	Boston Scientific	87
Iron Man	Guidewire with stiff body and soft tip designed for equipment delivery	Abbott Vascular	73
Mailman	Guidewire with stiff body and soft tip designed for equipment delivery	Boston Scientific	73
Mamba	Microcatheter	Boston Scientific	46
Mamba Flex	Microcatheter	Boston Scientific	46
Micro 14	Microcatheter (155 cm long)	Roxwood Medical	44
Micro 14 es	Microcatheter (extra support, 155 cm long)	Roxwood Medical	44
Micronester	0.018-inch coils	Cook Medical	89
Minnie	Microcatheter	Vascular Solutions	39

continued

Name	Type	Manufacturer	Pages
MiracleBros 3, 4.5, 6, 12	Stiff, nontapered tip guidewire	Asahi Intecc	59
Mizuki	Microcatheter	Kaneka Medix Corporation	47
Mizuki FX	Microcatheter (more flexible than Mizuki)	Kaneka Medix Corporation	47
MultiCross	Support catheter (self-expanding scaffold, with three separate lumens)	Roxwood Medical	54
Nhancer ProX	Microcatheter	Interventional Medical Device Solutions	47
Nhancer Rx	Dual lumen microcatheter	Interventional Medical Device Solutions	50
No Brainer	Radiation protection hat	Worldwide Innovations & Technologies	89
NovaCross	Support catheter (with elastic outer helical struts)	Nitiloop	55
NovaCross Xtreme	Support catheter (with increased intraocclusion crossing ability)	Nitiloop	56
Ostial Flash	Specialized balloon for treatment of ostial lesions	Cardinal Health	90
Papyrus	Covered stent	Biotronik	86
Persuader 3, 6, 9	Guidewire (stiff)	Medtronic	59
Pilot 50, 150, 200	Guidewire (polymer-jacketed, nontapered with increasing tip stiffness)	Abbott Vascular	58, 161
Prodigy	Support catheter (with an atraumatic balloon at the distal tip)	Radius Medical	55
Progreat	Microcatheter (0.018 inch, usually used for coil delivery)	Terumo	38
Progress 40, 80, 120, 140T, 200T	Guidewire (stiff, tapered tip)	Abbott Vascular	58
ProVia 3, 6, 9	Guidewire (stiff)	Medtronic	59,60
Prowler	Microcatheter	Cordis	38
PT Graphix Intermediate	Guidewire (polymer-jacketed)	Boston Scientific	58
PT2 Moderate Support	Guidewire (polymer-jacketed)	Boston Scientific	58
Quick Cross	Microcatheter	Spectranetics	38
Quick-Access Needle Holder	Needle holder for obtaining vascular access under X-ray guidance	Spectranetics	346
R350	Guidewire (350 cm long for externalization)	Vascular Solutions	25
RadPad	Disposable radiation shield	Worldwide Innovations & Technologies	89
Renegade	Microcatheter (0.018-inch, usually used for coil delivery)	Boston Scientific	42, 89
Revolution	Intravascular ultrasound catheter (rotational)	Volcano	85
RG3	Guidewire (330 cm long for externalization)	Asahi Intecc	25
Rotablator	Rotational atherectomy system	Boston Scientific	428

Name	Description	Manufacturer	Page
Rotaglide	Lubricant solution for rotational atherectomy	Boston Scientific	24
RotaWire Floppy and Extra support	Guidewire (for rotational atherectomy)	Boston Scientific	199
Runthrough	Guidewire (workhorse)	Terumo	59
Samurai RC	Guidewire (for retrograde crossing)	Boston Scientific	69
Shinobi and Shinobi Plus	Guidewire (stiff)	Cordis	58
Shuttle	Sheath	Cook Medical	24
Sion	Guidewire (composite core, excellent for collateral crossing in the retrograde approach)	Asahi Intecc	69
Sion Black	Guidewire (polymer-jacketed, composite core, excellent for collateral crossing)	Asahi Intecc	69
Sion Blue	Guidewire (workhorse)	Asahi Intecc	59
Stingray balloon	Balloon (for reentry into true lumen)	Boston Scientific	181
Stingray LP balloon	Low-profile Stingray balloon (for reentry into true lumen)	Boston Scientific	74
Stingray wire	Guidewire (for reentry through Stingray balloon)	Boston Scientific	186
Suoh 03	Guidewire (soft 0.3 g tip, for epicardial collateral crossing)	Asahi Intecc	69
Supercross	Microcatheter (with angled tip)	Vascular Solutions	52
Tandem Heart	Left ventricular assist device	Cardiac Assist Inc.	93
Tegaderm	Sterile cover	3M	25
Threader	Microdilation catheter	Boston Scientific	78
Tornus	Microcatheter (for balloon-uncrossable lesions)	Asahi Intecc	79
Transit	0.021-inch microcatheter (usually used for coil delivery)	Cordis	38
Trap it	Trapping balloon	Interventional Medical Device Solutions	91
Trapliner	Guide catheter extension that incorporates a trapping balloon	Vacular Solutions	28
Trapper	Trapping balloon	Boston Scientific	91
Turbo Elite	Laser catheter (without lasing duration limit)	Spectranetics	82
Turnpike	Microcatheter	Vascular Solutions	45
Turnpike Gold	Microcatheter (gold-plated threaded tip, for antegrade approach)	Vascular Solutions	45
Turnpike LP	Microcatheter (lower profile of the distal tip and shaft)	Vascular Solutions	45
Turnpike Spiral	Microcatheter (nylon coil distally, for antegrade crossing)	Vascular Solutions	45

continued

Name	Type	Manufacturer	Pages
TVC	Combined intravascular ultrasound and near-infrared spectroscopy catheter	InfraRedx	85
Twin Pass 5200	Dual lumen microcatheter	Vascular Solutions	50
Twin Pass Torque	Braided dual lumen microcatheter	Vascular Solutions	50
Ultimate 3	Guidewire (nontapered tip, composite core technology)	Asahi Intecc	63
Valet	Microcatheter	Volcano	39
Venture	Microcatheter (with deflectable distal tip)	Vascular Solutions	48
Viper	Guidewire (for orbital atherectomy)	CSI	251
Whisper LS, MS, ES	Guidewire (polymer-jacketed, soft)	Abbott Vascular	58
Wiggle wire	Guidewire (with curved distal shaft)	Asahi Intecc	73
Wolverine	Cutting balloon	Boston Scientific	83
Zero Gravity	Radiation protection system	Biotronik	89

Manufacturer	Location
Abbott Vascular	Santa Clara, CA
Abiomed	Danver, MA
Angioscore	Fremont, CA
Angiotech	Vancouver, BC, Canada
Asahi Intecc	Nagoya, Japan
Biotronik	Bülach, Switzerland
Boston Scientific	Natick, MA
Cardinal Health	Dublin, OH
Cook Medical	Bloomington, IN
Cordis	Warren, New Jersey
Covidien	Plymouth, MN
CSI (Cardiovascular Systems Inc.)	St Paul, MN
IMDS (Interventional Medical Device Solutions)	Roden, The Netherlands
Kaneka Medix Corporation	Osaka, Japan
Medtronic	Santa Rosa, CA
Merit Medical	South Jordan, UT
Nitiloop	Netanya, Israel
Philips Volcano	San Diego, CA
Radius Medical	Hudson, MA
Roxwood Medical	Redwood City, CA
Spectranetics	Colorado Springs, CO
Terumo	Somerset, NJ
Vascular Solutions	Minneapolis, MN
Worldwide Innovations & Technologies	Kansas City, KS
3M	St Paul, Minnesota

Appendix 2

Commonly Used
Acronyms in CTO Interventions

Acronym	Full Name	Description	Page (Main Explanation)	Pages (Mentioned)
	Anchor-Tornus technique	Combined use of the Tornus microcatheter and side-branch anchoring to cross a balloon-uncrossable CTO.	277	
	Antegrade balloon puncture	Variation of the reverse CART technique: the antegrade balloon remains inflated during retrograde crossing attempts and is punctured by the retrograde wire, which is then advanced while the antegrade balloon is retracted under fluoroscopy (the latter technique is also called the transit balloon technique).	229	
	Antegrade microcatheter probing	After retrograde guidewire crossing the retrograde microcatheter is advanced into the antegrade guide catheter, followed by the removal of the retrograde guidewire and intubation of the microcatheter with an antegrade wire.	240	
	Back bleeding sign	Blood coming out of the microcatheter after aspirating for at least 30 s (after guidewire withdrawal); it suggests (but does not prove) distal true lumen position.	384	416
BAM	Balloon-assisted microdissection (also called grenadoplasty)	Technique for crossing balloon-uncrossable lesions: a small (1.20–1.50 mm in diameter) balloon is advanced into the lesion as far as possible and inflated to high pressure until it ruptures. Balloon rupture can modify the proximal cap and facilitate crossing with another balloon.	269	
BASE	Balloon-Assisted Subintimal Entry	One of the "move-the-cap" techniques used in cases of proximal cap ambiguity: a slightly oversized balloon is inflated proximal to the CTO, to create a dissection, through which a knuckled guidewire is advanced subintimally around the proximal cap.	295	
BAT	Balloon-Assisted Tracking technique	Technique for advancing catheters through tortuous radial arteries: a balloon is advanced halfway in and halfway out the tip of the guide catheter and inflated. The guide/inflated balloon assembly is then advanced through the area of tortuosity. The BAT technique can also be used for sheathless insertion of guide catheters.	24	

Term	Description	Pages
Block and deliver	Technique for managing coronary perforations. A balloon is advanced proximal to or at the site of perforation and inflated to prevent continued bleeding into the pericardium. A covered stent or a microcatheter (for coil delivery) is advanced proximal to this blocking balloon. The blocking balloon is transiently deflated, followed by advancement of the covered stent or microcatheter for sealing the perforation. Coil delivery is done while the blocking balloon is inflated, minimizing bleeding into the pericardium. A large (8 Fr) guide catheter is needed for delivering covered stents, whereas smaller guide catheters suffice for delivering coils.	396, 403
Bobsled	Bobsled refers to changing the location of reentry attempts during antegrade dissection and reentry, using the Stingray balloon. Reentry is usually attempted at a healthier, straighter, and larger vessel segment.	189 191, 332
Bridge or rendezvous	Technique for inserting an antegrade wire through a CTO after successful retrograde crossing. After retrograde guidewire crossing the retrograde microcatheter is inserted into the antegrade guide catheter and aligned with an antegrade microcatheter, followed by insertion of an antegrade guidewire into the retrograde microcatheter.	240
Buddy wire stent anchor	Technique for increasing guide catheter support that can be used if the proximal vessel needs stenting: a buddy wire can be inserted and a stent deployed over the original guidewire, effectively trapping the buddy wire, which then provides strong guide catheter support.	133 131, 273
Confluent balloon	Variation of the CART technique in which both antegrade and retrograde balloons are inflated simultaneously in a kissing fashion to cause the subintimal space to become confluent, allowing wire passage through the CTO.	228 230
CART	Controlled Antegrade and Retrograde Tracking and dissection	Technique for reentry into the true lumen after subintimal CTO crossing during the retrograde approach: a balloon is inflated over the retrograde guidewire creating a space into which an antegrade guidewire is advanced.	221–222

Continued

Acronym	Full Name	Description	Page (Main Explanation)	Pages (Mentioned)
Contemporary reverse CART	Contemporary reverse Controlled Antegrade and Retrograde Tracking and dissection	Variation of reverse CART technique: a small (2.0–2.5 mm) antegrade balloon is used to facilitate crossing of the retrograde guidewire into the antegrade true lumen. Using a small balloon (instead of a larger one as is done in the standard reverse CART technique) minimizes the size of dissection and vessel injury.	229	220
Contrast-guided STAR	Contrast-guided Subintimal Tracking And Reentry; also called Carlino technique	Variation of the STAR technique in which contrast is injected through a microcatheter inserted into the proximal cap or within the subintimal space to create/visualize a dissection plane, thus facilitating guidewire advancement.	162	161, 193
Deflecting balloon or blocking-balloon technique		Technique for facilitating crossing when there is a side branch near the proximal or distal cap. A balloon is inflated at the ostium of the side branch, blocking entry of the guidewire into it and facilitating advancement of the guidewire through the target lesion.	316	
Double-blind stick-and-swap		Technique for reentering into the distal true lumen using the Stingray balloon after subintimal guidewire crossing. It is similar to the stick-and-swap technique, in which a stiff guidewire (such as the Stingray guidewire) is used to create an exit channel toward the distal true lumen, followed by exchange for a polymer-jacketed guidewire for completing the reentry. In stick-and-swap, contralateral contrast injection is used to determine the location of the distal true lumen relative to the Stingray balloon. In the double-blind stick-and-swap technique there is no contrast injection; instead a puncture is performed using a stiff wire on both sides of the Stingray balloon, followed by advancement of a polymer-jacketed guidewire on both sides of the Stingray balloon until reentry is achieved.	186	75, 183

Acronym	Name	Description		
DRAFT	Deflate, Retract, and Advance into the Fenestration technique	Variation of the reverse CART technique that requires two operators: the antegrade balloon is deflated and withdrawn by one operator while the other operator advances the retrograde guidewire through the space created by the balloon being retracted until the retrograde guidewire enters the antegrade guide catheter.	227	
e-CART	ElectroCautery-Assisted Re-enTry	Variation of the reverse CART technique used when the retrograde guidewire cannot penetrate the proximal cap and an antegrade wire cannot be advanced (usually in flush aortoostial lesions). A retrograde stiff guidewire (usually a Confianza Pro 12) is advanced as far as possible into the occlusion over a microcatheter, followed by cautery activation (for 1 sec) to burn through the impenetrable tissue into the aorta.	309	312, 313
	Fast-spin CrossBoss technique	Technique used for advancing a CrossBoss microcatheter: the catheter is rotated rapidly using the proximal torque device until it advances through the occlusion.	166	73, 257, 327
	Finish with the boss	Technique for minimizing the extent of subintimal dissection performed using a knuckled guidewire. Subintimal advancement of the knuckled guidewire is stopped proximal to the distal cap; followed by exchange of the knuckled wire for a CrossBoss catheter to complete the last part of subintimal crossing. The CrossBoss catheter has smaller profile than a knuckled guidewire, decreasing the likelihood of subintimal hematoma formation that can hinder reentry into the distal true lumen.	189	75, 178
Guideliner-assisted reverse CART	Guideliner-assisted reverse Controlled Antegrade and Retrograde Tracking and dissection	Variation of the reverse CART technique: a guide catheter extension is advanced over the proximal guidewire to form a proximal target for the retrograde guidewire to enter.	226	220

Continued

Acronym	Full Name	Description	Page (Main Explanation)	Pages (Mentioned)
	Hairpin technique; also called "reversed guidewire" technique	Technique for wiring highly angulated vessels. A polymer-jacketed guidewire is bent approximately 3 cm from the wire tip and the knuckle is advanced through the introducer into the coronary artery. Upon withdrawal the guidewire tip enters into the angulated side branch.	324	52, 214, 332
IVUS-guided CART	Intravascular ultrasound-guided Controlled Antegrade and Retrograde Tracking and dissection	Variation of the reverse CART technique: intravascular ultrasound is used to determine the location of the antegrade and retrograde guidewire and to allow precise sizing of the antegrade balloon. Intravascular ultrasound allows safe use of larger balloons, which in turn increase the likelihood of successful retrograde wire crossing. IVUS can also help determine whether significant recoil occurs after antegrade balloon inflation.	226	220
J-CTO score	Japan Chronic Total Occlusion Score	Five-point score for prediction of the likelihood of successful guidewire crossing within the first 30 min of crossing attempts. It was developed from the Multicenter Chronic Total Occlusion Registry in Japan. The five variables are: blunt stump, CTO calcification, within CTO tortuosity, occlusion length ≥20 mm, and prior failed attempt.	12	10, 126, 332
	Jet exchange, also called hydraulic exchange, or Nanto technique	Technique for removing a microcatheter while maintaining guidewire position. It is performed by connecting an inflating device over the back end of the microcatheter, inflating it at high pressure, and removing the microcatheter, while the antegrade flow keeps the guidewire in position. However, the trapping technique is more reliable and is preferred for over-the-wire system exchanges.	136	
	Just-marker technique	Variation of the retrograde technique. The retrograde wire is advanced to the distal cap and acts as a marker of the distal true lumen position, facilitating antegrade crossing attempts.	231	202, 218, 219, 258
	Kissing-wire technique	Variation of the retrograde technique that involves manipulation of both antegrade and retrograde wires within the occluded segment until crossing is achieved.	231	203, 219

Term	Description		References
Knuckle-boss technique	Technique for preventing entry of the CrossBoss catheter into side branches. The CrossBoss catheter is withdrawn proximal to the origin of the side branch, and a knuckled wire is advanced past the side branch (the larger size of the knuckle often prevents it from entering the side branch).	173	
LAST Limited Antegrade Subintimal Tracking	Wire-based technique for reentering into the distal true lumen after subintimal guidewire crossing. Usually a stiff guidewire with a 90 degrees angle is manipulated until it enters the distal true lumen. Because it is unpredictable, the LAST technique is currently used infrequently. The Stingray system is preferred for achieving distal true lumen reentry.	193	159, 162, 194, 384
Mini-STAR Mini Subintimal Tracking And Reentry	Variation of the STAR technique in which a polymer-jacketed guidewire (such as the Fielder FC or XT) is used to reenter into the distal true lumen immediately after the occlusion rather than further down the distal vessel. Similar with the LAST technique, mini-STAR is currently used infrequently; the Stingray system is the preferred strategy for distal true lumen reentry.	193	162, 384
Modified Carlino technique	Modification of the Carlino technique (subintimal contrast injection) in which a small amount of contrast (0.5–1.0mL) is injected into the subintimal space during cineangiography. The modified Carlino technique is often used to resolve proximal cap ambiguity and also for treating balloon uncrossable lesions.	195	
Mother–daughter–granddaughter technique	Simultaneous use of two guide catheter extensions (i.e., a 6 Fr extension through an 8 Fr extension) when multiple extreme vessel bends need to be navigated.	36	
Move-the-cap techniques	These techniques use antegrade dissection and reentry to clarify the course of the occluded vessel in case of proximal cap ambiguity and facilitate crossing. The following techniques are included in this category: balloon-assisted subintimal entry (BASE), scratch-and-go, and the Carlino technique.	295	164, 308, 318

Continued

Acronym	Full Name	Description	Page (Main Explanation)	Pages (Mentioned)
	Open-sesame technique	Technique for facilitating crossing when there is a side branch at the proximal cap. Balloon inflation is performed in the side branch inducing a geometrical shift of the proximal cap plaque, which in turn enables guidewire entry into the CTO.	314	
	Parallel-wire technique	During antegrade wire escalation if the initial guidewire enters the subintimal space, it is left in place and a new guidewire is inserted next to the original guidewire to facilitate crossing into the distal true lumen.	154	156, 173, 179, 242
	Ping-pong guide	Two guide catheters are used to simultaneously engage the same target vessel. One guide is pulled back to enable engagement with the other guide catheter and vice versa.	395–398	86, 205, 207, 294, 307, 387, 395
PROGRESS-CTO score	PROspective Global REgiStry for the Study of Chronic Total Occlusion Intervention score	Scoring system that uses four variables (proximal cap ambiguity, moderate/severe tortuosity, circumflex artery CTO, and absence of interventional collaterals) to create a four-point score that helps predict technical success.	13	10
	Proxis-Tornus technique	Insertion of a Tornus microcatheter through a Proxis device (increased support) to cross balloon-uncrossable CTOs. Since the Proxis device is no longer commercially available, a guide catheter extension can be used instead of the Proxis catheter.	277	
RASER technique	Rotablation and laser technique	Technique for treating balloon-uncrossable and balloon-undilatable lesions in which laser is used first to facilitate advancement of a rotational atherectomy wire (either directly or using a microcatheter for exchange), followed by rotational atherectomy.	282	
Reverse CART	Reverse Controlled Antegrade and Retrograde Tracking and dissection	Reverse CART is the opposite of the CART technique: a balloon is inflated over the antegrade guidewire creating a space into which the retrograde guidewire is advanced. At present, reverse CART is the most commonly used technique for retrograde reentry.	222–223	204, 219, 220, 227, 228, 309, 314

Technique	Description	Page(s)
Reverse wire trapping technique	Technique for delivering an antegrade guidewire through a CTO after successful retrograde guidewire crossing. The retrograde guidewire is snared with a small snare, followed by withdrawal of the retrograde guidewire by pulling the antegrade snare through the CTO into the distal true lumen. This technique is used very rarely.	240
Scratch-and-go technique	This is one of the move-the-cap techniques. A stiff guidewire is advanced toward the vessel wall into the subintimal space proximal to the CTO (scratching the wall). A microcatheter follows the wire into the subintimal space. The stiff guidewire is exchanged for a polymer-jacketed guidewire that is advanced to form a knuckle that then crosses the occlusion.	299
See-saw wire-cutting technique	Modified version of the wire-cutting technique for treating balloon-uncrossable lesions. Two balloons are advanced over two guidewires to the proximal cap of the CTO. The first balloon is inserted as far as possible into the lesion and inflated, effectively cutting the proximal cap with the second guidewire. The process is repeated with the second balloon, modifying the cap on the other side until the lesion is successfully crossed with a balloon.	271–272
See-saw technique	This is a variation of the parallel wire technique: during antegrade wire escalation if the guidewire enters into the subintimal space it is left in place and a second guidewire is advanced over a second microcatheter next to the first guidewire to cross the occlusion. Two microcatheters are used in the see-saw technique versus only one in the parallel-wire technique.	154–155
Septal surfing	Technique for retrograde guidewire crossing through septal collaterals. In contrast to the contrast-guided technique, in which contrast injection is used to determine the course of the septal collaterals and help guide the guidewire, in the surfing technique the guidewire is advanced blindly back and forth through the collateral until it crosses into the distal true lumen.	210

164, 222, 295

156

Continued

Acronym	Full Name	Description	Page (Main Explanation)	Pages (Mentioned)
	Side-branch anchor technique	Technique for increasing guide catheter support. A workhorse guidewire is advanced into a side branch proximal to the target lesion, followed by inflation of a small balloon into the side branch. The size of the balloon is selected to match the size of the collateral vessel. The side branch balloon is inflated to 6–8 atm, anchoring the guide into the vessel.	130–131	166, 273, 448
STAR	Subintimal Tracking And Reentry	STAR is the original antegrade dissection/reentry technique that was described by Antonio Colombo. A knuckled polymer-jacketed guidewire is advanced through the subintimal space until it spontaneously reenters into the distal true lumen, usually at a bifurcation.	193	159, 359
Stent Reverse CART	Stent Reverse Controlled Antegrade and Retrograde Tracking and dissection	This is a variation of the reverse CART technique: a stent is deployed from the proximal true lumen into the subintimal space to facilitate retrograde wiring into the stent. It is used infrequently as a last resort because it is irreversible and may sometimes hinder reentry, for example when the retrograde guidewire crosses through the stent struts.	227	220, 229
	Stick-and-drive technique	This is the classic technique for reentering into the distal true lumen using the Stingray balloon. The Stingray guidewire is advanced without rotation through the side port of the Stingray balloon so as to puncture back into the true lumen. After confirmation of distal true lumen position with contralateral injection the Stingray guidewire is rotated 180 degrees and advanced further down into the vessel. The stick-and-swap technique is preferred in most cases, especially when the distal vessel is of small caliber, tortuous, of diffusely diseased, as the stiff Stingray guidewire may cause injury of the distal vessel.	183	

	Stick-and-swap technique	Technique for reentry into the true lumen using the Stingray balloon: an initial puncture is performed using the Stingray wire to create a connection with the distal true lumen. The Stingray wire is removed and a Pilot 200 (or similar polymer-jacketed) guidewire is advanced through the same side port into the tunnel created by the Stingray wire to enter the distal true lumen.	186, 188	75, 183, 189
STRAW	Subintimal TRAnscatheter Withdrawal technique	Aspiration of hematoma that develops during antegrade dissection/reentry crossing to facilitate reentry. STRAW can be performed either through the Stingray balloon itself, or ideally through another microcatheter or over-the-wire balloon advanced proximal to the Stingray balloon.	189–190	191, 194, 389
	Subintimal distal anchoring technique	Technique for treating balloon-uncrossable lesions after successful guidewire crossing. A second guidewire is advanced subintimally through the occlusion and a balloon is advanced over the subintimal guidewire to anchor the true lumen guidewire, over which a balloon can then be advanced through the occlusion.	276–277	267, 272
	Tip-in technique	After retrograde guidewire crossing an antegrade microcatheter is advanced over the retrograde guidewire through the occlusion, followed by insertion of an antegrade guidewire and antegrade delivery of balloons and stents. This technique results in less strain on the collaterals; however, unlike guidewire externalization, loss of guidewire position may occur.	240–241	242, 372, 417
	Wire-cutting technique	Technique for treating balloon-uncrossable or balloon-undilatable lesions. A second guidewire is advanced through the occlusion and a balloon is inflated over the original guidewire while pulling the second guidewire, scoring and modifying the lesion.	271	

Index

'Note: Page numbers followed by "f" indicate figures, "t" indicate tables and "b" indicate boxes.'

Printed in the United States
By Bookmasters